Industrial X-Ray Compute

MW00563191

Simone Carmignato · Wim Dewulf
Richard Leach
Editors

Industrial X-Ray Computed Tomography

 Springer

Editors
Simone Carmignato
Department of Management and
  Engineering
University of Padova
Vicenza
Italy

Richard Leach
University of Nottingham
Nottingham
UK

Wim Dewulf
Department of Mechanical Engineering
KU Leuven
Leuven
Belgium

ISBN 978-3-319-86653-6   ISBN 978-3-319-59573-3   (eBook)
https://doi.org/10.1007/978-3-319-59573-3

© Springer International Publishing AG 2018
Softcover reprint of the hardcover 1st edition 2017
This work is subject to copyright. All rights are reserved by the Publisher, whether the whole or part of the material is concerned, specifically the rights of translation, reprinting, reuse of illustrations, recitation, broadcasting, reproduction on microfilms or in any other physical way, and transmission or information storage and retrieval, electronic adaptation, computer software, or by similar or dissimilar methodology now known or hereafter developed.
The use of general descriptive names, registered names, trademarks, service marks, etc. in this publication does not imply, even in the absence of a specific statement, that such names are exempt from the relevant protective laws and regulations and therefore free for general use.
The publisher, the authors and the editors are safe to assume that the advice and information in this book are believed to be true and accurate at the date of publication. Neither the publisher nor the authors or the editors give a warranty, express or implied, with respect to the material contained herein or for any errors or omissions that may have been made. The publisher remains neutral with regard to jurisdictional claims in published maps and institutional affiliations.

This Springer imprint is published by Springer Nature
The registered company is Springer International Publishing AG
The registered company address is: Gewerbestrasse 11, 6330 Cham, Switzerland

Acknowledgements

This book was a result of funding from the European Union's Seventh Framework Programme under grant agreement No. 607817, INTERAQCT project. Many of the chapter authors (with the exception of chapter 1) were early stage researchers from the INTERAQCT project. The authors of chapter 1 would like to thank the UK Engineering and Physical Sciences Research Council (grants EP/M008983/1 and EP/L01534X/1) and 3TRPD Ltd. for funding.

Contents

Chapter 1
Introduction to Industrial X-ray Computed Tomography

Adam Thompson and Richard Leach

Abstract The concept of industrial computed tomography (CT) is introduced in this chapter. CT is defined, and the history of CT scanning is outlined. As part of the history, the conventional tomography technique is explained. The development of CT is documented, outlining the initial experiments that lead to the development of modern CT. An overview of the current state of CT scanning is then presented, outlining the five main generations of clinical CT scanner and the two main types of scanner used in industrial CT. The industrial requirements of CT are then discussed, and the content of the following chapters is briefly outlined.

Computed tomography is defined as an imaging method in which the object is irradiated with X-rays or gamma rays and mathematical algorithms are used to create a cross-sectional image or a sequence of such images (VDI/VDE 2009). The word tomography itself comes from the Greek *tomos*, meaning a 'slice' or 'section', and *graphien*, meaning 'to write'. In the process of tomography, X-ray radiography is used to take a large number of radiographic projections, which are then reconstructed using a mathematical algorithm to form a slice image of the object being scanned (Hsieh 2009). These reconstructed slices can then be stacked to form a three-dimensional (3D) representation of the object that can be used in a wide array of applications. The tomographic technique was originally developed as a novel method of clinical visualisation, due to the advantage presented by the technology when compared to traditional two-dimensional (2D) radiography, which compresses a 3D volume into a 2D projection. 2D radiography, therefore, carries the caveat that, due to overlapping structures, locating a feature within a wider object becomes difficult due to the ambiguity regarding the depth of the feature. It is also the case that more dense structures overlapping a feature of interest may obscure that feature entirely, for example in the case of patient jewellery (see Fig. 1.1), or when a bone lies along the same X-ray path as a cancerous tumor, the tumor can be completely obscured by the presence of the bone.

A. Thompson (✉) · R. Leach
Faculty of Engineering, University of Nottingham, NG7 2RD Nottingham, UK
e-mail: Adam.Thompson@nottingham.ac.uk

© Springer International Publishing AG 2018
S. Carmignato et al. (eds.), *Industrial X-Ray Computed Tomography*,
https://doi.org/10.1007/978-3-319-59573-3_1

1

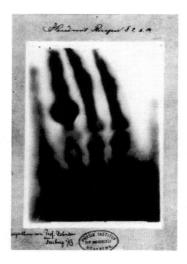 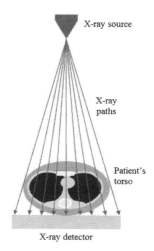

Fig. 1.1 *Left* An example X-ray radiograph taken by William Röntgen of his wife's hand, showing the finger bone concealed beneath the subject's wedding ring (NASA 2007). *Right* The basic X-ray acquisition setup (Hsieh 2009)

1.1 History of X-ray Computed Tomography

Since the discovery of X-rays and the inception of X-ray imaging, there have been many developments in X-ray technologies that have eventually resulted in the industrial CT scanners available today. This section provides a concise history of conventional and computed tomography, and examines the progress of the technologies that have evolved into that which is seen in modern scanners. The initial experiments that defined modern CT are explored, and key developments over time are documented.

1.1.1 X-ray Tomography

Prior to the invention of modern computing, there existed a version of tomographic technology that did not require the complex computation that allows today's computed tomography (CT) scanners to function. Conventional tomography was conceived initially in 1921 by André Bocage (Bocage 1921; Hsieh 2009), in a patent describing a method of attaining slice images of an object by blurring structures above and below the plane of interest. Bocage used the simple setup of an X-ray source, an object and a detector, in this case a piece of X-ray film. For his experiment, Bocage synchronously translated the X-ray source and the detector linearly in opposite directions, in order to blur the structures that lay outside of the focal plane of the system. To understand how Bocage's method works, two points

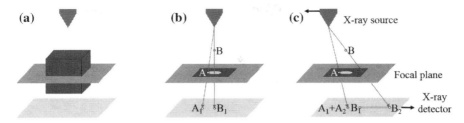

Fig. 1.2 Diagrammatic representation of the principle of conventional tomography: **a** a plane is selected from the object being scanned, **b** the initial position of the shadows on the detector from points *A* and *B*, **c** the position of the shadows of points *A* and *B* following linear translation in opposite directions. Subscripts *1* and *2* represent the shadows before and after translation respectively (Hsieh 2009)

labelled A and B inside the object being scanned should be considered, where A is in the focal plane and B lies above the focal plane. When the X-ray source and the detector are translated, the shadow of point A remains in the same position on the detector and so the image does not blur. The shadow of point B, however, moves across the detector and the image blurs into a line segment. Any other point not in the focal plane of the X-ray beam will, therefore, similarly blur, whilst any other point lying in this plane will not blur, similarly to point A. The result is a relatively clear image of a single slice, relating to the plane at the focal point of the X-ray beam. This process is explained diagrammatically in Fig. 1.2.

There are many significant caveats to the conventional tomography technique, the first of course being the blurred shadows present in the final image due to the many points not in the focal plane, which result in a general reduction in shadow intensity across the image. It should also be noted that planes in close proximity to the focal plane will experience very low levels of blurring compared to planes further away. This means that, while the image correlating to the true plane formed by the X-ray beam will have the highest clarity, the final image will actually represent a slice of definite thickness as opposed to a true plane. The amount of blurring can be used for thresholding the slice thickness, which depends on the angle swept out by the X-ray beam, as shown in Fig. 1.3. The slice thickness is inversely proportional to $\tan(1/\alpha)$, so α must by sufficiently large to produce a reasonably thin slice.

The other significant caveat to the conventional tomography technique is that the blurring effect is generally present only in the direction parallel to motion of the X-ray source and detector, so little to no blurring occurs perpendicular to this direction. The effect of this direction-dependent blurring is that, for structures in the object being scanned that lie parallel to the direction of source and detector motion, blurring occurs only along the direction of motion. Figure 1.4 explains the phenomenon of direction-dependent blurring in conventional tomography. In Fig. 1.4, the reference artefact (right) is made from two parallel cylinders, each topped with a sphere of greater density than the cylinders. The two spheres are placed so that they lie above the cylinders in a plane parallel to the plane formed by the cylinders.

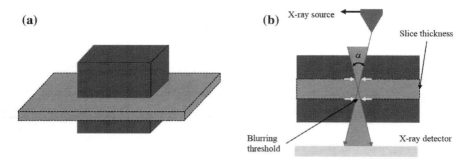

Fig. 1.3 **a** Representation of the finite size slice through the object being scanned, **b** slice thickness as a function of the angle α (Hsieh 2009)

Figure 1.4a, b represent the case in which the long axis of the artefact is placed perpendicular to the direction of source and detector motion, while Fig. 1.4c, d represent the case in which the long axis of the artefact is placed parallel to the direction of source and detector motion. In the perpendicular case, Fig. 1.4b clearly shows blurring of the cylinders which are placed out of the focal plane, and shows clearer images of the spheres (which, for this example, can be considered as the objects of interest within the structure). In Fig. 1.4d, all blurring of the cylinders occurs parallel to their direction and so the image produced does not differ from the conventional radiograph. This case also carries with it the problem that the images of the spheres are not enhanced as the blurring effect is negated by the orientation of the reference artefact.

It is possible to account in part for the direction-dependent blurring problem using a more complicated source and detector motion path than the simple linear translation demonstrated here, in a process known as 'pluridirectional' tomography. Using, for example, a circular or elliptical trajectory, the X-ray source and detector can be moved synchronously in opposite directions to achieve uniform blurring in all directions. Figure 1.5 shows an example of this form of conventional tomography. It should be noted that a number of disadvantages come from using the pluridirectional method, relating to increased scan cost and time, as well as larger X-ray dosage, which is relevant especially in the case of a living scan subject.

Although pluridirectional tomography is successful in accounting for the directional blurring that causes imaging problems in linear conventional tomography, the pluridirectional technique is still incapable of completely removing objects outside the focal plane from the image. This problem results in a reduction of contrast between different structures within the focal plane of the image. Conventional tomography (in both linear and pluridirectional forms) is, therefore, severely limited in application and so initially experienced little clinical or industrial use on its inception. As image processing techniques have been developed over time, however, in addition to the invention of digital flat-panel detectors (as opposed to radiographic film), conventional tomography has experienced a renewed interest in clinical applications where a lower X-ray dose than from CT is required

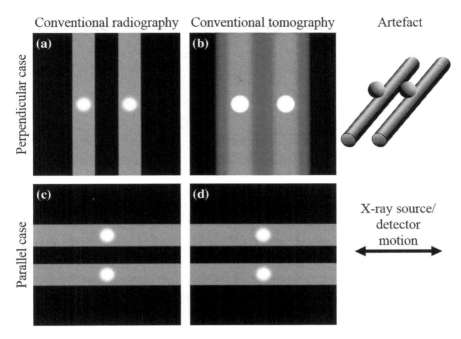

Fig. 1.4 Simulations of images produced using reference artefacts for conventional tomography, showing direction-dependent blurring. The focal plane for each tomographic image is at the centre of the spheres (Hsieh 2009)

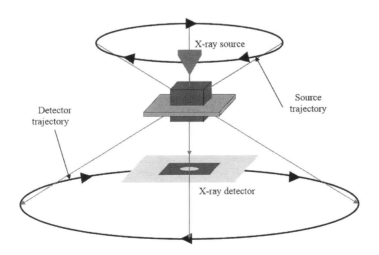

Fig. 1.5 An example of a method of pluridirectional tomography (Hsieh 2009)

(Hsieh 2009; Dobbins III and Godfrey 2003; Warp et al. 2000; Stevens et al. 2001; Nett et al. 2007; Deller et al. 2007), as well as in industrial applications where planar objects or parts with high dimensional aspect ratios have to be analysed (Kruth et al. 2011; De Chiffre et al. 2014). The modern version of the conventional tomography technique (known as *laminography* or *tomosynthesis*) can acquire several images of different focal planes using a single data acquisition and offers improved image quality using modern image processing.

1.1.2 X-ray Computed Tomography

The mathematical theory behind CT scanning dates back as far as 1917, when Johann Radon showed that a function can be reconstructed from an infinite set of its projections using the 'Radon transform' (Radon 1986). The Radon transform in two dimensions is the integral transform (Arfken and Weber 1985), consisting of the integral of a function over straight lines. Radon transform data is commonly referred to as a 'sinogram' (see Chap. 2), as the Radon transform of a Dirac delta function can be represented by a sine wave. The accepted invention of X-ray CT was by Gabriel Frank in 1940, as described in a patent (Frank 1940; Webb 1992; Hsieh 2009). Considering the fact that modern computation did not exist in 1940, this can be considered to have been no simple feat. In his patent, Frank described equipment to form linear representations of an object being scanned (i.e. sinograms), as well as describing optical back projection techniques used in image reconstruction (see Chap. 2).

Frank's patent laid the foundations for today's CT, but the method at the time was heavily hindered by a lack of modern computational technology and so no progress was made for some time following the initial publication of Frank's patent. In 1961, however, a neurologist from the University of California, Los Angeles named William H. Oldendorf published his experiments using gamma radiation to examine a nail surrounded by a ring of other nails (i.e. preventing examination of the nail by conventional radiography) (Oldendorf 1961). Olendorf's experiment was designed to simulate a human head and to investigate whether it was possible to gain information about a dense internal structure surrounded by other dense structures using transmission measurements. An image of Oldendorf's experimental setup is shown in Fig. 1.6. The test sample was translated linearly at 88 mm per hour, between a collimated I^{131} gamma source and a NaI scintillation detector, whilst simultaneously being rotated at a rate of 16 rpm. This setup is shown schematically in Fig. 1.7.

Oldendorf set up this experiment so that the gamma ray beam always passed through the centre of rotation of the system, allowing each nail in the rings to pass through the gamma ray beam twice per rotation, while the two central nails passed comparatively slowly through the beam owing to the much slower linear translation. Oldendorf then recorded the intensity of the measured beam against time. As the nails in the ring passed relatively quickly in and out of the beam, the resultant

Fig. 1.6 The sample used by William H. Oldendorf, comprising two concentric rings of iron nails surrounding an aluminium nail and another iron nail separated by approximately 15 mm, set into a plastic block of 100 mm × 100 mm × 4 mm. The gamma source used is shown in the *top right* of the image (Oldendorf 1961)

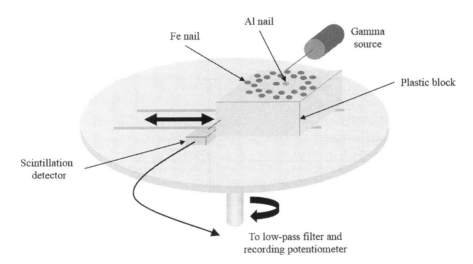

Fig. 1.7 Schematic diagram of Oldendorf's experiment

intensity variations could be considered as high frequency signals and, therefore, removed from the intensity profile using a low-pass filter. Due to the slower linear translation, the two central nails remained in the beam for a much longer period of time and so represented relatively low frequency signals compared to the nails in the rings. These signals were, therefore, preserved by the low-pass filter and Oldendorf was able to detect the presence of the two central nails through reconstruction of a single line through the centre of rotation. In order to reconstruct a full 2D slice, Oldendorf would have to have shifted the sample relative to the centre of rotation, and with no means to store the line data at the time, no attempt was made to reconstruct a slice. The method devised by Oldendorf does, however, represent as example of the basic principles used in today's CT scanners.

The next advancement in tomographic technology came in 1963, when the idea of transverse tomography was introduced by David Kuhl and Roy Edwards (Kuhl and Edwards 1968). In their study, Kuhl and Edwards used radioisotopes to acquire a section from twenty-four separate exposures each taken at twenty-four regularly-spaced angles around a patient's brain. At each step, two opposed detector films, placed tangentially to the axis of rotation, were exposed to a thin line of light moving across a cathode ray tube screen orientated and speed-matched to the detectors' line of sight. By following this procedure, Kuhl and Edwards were in fact performing a backprojection operation (Hsieh 2009), which they later performed digitally by replacing the detector films with a computational backprojection operation. Kuhl and Edwards did not at the time have a method of reconstruction, but this particular experiment has been developed in the time since Kuhl and Edwards initially performed it, and now forms the basis of modern emission CT technology.

The first actual CT scanner was also detailed in 1963 and reported by Cormack (1963) following work performed by himself over a number of years prior to 1963 (Hsieh 2009). In his paper, Cormack published his method of finding a function in a finite region of a plane through the use of the line integrals along all straight lines intersecting that region, and applied that method to his interest in the reconstruction of tissue attenuation coefficients for improvement of radiotherapy treatments. After deriving a mathematical theory of image reconstruction, Cormack tested his theory using test artefact in the form of an 11.3 mm diameter aluminium disk surrounded by an aluminium alloy annulus, which was then surrounded by an oak annulus; totalling 200 mm in diameter. Using a C^{60} gamma ray source and a Geiger counter detector, Cormack scanned the test artefact in 5 mm increments to a form linear scan at a single angle. As all other scans would have been identical due to the circular symmetry of the artefact, Cormack was able to use the single line scan to calculate attenuation coefficients for both aluminium and wood using his reconstruction theory.

In his aforementioned 1963 paper and its sequel in 1964 (Cormack 1963, 1964), Cormack repeated his experiments using a non-symmetrical test artefact, consisting of two aluminium disks surrounded by plastic and encased in an outer aluminium ring, to represent tumours within a patient's skull. In this case, Cormack once again used a gamma ray source and a Geiger counter detector, but this time acquired linear scans at twenty-four increments around a 180 angle.

Concurrently to Cormack's discoveries in the early 1960s, Godfrey Hounsfield of EMI Ltd similarly postulated that it would be possible to reconstruct internal

Fig. 1.8 The original X-ray CT scanner developed by Godfrey Hounsfield at EMI in 1967 (Bioclinica 2011)

structure by taking X-ray measurements along all lines through the object being examined (Hounsfield 1976). As a result of his ideas, Hounsfield began to develop the first clinical CT scanner in 1967 at EMI, producing the prototype shown in Fig. 1.8 using an americium gamma ray source of relatively low intensity. Hounsfield's prototype took nine days to acquire a full data set and reconstruct a picture, including the 2.5 h of computation required to perform the 28,000 simultaneous equations that Hounsfield's equipment needed to form a reconstruction, given his lack of access to Cormack's algorithm. Hounsfield modified later his initial prototype to use an X-ray source of much higher radiation intensity than the americium alongside a scintillation detector, greatly decreasing scan time and increasing the accuracy of the measured attenuation coefficients (Hsieh 2009).

Hounsfield's CT scanner was continually developed until its release in 1971, where it was installed in the Atkinson-Morley Hospital in London. The final version of the scanner was capable of scanning a single layer in less than five minutes at a resolution of 80 × 80 pixels, with a reconstruction time of approximately seven minutes. The scanner itself was limited in application to head scans only and required a water tank to surround the patient's head in order to reduce the dynamic range of the X-rays onto the detector. A scan of the first patient successfully

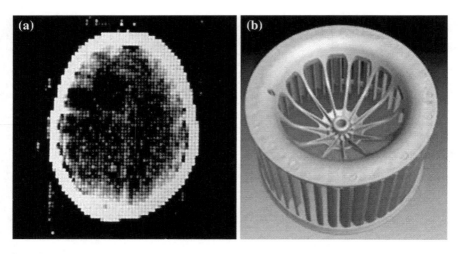

Fig. 1.9 **a** An image from the first clinical scan of a patient with a frontal lobe tumor, taken on the 1st October 1971 (Impactscan 2013), **b** an image of a reconstructed model of a car part for industrial quality checks (Kruth et al. 2011)

identified a frontal lobe tumor in the patient's head; an image from this scan is shown in Fig. 1.9 (Paxton and Ambrose 1974). The invention of the CT technology eventually earned the 1979 Nobel Prize in Physiology and Medicine for both Cormack and Hounsfield.

Since these early scans, a great deal of development has occurred in the field of CT scanning, initially with the development in 1974 of Robert Ledley's Automatic Computerized Transverse Axial (ACTA) scanner, which was the first scanner capable of full body scans of patients (Ledley et al. 1974). Beyond the obvious improvement in versatility, the ACTA scanner also offered improved resolution and faster scan times than the EMI scanner released three years earlier.

Speed, spatial and low-contrast resolution, as well as slice count improvements have continued over the years since the ACTA scanner was released with required scan times per slice in clinical CT halving approximately every 2.3 years, in line with Moore's law for the exponential increase in the number of transistors in a dense integrated circuit (Hsieh 2009). CT scanning technology has also been adapted to a wide array of applications, from veterinarian use (Hsieh 2009), to more recent advances in high resolution industrial CT (as shown in Fig. 1.8). In terms of industrial developments, CT is now increasingly used in the fields of materials characterization and non-destructive testing for measurements of internal structures and detection of defects, as well as in dimensional metrology for direct measurement of parts in reference to dimensional and geometrical product specifications (Kruth et al. 2011; De Chiffre et al. 2014).

1.2 Evolution of CT Scanners

As discussed, there have been several developments of CT technology over time, many of which have been driven by the improvements required in the field of clinical imaging. This section provides an overview of the various scanner types developed since Hounsfield's original prototype; a more in depth discussion of scanning modes and their effect on scan quality is presented in Chap. 3. Similarly, the processing of CT datasets for metrology, non-destructive testing and other specialised analyses, is outlined in Chap. 4.

1.2.1 Clinical CT Scanners

As with many aspects of medical physics, clinical CT scanner technology has been driven by the needs of the patient. Improvements to scanners were generally made to decrease scan times, in order to reduce the X-ray dosage to the patient and to decrease blurring caused by involuntary patient movement. Clinical scanners are, therefore, generally constructed so that the scanner rotates at high speed around the patient, as rotating the patient is not feasible, although the patient table is commonly translated through the scanner axis. These constraints, therefore, place limitations on the accuracy and precision of scanners, but it is interesting to note how clinical scanners have evolved over time in accordance with the needs of the application. The scanning methods described here are each used to gain single slices of CT data; in order to gain volumetric scans, the patient is usually translated through the axis of the scanner.

1.2.1.1 First and Second Generation Clinical CT Scanners

Hounsfield's original 1971 scanner was known as the first generation CT scanner, and used a collimated pencil beam of X-rays (3 mm wide along the scanning plane and 13 mm long across the plane) to acquire a single data point at a time. The X-ray source and detector moved linearly across the sample to acquire data points across the scan plane. The source and detector were then rotated by a single degree and the process repeated in order to gather data about the object, as shown schematically in Fig. 1.10.

The first generation scanner represented a good proof of concept for CT as a technology but, as demonstrated by the poor quality shown in the brain scan in Fig. 1.8, the scanner was limited in its ability to provide a representation of the patient's internal geometry. The low quality of this scan was attributed to the patient's movement during the 4.5 min scan time. To reduce the scan time, a second generation of CT scanner was built; presenting a number of improvements over the first design. The second generation scanners operated using essentially the same

Fig. 1.10 Hounsfield's first
generation CT scanner
showing the
translation-rotation system

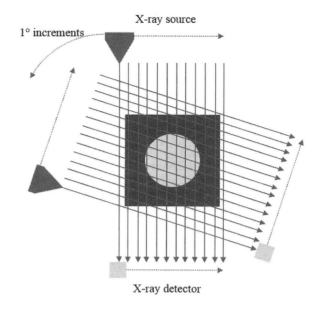

Fig. 1.11 The second
generation CT scanner,
showing the improved
translation-rotation system

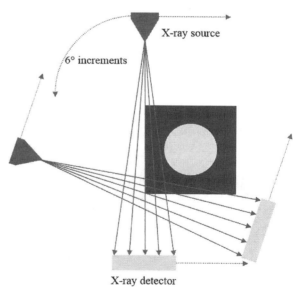

design as the first scanner, still utilising the translation-rotation format, but multiple
pencil beams orientated at different angles allowed a reduction in the number of
rotation steps required, thus speeding up the scan. The second scanner generation is
shown schematically in Fig. 1.11.

Improvements made to the second generation scanner allowed much faster scan acquisition and in 1975 EMI released a second generation scanner that was able to acquire a scan slice in less than 20 s through the use of thirty separate pencil beams (Hsieh 2009). The 20 s threshold was particularly important in clinical CT, as 20 s is a small enough amount of time for a patient to be able to hold their breath, and so problems related to patient motion could be considerably reduced.

1.2.1.2 Third Generation Clinical CT Scanners

As discussed, in medicine the eventual aim of CT scanners is to produce a snapshot of patient geometry at a single instant, and the translation-rotation type scanners are hampered by a ceiling in terms of the fastest achievable speed. As a result, a third generation of CT scanner was developed, capable of faster data capture than the previous two generations by removing the linear motion component required by older designs. The third generation scanner design has endured and remains the most popular design for clinical scanners.

In third generation clinical scanners, an array of detector cells is placed on an arc opposite an X-ray source emitting a fan of X-rays that encompasses the whole patient within the field of view of the fan.

In the third generation setup, the source and detector remain stationary with respect to each other and scan by rotating around the patient. Figure 1.12 shows this scanner design schematically.

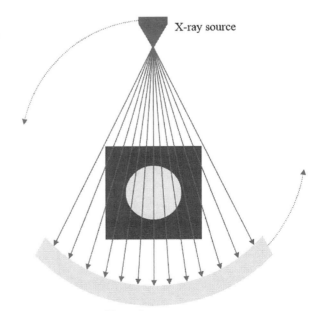

Fig. 1.12 The third generation of CT scanner, the first fan-beam type scanner

X-ray source

X-ray detector array

   Third generation scanners have been developed substantially since their inception, as the first designs required cabling to transfer signals and so significantly
limited the rotation of the scanners. This limitation, therefore, required scanners to
rotate alternately clockwise and anticlockwise to acquire slice data, which then
limited the lowest slice time to approximately 2 s (Hsieh 2009). Later designs
incorporated slip rings to allow a constant rotation of the source and detector gantry,
therefore, allowing a reduction in slice capture time to 0.5 s by removing the time
lost to repeated acceleration and deceleration of the scanner. The third generation of
scanners was also the first to allow for helical as opposed to step-wise scanning
(more information on helical and step-wise scanning is presented in Chap. 3).

1.2.1.3   Fourth Generation Clinical CT Scanners

Third generation scanners are not without flaws, particularly regarding the stability
of the X-ray detector and aliasing that occurs in scan data. To combat these flaws,
the third generation design was modified to use a stationary detector in the form of
an enclosed ring surrounding the entire patient. The X-ray source, in contrast, still
rotates around the patient, projecting a fan of X-rays that sweeps the detector ring as
the source rotates (Hsieh 2009). This configuration, known as the fourth generation
CT scanner and shown in Fig. 1.13, allows a higher number of projections per
rotation as, unlike in third generation scanners, the sample spacing is not dependent
on the detector element size. This advantage allows a reduction of aliasing effects
and allows the detector to be recalibrated dynamically, as each element is exposed

Fig. 1.13 A fourth
generation CT scanner,
showing the full ring of
detector elements

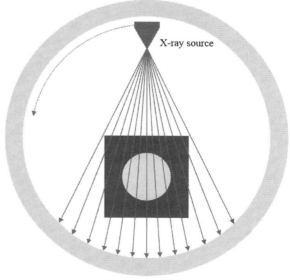

X-ray detector ring

at some point during each rotation to an unattenuated X-ray beam. The dynamic recalibration can be used to account for detector instability, solving the stability problem presented by third generation scanners.

The primary concern with fourth generation scanners in the amount of hardware required for the setup—a very large number of detectors are required to complete a ring large enough to accommodate patients (Hsieh 2009). Increasing the number of detectors carries a substantial cost increase and so in medical practice (where instrument cost is often a prohibitive factor), it is difficult to justify purchase of fourth generation scanners given the limited associated advantages. Fourth generation scanners are, therefore, less popular than their older counterparts.

Another significant problem relating to fourth (as opposed to third) generation scanners is in relation to scattering effects, as each detector element must be able to detect X-rays from a relatively wide angle (Hsieh 2009). This requirement allows for scattered X-ray to reach detector elements and distort the scanned data, with no simple method of reducing this problem. Scatter can be corrected for using post-processing algorithms or reference detectors, but both of these corrections carry the caveat of increasing the time taken in reconstructing images.

1.2.1.4 Fifth Generation Clinical CT Scanners

Both the third and fourth generation scanners have now developed to the point that they can acquire a full slice of scan data in less than a second, but applications in medicine exist where the human body moves at rates comparable or faster than this time period. Third and fourth generation scanners are ultimately limited in their ability to decrease scan time, as the maximum rotation speed of the gantry is limited by the maximum centripetal forces it is possible for the equipment to experience. This limitation necessitated a further technological development in order to decrease scan times, and so in the early 1980s, the fifth generation CT scanner was invented specifically for inspection of patient's hearts (Boyd et al. 1979; Hsieh 2009). The required scan time that is fast enough to accurately provide a snapshot of cardiac motion is below 50 ms. The fifth generation CT scanner was designed in such a way that no mechanical movement of the source, detector or patient is required; instead magnetically sweeping the electron beam back and forth along a cylindrical X-ray target to produce the X-ray fan beam. Figure 1.14 shows a fifth generation scanner diagrammatically.

Because of the use of an electron beam, the entire fifth generation scanner must be sealed inside a vacuum. The presence of the much larger vacuum presents obvious associated problems, not least in limiting the size of the scanner compared to other designs. The scanner was designed in order to specifically image the heart, and so allows 80 mm of translation through the direction of the scanner; much less than possible using third or fourth generation scanners, but at the much faster rate available to the electron beam system.

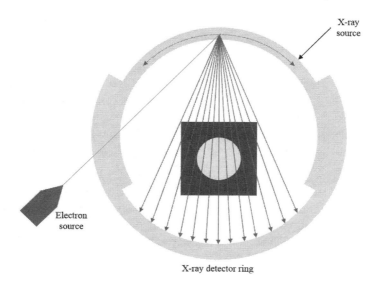

Fig. 1.14 The fifth generation CT scanner. The source and detector rings both sweep through angles of 210° and are arranged to be non-coplanar, so as to allow for the overlap

1.2.2 Industrial CT Scanners

In an industrial setting, the primary aim of performing CT scans differs greatly from the requirements of the medical field. In most of the cases, industrial CT scanning is not so concerned with X-ray dosage to the sample as in medicine, and while it is certainly an advantage to achieve fast scans, the requirement for ultra-fast scan times is no longer present in an industrial setting. Industrial CT is, therefore, capable of using higher intensity X-ray sources, and of increasing scan times to long periods when useful to achieve high precision in scans. As discussed, industrial CT is mostly used for materials characterization, non-destructive testing and metrology applications, and so the focus is generally geared more towards achieving the maximum possible scan resolution, accuracy and precision. Further industrial CT applications also exist in areas such as in the examination of fibre reinforced composites, as well as multimodal data analysis. Discussions of the particular areas of focus for industrial CT will be given later in this book (see Chaps. 8 and 9), but this section will explain the particular modifications made to clinical CT scanning devices in recent years, in order to suit industrial needs (De Chiffre et al. 2014).

For both material analysis and metrology applications, the fundamental difference in industrial CT scanners is the rotational component of the scanner. As opposed to in medical applications, where the scanning apparatus is rotated around the patient at high velocity, in most industrial systems the scanning apparatus is fixed and the sample rotated. This modification allows the construction of CT systems with higher accuracy and stability, features that are crucial to the applications for which they are produced (Kruth et al. 2011). Another major difference between the clinical and industrial type scanners is the input parameters used. Parameters differ significantly because the material being scanned (human tissue in medicine, mainly metals and polymers in industry), as well as the desired output and the size of the object being scanned, differ greatly between applications (De Chiffre et al. 2014). Most industrial CT scanners are based on the third generation CT scanner design, split into two categories each using a fan or cone beam of X-rays respectively; dependent on the specific application of the scanner.

A number of other additions and modifications are made particularly in the case of metrology CT systems, designed in reference to touch-probe co-ordinate measuring systems. For example, metrological CT systems commonly have high precision mechanical setups for modifying the relative positions of sample, detector and source, as well as thermally stable structures in their construction (De Chiffre et al. 2014). These systems also commonly contain high precision temperature stabilising hardware, and are kept in temperature-controlled laboratories.

1.2.2.1 Industrial Fan Beam CT Scanners

In terms of their setup, fan beam CT scanners are essentially the same as third generation scanners, except for the previously-stated difference of rotating the sample as opposed to the scanning gantry. Figure 1.15 shows a fan beam setup schematically. As with conventional third generation scanners, the X-ray source outputs a 2D fan of X-rays, which pass through the object being scanned and onto a detector. The detectors used in industrial CT commonly utilise scintillators with modern charge-coupled devices (CCDs), and may be curved or straight, line or flat panel in construction; see Chap. 3 for a more in-depth discussion of detector

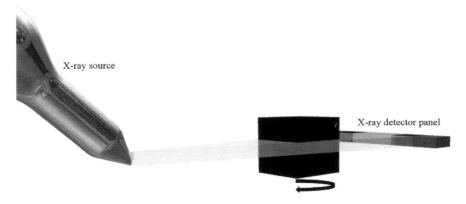

Fig. 1.15 Schematic diagram of a fan beam scanning setup, operating similarly to a conventional third generation scanner. Note how the sample rotates while the scanning gantry remains stationary

characteristics. Fan beam systems can acquire slice data in either a helical or step-wise manner, and are, therefore, much slower than the cone-beam scanner type (which will be discussed in the next paragraph), when used to produce a full 3D reconstruction of the object being measured. Fan beam scanning, however, does not suffer from some of the imaging artefacts (see Chap. 5) that cone-beam CT experiences, and so is capable of producing scan data of higher accuracy than the cone-beam counterpart, especially in case of high-energy sources. This makes fan beam scanning a useful option in dimensional metrology, where dimensional accuracy is of the utmost importance. Fan beam scanners are occasionally used also in material analysis applications, when higher X-ray energies and precision are required.

1.2.2.2 Industrial Cone Beam CT Scanners

Unlike in other CT systems discussed previously, cone beam CT scanners are capable of acquiring 3D volumetric data in a single rotation. Cone beam scanners utilise a 3D cone of X-rays to scan an entire object (or part) in one go, and so represent a very fast acquisition compared to fan beam systems. Similarly to the fan beam scanners, these systems operate by rotating the object being scanned between a stationary source and detector. For applications where the object being scanned is larger than the field of view, it is possible to move the object through the X-ray beam in either a step-wise or helical cone-beam manner (De Chiffre et al. 2014). Figure 1.16 shows the cone beam scanner geometry schematically.

As discussed, compared to fan beam, cone beam CT is subject to a number of additional imaging artefacts (see Chap. 5) and so cone-beam scans will typically be

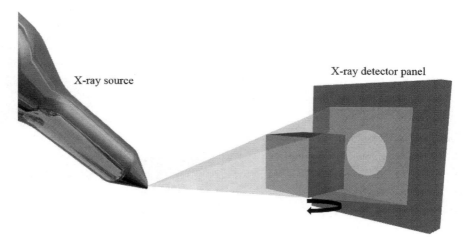

Fig. 1.16 Schematic diagram of a cone beam scan setup, showing how these systems acquire 3D volumetric data in a single rotation

of lower quality than scans taken by fan beam systems. Cone beam CT is, therefore, less commonly used when large parts and/or difficult-to-penetrate materials are scanned using high-energy X-rays. In such cases, the use of cone beam CT would result in substantial imaging artefacts and, therefore, inaccurate results (see Chap. 5). In these situations, fan beam scans are preferred, though in general, for relatively small parts that can be relatively easily penetrated, a majority of applications today utilise cone beam scanning for the speed advantages offered by the technology.

1.2.2.3 Other Advanced CT Setups

In recent years, specific industrial requirements have necessitated developments in CT technology to suit individual requirements (De Chiffre et al. 2014). For example in the case of in-line inspection, robot integration is now beginning to be used to load parts into in-line systems to reduce handling time (see Fig. 1.17). CT systems can also be integrated into scanning electron microscopes (SEMs) for high resolution examination of small samples, using the electron beam in the SEM to generate X-rays (Bruker 2016). Large scale CT for evaluation of, for example, large assemblies or shipping containers is also now possible using very high voltage X-ray tubes or radiation from particle accelerators (Salamon et al. 2015). Linear accelerator and synchrotron CT is also now increasingly being performed for specific industrial applications (De Chiffre et al. 2014). Linear accelerators can be used to produce electron beams of very high energy, which can then produce highly

Fig. 1.17 Automated robotic loading into a CT system for in-line inspection (Zeiss 2016)

Fig. 1.18 Aerial view of the Diamond Light Source in Oxfordshire, UK (Diamond Light Source 2006)

penetrating X-ray beams for inspection of dense or large parts as outlined previously. Synchrotrons are also of interest, as these particle accelerators are capable of producing highly collimated, monochromatic X-ray beams that are also coherent, bright and easily tuneable. Synchrotrons do, however, carry the disadvantage of requiring a large footprint, often requiring setups of hundreds of metres or even kilometres in diameter. Most synchrotron facilities do now contain CT devices available for use, although they often suffer from long booking lists and carry a large associated cost (De Chiffre et al. 2014). Figure 1.18 shows the UK's national synchrotron facility, the Diamond Light Source in Oxfordshire.

1.3 Industrial Requirements

Recent review work has identified a number of industrial requirements as CT technology moves forward into the future (De Chiffre et al. 2014; Kruth et al. 2011). The review by Kruth et al., specifically focusing on CT metrology, concludes that "CT metrology has a high potential for dimensional quality control of components with internal cavities that are not accessible with other measuring devices. It also allows for a holistic measurement of mechanical components: i.e. full assessment of inner and outer part surface instead of a limited set of points". The Kruth et al. review also discusses how CT measurements represent the only currently available method of performing dimensional metrology and material analysis in a single inspection process, although identifies that there remains a lot of work required to establish CT as a viable method of inspection. Particularly for metrological applications, work needs to be undertaken in the areas of system qualification and performance verification, as well as in the establishment of traceability for CT measurements; both of which will be covered in detail in

Chaps. 6 and 7. Kruth et al. also noted that the possibility of using data fusion to merge data from different measurement systems in the future represents a way of further unlocking the benefits of CT for industrial applications.

In the later review by De Chiffre et al., the authors further noted future industrial demands for both metrological and materials analysis applications, citing a reduction in the inspection cycle time and cost of CT instruments, as well as an improvement in user-friendliness and an enlargement of the range of applications as the primary goals for the future of industrial CT. In addition to these goals, De Chiffre et al. discussed the requirement for improvement in scan resolution and accuracy through the reduction of the focal spot size, the option for measurement of multi-material assemblies and development in techniques for the measurement of larger parts.

In another recent review (Thompson et al. 2016), the authors particularly noted the applicability of CT to the measurement of novel additive manufactured (AM) parts, because of the inherent ability of AM to produce parts with complex internal geometries. These geometries are often unable to be measured by conventional methods because of the access requirements of optical or contact probes. The volumetric nature of CT allows manufacturers to measure AM parts without this requirement, and so measurement of the complex internal geometries particular to AM parts becomes possible. Thompson et al. did however conclude that significant work remains in calibration and verification of CT systems for dimensional metrology, as well as the requirement for greater scan resolution for the purpose of ever-increasingly precise non-destructive testing. The application of CT measurements to AM will be discussed further in Chap. 9.

As previously mentioned, in Chap. 2, the authors detail the principles of computed tomography in greater detail, examining the fundamental physics of X-ray generation and detection, as well as the concept of image reconstruction and the various reconstruction algorithms. Chapter 3 then includes a discussion on the components of CT devices, while Chap. 4 includes an explanation of the steps required in processing, analysing and visualisation of CT data. Later in the book, advances made since the aforementioned technology reviews were performed will be summarised, particularly looking at the sources of error in CT measurements (Chap. 5), improvements in system qualification and set-up (Chap. 6) and advances in traceability establishment for CT measurements (Chap. 7). State-of-the-art applications of CT for non-destructive testing and materials characterisation, as well as for dimensional metrology, will be presented in Chaps. 8 and 9.

References

Arfken G, Weber HJ (1985) Integral transforms. Mathematical methods for physicists, 3rd edn. Academic Press, Orlando, pp 794–864

Bioclinica (2011) The evolution of CT scan clinical trials. http://www.bioclinica.com/blog/evolution-ct-scan-clinical-trials. Accessed 12th May 2016

Bocage AEM (1921) Procédé et dispositifs de radiographie sur plaque en mouvement. French Patent No. 536464

Boyd D, Gould R, Quinn J, Sparks R, Stanley J, Herrmmannsfeldt W (1979) A proposed dynamic cardiac 3-D densitometer for early detection and evaluation of heart disease. IEEE Trans Nucl Sci 26(2):2724–2727

Bruker (2016) Micro-CT for SEM: true 3D imager for any scanning electron microscope (SEM). Bruker microCT. http://bruker-microct.com/products/SEM_microCT.htm. Accessed 16th May 2016 2016

Cormack AM (1963) Representation of a function by its line integrals, with some radiological applications. J Appl Phys 34(9):2722–2727

Cormack AM (1964) Representation of a function by its line integrals, with some radiological applications II. J Appl Phys 35(10):2908–2913

De Chiffre L, Carmignato S, Kruth J-P, Schmitt R, Weckenmann A (2014) Industrial applications of computed tomography. Ann CIRP 63(2):655–677. doi:10.1016/j.cirp.2014.05.011

Deller T, Jabri KN, Sabol JM, Ni X, Avinash G, Saunders R, Uppaluri R (2007) Effect of acquisition parameters on image quality in digital tomosynthesis. Proc SPIE 6510:65101L

Diamond Light Source (2006) Diamond about us. Diamond Light Source. http://www.diamond.ac. uk/Home/About.html. Accessed 5th Apr 2016

Dobbins JT III, Godfrey DJ (2003) Digital x-ray tomosynthesis: current state of the art and clinical potential. Phys Med Biol 48(19):65–106

Frank G (1940) Verfahren zur herstellung von körperschnittbildern mittels röntgenstrahlen. German Patent 693374

Hounsfield G (1976) Historical notes on computerized axial tomography. J Can Assoc Radiol 27(3):135–142

Hsieh J (2009) Computed tomography: principles, design, artifacts, and recent advances. SPIE Press, Bellingham, WA, USA

Impactscan (2013) CT History. http://www.impactscan.org/CThistory.htm. Accessed 5th April 2016

Kruth J-P, Bartscher M, Carmignato S, Schmitt R, De Chiffre L, Weckenmann A (2011) Computed tomography for dimensional metrology. Ann CIRP 60(2):821–842. doi:10.1016/j. cirp.2011.05.006

Kuhl DE, Edwards RQ (1968) Reorganizing data from transverse section scans of the brain using digital processing. Radiology 91(5):975–983

Ledley R, Wilson J, Huang H (1974) ACTA (Automatic Computerized Transverse Axial)—the whole body tomographic X-ray scanner. Proc SPIE 0057:94–107

NASA (2007) X-rays. NASA. http://science.hq.nasa.gov/kids/imagers/ems/xrays.html. Accessed 28th Sept 2016

Nett BE, Leng S, Chen G-H (2007) Planar tomosynthesis reconstruction in a parallel-beam framework via virtual object reconstruction. Proc SPIE 6510:651028

Oldendorf W (1961) Isolated flying spot detection of radiodensity dis-continuities-displaying the internal structural pattern of a complex object. IRE Trans Biomed Electron 8(1):68–72

Paxton R, Ambrose J (1974) The EMI scanner. A brief review of the first 650 patients. Br J Radiol 47(561):530–565

Radon J (1986) On the determination of functions from their integral values along certain manifolds. IEEE Trans Med Imaging 5(4):170–176

Salamon M, Boehnel M, Riems N, Kasperl S, Voland V, Schmitt M, Uhlmann N, Hanke R (2015) Computed tomography of large objects. Paper presented at the 5th CIA-CT conference and InteraqCT event on "industrial applications of computed tomography", Lyngby, Denmark

Stevens GM, Fahrig R, Pelc NJ (2001) Filtered backprojection for modifying the impulse response of circular tomosynthesis. Med Phys 28(3):372–380

Thompson A, Maskery I, Leach RK (2016) X-ray computed tomography for additive manufacturing: a review. Meas Sci Technol 27(7):072001

VDI/VDE 2360 Part 1.1 (2009) Computed tomography in dimensional measurement, basics and definitions

Warp RJ, Godfrey DJ, Dobbins JT III (2000) Applications of matrix inversion tomosynthesis. Proc SPIE 3977:376–383

Webb S (1992) Historical experiments predating commercially available computed tomography. Br J Radiol 65(777):835–837

Zeiss (2016) Inline Computer Tomography. Zeiss. http://www.zeiss.co.uk/industrial-metrology/en_gb/products/systems/process-control-and-inspection/volumax.html. Accessed 12th May 2016

Chapter 2
Principles of X-ray Computed Tomography

Petr Hermanek, Jitendra Singh Rathore, Valentina Aloisi
and Simone Carmignato

Abstract In this chapter, the physical and mathematical principles of X-ray computed tomography are summarised. First, the fundamentals of X-ray physics are covered, with details on generation, propagation and attenuation of X-rays, including a brief introduction to phase-contrast and dark-field imaging. Then, the basics of detection, digitisation and processing of X-ray images are discussed. Finally, the chapter focuses on the fundamentals of tomographic reconstruction and introduces the main reconstruction algorithms.

X-ray computed tomography (CT) is an imaging technique that originally found its application in the medical field and over many years of technological advancements has been extended to several industrial applications (see Chap. 1 for historical details). X-ray CT technology has been used extensively for non-invasive human body investigation and non-destructive analyses of industrial parts and materials. The most notable difference between current clinical and industrial CT systems is their source-object-detector configurations (see Sect. 1.2). In the case of clinical CT, the detector and the source rotate around the object (patient), whereas in the case of an industrial CT, the detector and the source are in most cases stable and the object under investigation performs rotatory movement (further details on different industrial CT configurations are given in Chap. 3). The basic configuration of cone beam CT scanning acquires two-dimensional (2D) projections (radiographs) from multiple angles and applies a reconstruction algorithm to obtain a three-dimensional (3D) volumetric representation of the scanned object (a schematic representation of a typical CT scanning procedure is shown in Fig. 2.1). The principle is explained using the cone beam CT configuration as an example, as it is often used in industry. Alternative configurations will be described in Chap. 3.

This chapter gives a brief description of the physical principles of X-ray CT and provides a general background to allow for a better understanding the topics

P. Hermanek · J.S. Rathore · V. Aloisi · S. Carmignato (✉)
Department of Management and Engineering, University of Padova,
Stradella San Nicola 3, Vicenza 36100, Italy
e-mail: simone.carmignato@unipd.it

© Springer International Publishing AG 2018
S. Carmignato et al. (eds.), *Industrial X-Ray Computed Tomography*,
https://doi.org/10.1007/978-3-319-59573-3_2

Fig. 2.1 Schematic representation of a typical X-ray CT scanning workflow

discussed in the subsequent chapters. Section 2.1 covers topics regarding generation, propagation and attenuation of X-rays, Sect. 2.2 describes the detection of
partially attenuated radiation, and Sect. 2.3 explains the principle of reconstruction
and summarises the main reconstruction algorithms.

2.1 Fundamentals of X-ray Physics

This section describes the physical principles of X-ray imaging, which is the first
step in the CT scanning chain. It is important to understand the basic principles of
X-ray physics in order to understand the phenomena that arise from the nature of
X-rays and which will be discussed in later chapters.

Firstly, the generation of X-rays is discussed, followed by a description of the
X-ray spectrum and the factors that influence it. Finally, the attenuation of X-rays is
discussed, supplemented with alternative evaluation techniques for studying the
interaction of X-rays with matter: phase-contrast and dark-field imaging.

2.1.1 X-ray Generation

X-rays are often referred to as Röntgen rays after the name of the discoverer,
Wilhelm Conrad Röntgen. Röntgen was awarded the first Nobel Prize for Physics in
1901 for the discovery of X-rays. These newly discovered rays had no charge and
much more penetrating power than the already discovered cathode rays (Curry et al.
1990). The revolutionary thing observed was that X-rays could pass through biological tissue, showing bones and other tissues in the body. The first X-ray image
was obtained in medicine by Michael Pupin in 1896 (Carroll 2011); it was a
surgical case where a man's hand was full with buckshot (see Fig. 2.2). More
details on the history of X-rays and CT are given in Chap. 1.

X-rays are electromagnetic waves. The energy of each X-ray photon is proportional to its frequency, which is expressed by the following equation

Fig. 2.2 First X-ray image obtained by Michael Pupin in 1896 (Tesla Memorial Society of New York)

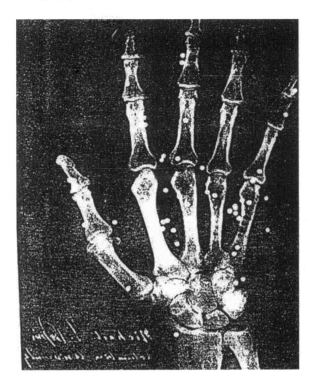

$$E = h\nu = \frac{hc}{\lambda} \tag{2.1}$$

where h is Planck's constant (6.63×10^{-34} J s), c is the speed of light (3×10^8 m s^{-1}), and λ is the wavelength of the X-ray.

As illustrated in Fig. 2.3, the characteristic wavelength of X-rays ranges from 0.01 to 10 nm. X-rays are classified as soft and hard X-rays based on the wavelength and on their ability to penetrate through materials. Longer wavelength X-rays ($\lambda > 0.1$ nm) are termed "soft" X-rays and those with shorter wavelengths are termed "hard" X-rays.

X-rays are generated by means of various sources (linear accelerators, synchrotrons, etc.); however, for sake of simplicity, the principles will be explained on the most commonly used source—an X-ray tube. Details of other source types are given in Chap. 3. The basic principle on which most X-ray tubes used in CT systems operate relies on the Coolidge tube, also known as a hot cathode tube, which dates back to 1913 (Hsieh 2009). In a Coolidge tube, the thermionic effect is exploited for electron production.

The main components of an X-ray tube are a cathode and an anode, which are inserted in an evacuated enclosure. There are different types of anodes (targets) used in industrial CT systems, e.g. transmissive, rotating or liquid. The principle of X-ray

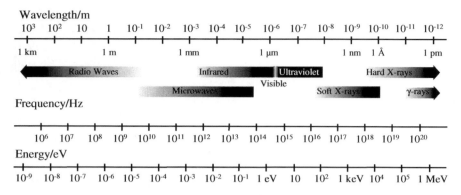

Fig. 2.3 Electromagnetic spectrum with wavelengths, frequencies and energies. Adapted from (Wittke 2015)

generation will be explained referring to the most frequently used reflective target. Other types of target and auxiliary CT components are described in detail in Chap. 3. The cathode usually consists of a thin tungsten filament from which the electrons are emitted. A low voltage generator is connected to the cathode, which is then heated by the Joule effect. As the cathode temperature increases, the kinetic energy of the electrons increases. When this thermally-induced kinetic energy becomes strong enough to overcome the binding energy of the electrons to the atoms of the filament, the electrons can escape the metal (Buzug 2008). This effect is called the thermionic effect. Due to the attractive force between opposite charges, free electrons remain in the proximity of the filament, forming an electron cloud. For thermionic emission, therefore, the filament temperature should be high enough so that the kinetic energy of the electrons is the minimum required to overcome the attractive force holding them in the metal. This minimum energy is called the work function. During thermionic emission, the temperature of the tungsten filament reaches 2400 K (the melting point of tungsten is 3695 K) (Buzug 2008).

The emitted electron current density is described by the Richardson-Dushman equation

$$J = AT^2 e^{-W/kT} \qquad (2.2)$$

where J is the current density, A is a constant, equal to 1.20173×10^{-6} A m^{-2} K^{-2}, T is the filament temperature, W is the work function and k is the Boltzmann constant, equal to 1.38×10^{-23} J K^{-1}.

The anode consists of a metal part (the target), usually tungsten or molybdenum, embedded in a thicker part, typically made from copper for good heat dissipation. The cathode and anode are connected respectively to the negative and positive poles of a high voltage generator. Due to the difference in electric potential, the electrons generated by thermionic effect at the cathode, are accelerated towards the anode, where they strike a target (Fig. 2.4).

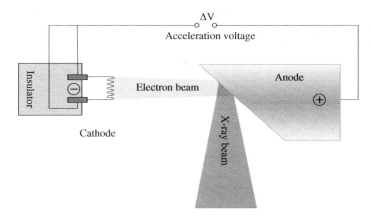

Fig. 2.4 Schematic representation of an X-ray tube. The electrons generated by thermionic emission at the cathode are accelerated by the applied voltage between the cathode and anode. When the electrons reach the anode, X-rays are generated

The trajectory of the electrons emitted from the filament, is usually controlled by means of a focus cup, in order to focus the electrons to a small region on the target. The energy of the electrons reaching the target depends on the applied acceleration voltage

$$E_{electrons} = e\Delta V \tag{2.3}$$

where e is the electric charge of an electron, equal to 1.6×10^{-19} C, and ΔV is the acceleration voltage between the cathode and anode. The acceleration energy is measured in electron volts (eV). 1 eV is the kinetic energy gained by an electron accelerated by a ΔV of 1 V.

The target material is bombarded by high-speed electrons and several types of interaction occur. Around 99% of the energy is transformed into heat as an outcome of the low energy transfer between the accelerated and anode electrons resulting in ionisation of target atoms. The remaining 1% of the energy is converted into X-rays by the following interactions (Hsieh 2009; Buzug 2008):

1. Deceleration of fast electrons in the atoms of the target material.
2. Emission of X-rays generated by re-filling a vacant place in the inner shell of the atom by an outer shell electron.
3. Collision of an accelerated electron with a nucleus.

Interaction 1 is described in Fig. 2.5b. The fast electron comes close to the nucleus and is deflected and decelerated by the Coulomb fields from the atom. The energy loss caused by the deceleration generates X-ray radiation. This radiation is generally called *bremsstrahlung*, which is a contraction of the German words *bremsen* "to brake" and *Strahlung* "radiation" (i.e. "braking radiation" or "deceleration radiation"). The *bremsstrahlung* process usually comprises several

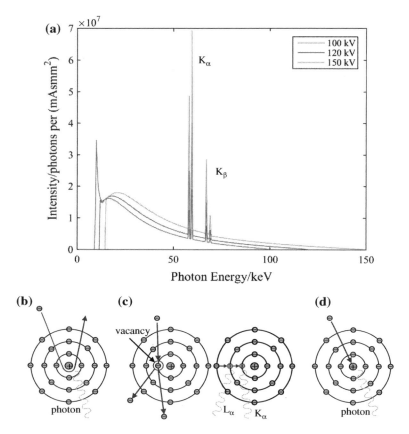

Fig. 2.5 X-ray generation principle: **a** X-ray spectrum at acceleration voltages 100, 120 and 150 kV for a tungsten target, in which the continuous curve is due to *bremsstrahlung*, while the spikes are characteristic K lines, **b** the principle of *bremsstrahlung*, **c** the principle of characteristic radiation, **d** direct interaction between the electron and the nucleus. X-ray spectra were obtained by software SpekCalc (Poludniovski et al. 2009)

deceleration steps as one electron emits, in most cases, several photons. The energy of a single photon is determined by the amount of the energy transferred during the cascade deceleration. Another possible, but very low probability situation, can happen when the total energy of a single electron is transferred to a single photon. Since the number of emitted photons and the transferred energy depend on the trajectory of the electron and the proximity to the nucleus, the distribution of energies is continuous (see Fig. 2.5a).

Interaction 2 occurs when accelerated electrons directly collide with the inner shell electrons of the target material. Figure 2.5c depicts the situation when an electron emitted from the filament knocks out a K-shell electron. As a consequence of this collision, the atom is ionised and a vacant place in the K-shell level is left.

The hole is filled by an electron from one of the outer shells (L, M, N, etc.) along with the emission of a photon. The most probable electron that fills the vacant hole is from the adjacent shell, i.e. in the case of a K-shell, the electron from the L-shell with the characteristic energy K_α. However, the transition from more distant shells can occur; M, N, O, etc. transitions to the K-shell are denoted by K_α, K_β, K_γ and similarly transitions to the L-shell are denoted by L_α, L_β, L_γ. The characteristic line spectrum generated by this direct electron interaction is shown in Fig. 2.5a, denoted by spikes in the spectrum. The energy of the characteristic radiation is determined by the binding energies of the two interacting shells and depends on the anode material.

Figure 2.5d shows the direct interaction between the electron and the nucleus of an anode atom. In this case, the entire energy of the electron is transferred to *bremsstrahlung* with the maximum energy of the X-ray spectrum. However, the probability of a direct electron-nucleus interaction is low, as demonstrated by the low intensity of high energy X-rays in Fig. 2.5a.

2.1.2   X-ray Radiation Spectra and Focus

An X-ray energy spectrum, which has a direct impact on the manner of the consequent X-ray attenuation, is affected by several factors. The following list the most important factors (Cierniak 2011; Stock 2008; Buzug 2008):

- acceleration voltage,
- tube current,
- filtration, and
- anode angle and material.

The acceleration voltage determines both the "quality" and the "quantity" of the X-ray beam, i.e. voltage affects both the energy interval and the amplitude, which can be observed in Fig. 2.5a. Therefore, increasing tube voltage shifts the spectrum to higher energies, which facilitates the penetration of more absorbing samples.

The tube current, on the contrary, generates a linear increase in X-ray intensity, whereas the distribution of X-ray energies remains unchanged. Thus, current affects only the "quantity" (number of emitted X-ray quanta) of the X-ray beam. The effect of current variation is schematically represented in Fig. 2.6. By increasing the current, lower levels of noise in the acquired images can be reached as a trade-off for an increase in tube power (which can increase the focal spot size, hence having a negative impact on the quality of the acquired images).

Filtration of an X-ray beam is another important way to modify and improve the X-ray spectra for a specific application. The unfiltered X-ray spectrum is composed of a wide range of emitted photons, including those with a large wavelength. These low-energy (soft) X-rays are more easily attenuated by matter than X-rays with high energy (hard). This phenomenon is responsible for beam hardening artefacts, which

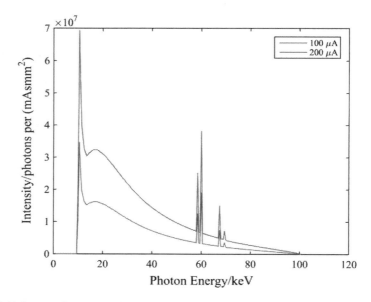

Fig. 2.6 Influence of current on X-ray spectra at 100 kV for a tungsten target. X-ray spectra were obtained by software SpekCalc (Poludniovski et al. 2009)

will be discussed in detail in Chaps. 3 and 5. As the soft X-rays are completely absorbed by the investigated object, they do not contribute to the detected signal, and the mean energy of the X-ray beam passed through the matter is increased, i.e. the beam is hardened. In order to reduce the effect of low-energy X-rays, physical filtration is applied. Most CT users employ flat metal filters, e.g. made of copper, aluminium or tin. Filters are usually from several tenths to several millimetres thick, and are commonly placed between the X-ray source and the scanned object. The effect of various filter materials and thicknesses is shown in Fig. 2.7. The pre-hardening effect of a physical filter results in an X-ray beam with reduced intensity, however, with a spectrum with higher average energy. By filtering out the soft X-rays, the beam hardening artefacts can be reduced (Hsieh 2009).

The intensity of *bremsstrahlung*, which is the main contributor to the whole spectrum, can be defined as (Agarwal 2013)

$$I \propto Zh(v_{max} - v), \qquad (2.4)$$

where Z is the atomic number, h is Planck's constant and v is the electron velocity. Equation 2.4 shows that the intensity is directly dependent on the atomic number of the anode material and a higher intensity of X-ray beam can be reached with materials with high Z, e.g. tungsten ($Z = 74$) is used as a target material in common industrial CT systems. The same intensity improvement can be obtained by incrementing the anode angle (Banhart 2008).

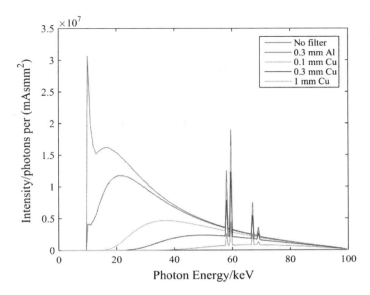

Fig. 2.7 Influence of X-ray filtration by different filter materials and thicknesses on X-ray spectra at an acceleration voltage 100 kV and a tungsten target material. The X-ray spectra were obtained by software SpekCalc (Poludniovski et al. 2009)

The target material suffers from extreme heat load (as explained earlier in this chapter). Therefore, the spot where the electron beam is focused (called the focal spot) must have a finite size (i.e. it cannot be a single point, which would be a focal spot with an ideally null size). Not respecting this condition results in a local melt of the target. However, as a consequence of the increase of the focal spot size, the so-called penumbra effect occurs, which is shown in Fig. 2.8. The size of the focal spot limits the resolution of the acquired images and affects the image quality, producing edge blurring. Therefore, general good practice is to try to reduce the spot size to a minimum. Based on the achievable size of the focal spot, industrial CT systems can be divided into several groups:

- conventional or macro CT systems with focal spots larger than 0.1 mm,
- systems with a spot size in the range of a few micrometres, referred to as microfocus (μ-CT),
- nanofocus systems with focal spot sizes lower than 1 μm (down to 0.4 μm), and
- synchrotron CT where focal spot sizes can go down to 0.2 μm or to 0.04 μm using Kirkpatrick-Baez optics (De Chiffre et al. 2014; Requena et al. 2009).

There are two major factors determining the spot size: (i) the orientation of the anode and (ii) the tube power (to which the amount of the heat transmitted by the electron beam is proportional). As can be seen from Fig. 2.8, the effective size of the focal spot can be adjusted by changing the angle α, which is the angle between

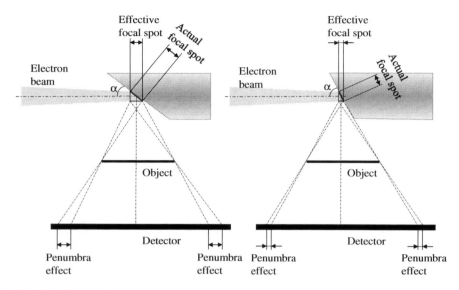

Fig. 2.8 Influence of target angle α on focal spot size and subsequently the influence of focal spot size on penumbra effect

the electron beam and the target surface. However, having this angle too large can cause backscattering of electrons and thus reduction of X-ray generation efficiency. The dependency of the focal spot size on the tube power is usually defined by manufacturers of common μ-CT systems as linear, i.e. 1 W ∼ 1 μm enlargement of the focal spot (Müller et al. 2013).

2.1.3 Interaction with the Object

When X-rays pass through an object, the intensity of the X-ray beam is reduced exponentially due to changes in the number, energy and direction of the incoming photons. This complex intensity reduction mechanism is called attenuation and is described by four primary interactions between the X-rays and the matter (Cierniak 2011; Buzug 2008; Stock 2008):

- the photoelectric effect,
- Compton (incoherent) scattering,
- Rayleigh/Thomson (coherent) scattering, and
- pair production.

Photoelectric effect

When the incident photon has a higher energy than the binding energy of the atomic electrons in the irradiated sample, the photon energy is sufficient to free an electron from a lower shell and to provide an additional kinetic energy to the ejected electron (which is often called the photoelectron). This energy transfer is described by

$$E_p = E_b + E_{pe} \qquad (2.5)$$

where $E_p = h\nu$ is the energy of the photon, E_b is the electron binding energy and E_{pe} is the kinetic energy of the free photoelectron. As the electron is ejected, the atom remains ionised until the hole in the lower shell is refilled by an outer shell electron. As a result of the energy difference between the two electron levels, a characteristic fluorescence photon is emitted. Furthermore, when the secondary energy is sufficient to eject another electron, the new free particle is called an Auger electron.

The photoelectric effect was discovered by Albert Einstein in 1905 and is a phenomenon where the incident photon is absorbed. The absorption is proportional to the atomic number Z of the irradiated material and the energy of the photon $E = h\nu$. There are different approaches to describe the relationship between the abovementioned factors (absorption, Z and E) in the literature. One approach (Buzug 2008) uses the following equation as a rule of thumb for estimating the absorption coefficient due to the photoelectric effect

$$\mu_{pe} \propto Z^4 \lambda^3 \qquad (2.6)$$

where λ is wavelength of the photon. The total cross-section, i.e. the intrinsic likelihood of the event, is described, e.g. by (Jackson and Hawkes 1981)

$$\sigma_{pe} \propto Z^5 E^{-3.5} \qquad (2.7)$$

or by (Jenkins and Snyder 1996)

$$\sigma_{pe} \propto A Z^4 E^{-3} \qquad (2.8)$$

where A is the atomic mass number. There are several other models for evaluating the absorption coefficient, however, in most models the atomic number Z has the predominant effect. Because of this strong dependency on Z, high atomic number materials are used for shielding X-ray radiation, e.g., lead ($Z = 82$) is often used.

Compton (incoherent) scattering

In case of Compton scattering, the colliding X-ray quanta lose energy and change their direction by interacting with atomic electrons. This kind of reaction occurs when a photon has considerably higher energy than the electron binding energy. The photon kicks the electron out of the atom; however, compared to the photoelectric effect, the photon is not absorbed and continues to travel with lower energy, shifted wavelength and different direction as it is deflected or scattered. The energy balance is defined as

$$E_p = E_e + E'_p \tag{2.9}$$

where E_e is the kinetic energy of the ejected electron and E'_p is the energy of the deflected photon. The recoil electron that carries part of the energy of the incident photon is called the Compton electron. The scattered photon and the Compton electron can have further collisions before leaving the matter.

The angle of scattering depends on the energy of the incoming X-ray photons. High-energy photons tend to forward scatter, whereas photons with lower energies have higher probability of backscattering, which can produce so-called ghost images on the detector. The change of wavelength is described as (Lifshin 1999)

$$\Delta\lambda = \frac{2\pi h}{m_e c} + (1 - \cos\theta) \tag{2.10}$$

where m_e is the electron mass, c is the speed of light in a vacuum and θ is the scatter angle.

The probability of Compton scattering depends, in contrast to the photoelectric effect, on the electron density and not on the atomic number of the irradiated object. The attenuation coefficient caused by Compton scattering is defined by

$$\mu_{compt} = n \cdot \sigma_{compt}, \tag{2.11}$$

where n is the electron density and σ_{compt} is the total cross-section for Compton scattering derived from the Klein-Nishina expression (Leroy and Rancoita 2004).

Rayleigh (coherent) scattering

Rayleigh (or Thompson) scattering occurs when low-energy photons interact with atomic electrons. The process of Rayleigh scattering is a consequence of the common property of electromagnetic waves, which mainly interact with electrons through their oscillating electric field. These oscillating electrons then, no matter whether they are tightly bound to an atom or not, can elastically scatter. The main difference compared to Compton scattering is that no energy is transferred and no ionisation takes place during the event. Furthermore, as Rayleigh scattering is elastic, both the incident and the scattered photons have the same wavelength; only the direction of the scattered X-ray changes. Rayleigh scattering is a strongly forward-directed process and the total cross-section is described by (Buzug 2008)

$$\sigma_{ray} = \frac{8\pi r_e^2}{3} \frac{\omega^4}{\left(\omega^2 - \omega_0^2\right)^2} \tag{2.12}$$

where r_e is the classical electron radius, ω is the oscillation frequency and ω_0 is the natural frequency of the bounded atom electrons. It is worth noting that elastic scattering is important only for low energies, whereas as the energy of X-rays is increased, the photoelectric effect, with contributions of Compton scattering, becomes predominant. Therefore, Rayleigh scattering is a relatively small effect for typical CT use.

Pair production

The effect occurring in high X-ray energy ranges (some mega-electron-volts), based on the formation of an electron-positron pair, is called pair production. The pair production process takes place inside the Coulomb field of an electron or a nucleus. The newly emerged pair mutually annihilate, and two γ-rays are emitted. The two newly emerged photons can travel in opposite directions, and be scattered or absorbed by the nucleus. However, as the range of energies for common industrial applications is well below 1 meV, the pair production contribution to the total attenuation can normally be considered negligible (Buzug 2008).

Total attenuation

The sum of the four contributions explained in this section determines the total attenuation of X-rays in the penetrated object. The total attenuation coefficient is defined as follows:

$$\mu = \mu_{pe} + \mu_{compt} + \mu_{ray} + \mu_{pp} \tag{2.13}$$

where μ_{ray} is the attenuation coefficient due to Rayleigh scattering and μ_{pp} is the attenuation coefficient due to pair production. As discussed earlier, the single attenuation contributors depend on the energy of the X-rays and the properties of the irradiated material. Therefore, it is difficult to determine generally the relative importance of the single attenuation mechanisms. The impact of the photoelectric effect, and Compton and Rayleigh scattering on the total attenuation is shown in Fig. 2.9 for different materials (carbon, aluminium and tungsten). Figure 2.9a shows that for carbon (which has the lowest atomic number among the three selected materials, $Z = 6$), Compton scattering becomes dominant for energies above 25 keV. In the case of aluminium ($Z = 13$), depicted in Fig. 2.9b, the main part of the attenuation is attributed to Compton scattering for energies higher than 50 keV. Finally, for tungsten ($Z = 74$) the photoelectric absorption is the dominant effect for a larger range of energies (Fig. 2.9c). Furthermore, the characteristic K-, L- and M-absorption edges for tungsten can be seen in Fig. 2.9c.

Beer-Lambert law of attenuation

When an X-ray beam travels through matter with an attenuation coefficient $\mu > 0$, it suffers an exponential intensity loss. Assuming the incident beam to be monochromatic and the irradiated object to be homogenous (i.e. with constant attenuation coefficient), the X-ray attenuation can be expressed by the relationship

$$\frac{\mathrm{d}I}{I(x)} = -\mu \mathrm{d}x \tag{2.14}$$

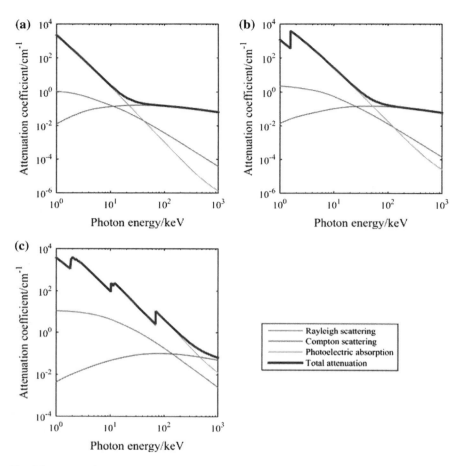

Fig. 2.9 Attenuation coefficients for **a** carbon, **b** aluminium and **c** tungsten, in relation to photon energy, with highlighted single contributors to the total attenuation. Values of attenuation coefficients were obtained from the National Institute of Standards and Technology (NIST) online database (Berger et al. 2010)

where I is the X-ray intensity and x is the distance travelled through the matter. Integrating both sides of Eq. 2.14, gives an exponential equation, known as the Beer-Lambert law:

$$I(x) = I_0 e^{-\mu x} \tag{2.15}$$

where I_0 is the incident X-ray intensity and I is the intensity after the interaction with the matter (Fig. 2.10). However, Eq. 2.15 suffers from the assumption of having a homogenous material and a monochromatic X-ray beam, which is not realistic in common industrial CT applications. Therefore, in order to consider a varying attenuation coefficient, the exponent of Eq. 2.15 must be supplemented

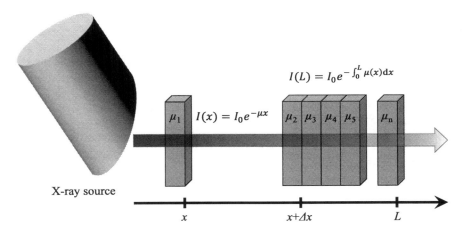

Fig. 2.10 Attenuation of X-rays by the Beer-Lambert law. Attenuation in the case of a homogenous object with an attenuation coefficient μ_1 and thickness x is depicted in the first part of the image (*left*). Attenuation for an object composed of different materials with different attenuation coefficients μ_2, μ_3... μ_n is shown in the *right part* of the figure

with the line integral along a path L in which μ varies in every point of the path (Fig. 2.10). The new equation is given by

$$I(L) = I_0 e^{-\int_0^L \mu(x)dx}. \tag{2.16}$$

In addition, to adapt the relation to a polychromatic X-ray beam spectrum, Eq. 2.16 must be modified as follows:

$$I(L) = \int_0^{E_{max}} I_0(E) e^{-\int_0^L \mu(E,x)dx} dE \tag{2.17}$$

where E is the energy of X-rays. However, for simplicity, Eq. 2.16 is usually implemented in practice, which is one of the reasons why there are beam hardening artefacts in CT reconstructions (see Chap. 5).

Phase-contrast and dark-field imaging

At the end of this section, it is worth mentioning that, while conventional transmission-based CT relies on X-ray attenuation, other CT methods, such as phase-contrast and dark-field imaging, rely on the different wavelength characteristics of X-rays. These non-conventional methods are especially useful when the attenuation contrast is not sufficient to identify structures of interest.

Phase-contrast imaging uses information on changes in the phase of an X-ray beam that travels through an object in order to produce images. The interaction with the object results in the refraction of the X-ray beam, which produces a phase shift of the incident beam, depending on the specific refraction index of the materials composing the object. The primary task of phase-contrast imaging is to convert the phase shift into a change in intensity. There are several techniques for this conversion (Pfeiffer et al. 2006; Banhart 2008):

- Interferometric methods, i.e. methods based on superposition of a reference wave and the wave that passed through the object.
- Diffraction-enhanced methods, i.e. methods using an optical element that is configured to reveal the phase gradient.
- Propagation-based methods, i.e. methods based on focused and defocused light caused by the curved phase of the beam yielding areas of higher or lower intensity.

Dark-field imaging is a scattering-based method used to enhance the contrast of tiny features within the material, such as sub-micrometre details that would be too small to be imaged by conventional transmission-based CT. Dark-field imaging takes advantage of the fact that homogenous substances cause negligible small-angle X-ray scattering, whereas specimens that have density fluctuations on the micro-scale cause strong small-angle scattering. The non-scattered signal is thereafter removed from the acquired image resulting in the dark field around the specimen (Pfeiffer et al. 2008).

Examples of X-ray images comparing conventional (attenuation-based) radiography to dark-field and phase-contrast imaging can be seen in Fig. 2.11.

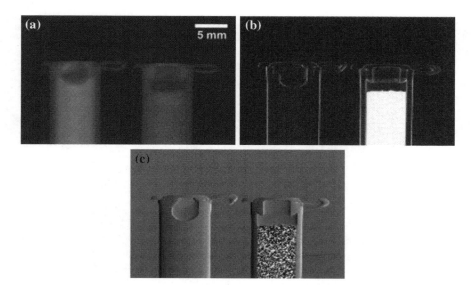

Fig. 2.11 Comparison of X-ray images of two plastic containers filled with water (*left*) and sugar (*right*) acquired by three different techniques: **a** conventional (attenuation-based) X-ray image, **b** dark-field image and **c** phase-contrast image (Pfeiffer et al. 2009)

2.2 Signal Detection and Processing

The workflow of signal detection and processing in CT is shown in a simplified form in Fig. 2.12. In the X-ray CT process chain, the detection step is carried out by means of physical detectors, which capture the attenuated X-ray beam and convert it into electrical signals, which are subsequently converted into binary coded information. As depicted in Fig. 2.12, after several processing steps, the results are available for final visualisation and interpretation on a computer. The first part of this section concerns the detectors, while the latter part includes some basics of image processing.

2.2.1 X-ray Detectors

As explained in the Sect. 2.1, in conventional attenuation-based CT, the X-rays are attenuated due to absorption or scattering during their travel through the material. The attenuation is then measured by capturing the attenuated X-rays using an X-ray detector. Over the years, there has been a considerable improvement in the detection technology. Interestingly, the various generations of CT systems are sometimes referred to by the different type/configuration of detectors (Hsieh 2009; Panetta 2016); more details about the different CT generations can be found in Chap. 1.

The detector captures the X-ray beam and converts it into electrical signals. This conversion of X-ray energy to electrical signals is primarily based on two principles: gas ionisation detectors and scintillation (solid-state) detectors. A schematic representation of both detection principles is shown in Fig. 2.13. Gas ionisation detectors convert X-ray energy directly into electrical energy, whereas scintillation detectors convert X-rays into visible light and then the light is converted into electrical energy; a detailed description of each type follows.

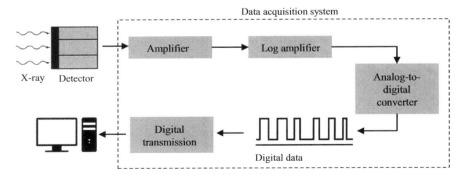

Fig. 2.12 Signal detection and processing in a CT system. Adapted from (Seeram 2015)

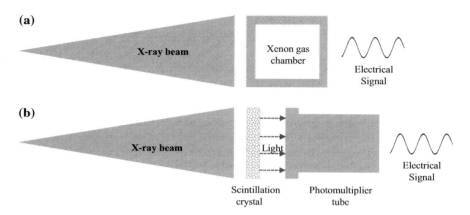

Fig. 2.13 Schematic representation of the capture of an X-ray beam by **a** gas detector and **b** solid-state detector

Gas ionisation detectors

Gas ionisation detectors (also termed gas detectors) were first introduced in the third generation scanners (Chap. 1 discusses the various generations). Gas ionisation detectors utilise the ability of X-rays to ionise gases (xenon). When the X-rays fall on the gas chamber, a positive ion (Xe^+) and an electron (e^-) are generated as a result of the photoelectric interaction; the positive ion and the electron are attracted to a cathode and anode respectively, see Fig. 2.14a. This migration of the positive ion and the electron introduces a small signal electrical current, which varies with the intensity of the X-ray beam, i.e. the number of photons. In the basic configuration of a gas detector, there is a series of individual gas chambers separated by tungsten plates, which act as a cathode as shown in Fig. 2.14b (Seeram 2015).

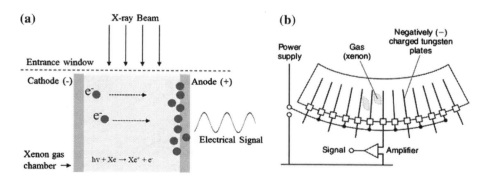

Fig. 2.14 **a** The principle of gas detector and **b** basic configuration of a gas ionisation detector in a series arrangement. Adapted from (Seeram 2015)

Scintillation (solid-state) detectors

Scintillation detectors, or solid-state detectors, are used in almost all modern industrial CT systems. A solid-state detector consists of scintillation crystals and a photon detector. When the X-rays strike the scintillator crystals, they are converted into long-wave radiation (visible light) within the scintillation media (see Fig. 2.15a). The light is then directed to an electronic light sensor such as a photomultiplier tube (Fig. 2.15b). The photomultiplier absorbs the light emitted by the scintillator and ejects electrons via the photoelectric effect. As illustrated in Fig. 2.15b, electrons are released when the light strikes the photocathode and then the electrons cascade through a series of dynodes, maintained at different potentials, to result in an output signal (Seeram 2015).

The choice of scintillator material is critical and depends on the desired quantum efficiency for the X-ray-to-light conversion, and on the time constant for the conversion process, which determines the "afterglow" of the detector (i.e. the persistence of the image even after turning off the radiation, as described in the following section). Modern sub-second scanners require a very small time constant (or very fast fluorescence decay); for such needs ceramic materials made of rare earth oxides are used. Typical materials for scintillators are sodium iodide doped with thallium caesium iodide cadmium tungstate ($CdWO_4$), zinc sulphide (ZnS), naphthalene, anthracene and other compounds based on rare earth elements, or gadolinium oxysulfide (Gd_2O_2S) (Buzug 2008; Fuchs et al. 2000; Cierniak 2011).

Detector characteristics

As with any interaction of X-rays with matter, the previously-discussed phenomena occur when the X-rays come into contact with the detector material. The mass attenuation coefficient of the detector material must be known in order to assess the quality of the detector material with respect to the desired behaviour (high quantum efficiency). As illustrated in Fig. 2.16, Gd_2O_2S (a scintillation

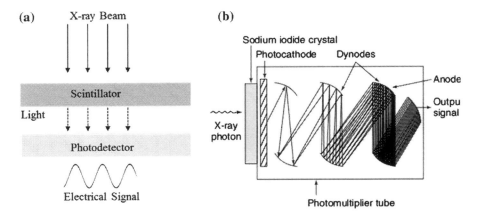

Fig. 2.15 **a** The principle of a scintillator detector and **b** schematic view of a photomultiplier tube. Adapted from (Seeram 2015)

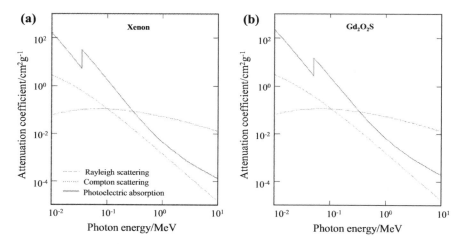

Fig. 2.16 Mass attenuation coefficient for the main photon-matter interaction principles in detector material; values of attenuation coefficients were obtained from the NIST online database (Berger et al. 2010)

detector) has a higher probability of energy conversion through photoelectric absorption than xenon gas; thus, it possesses higher quantum efficiency (defined later).

Several characteristics describe the quality of the results achieved by a detector. The most important of these characteristics are: detection efficiency, stability over time, energy resolution, afterglow, response time and dynamic range (Cierniak 2011).

The overall detection efficiency is primarily determined by the geometric efficiency (fill factor) and the quantum efficiency (capture efficiency). The geometric efficiency refers to the X-ray sensitive area of the detector as a fraction of the total exposed area, and the quantum efficiency refers to the ratio of the incident quanta that are absorbed to the quanta that contribute to the signal (Cunningham 2000). Scintillator detectors have higher quantum efficiencies than gas (xenon) detectors due to the fact that the atomic number of the materials used for scintillators are much larger than for gases. However, this advantage can be somewhat balanced out by the losses due to the two steps (X-ray-to-light and light-to-electrical signal) conversion in solid-state detectors.

The stability over time refers to the steadiness of the detector response. Scintillation detectors tend to vary their properties during irradiation, which subsequently results in a change in transfer characteristics. However, once the irradiation is stopped, the semiconductor crystals can return to their original state after some time, which varies from minutes to hours corresponding to the type of material. In contrast to this, gas detectors are insensitive to the irradiation.

Energy resolution is the ability of the detector to resolve the energy of the incoming radiation, which is described as the full width at half maximum (FWHM)

Fig. 2.17 Determination of full width at half maximum (FWHM)

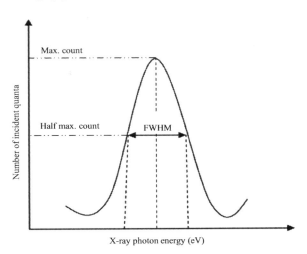

of the detector's transfer characteristic as a function of the incident X-ray photon energy (see Fig. 2.17). The X-ray frequency at which the maximum detection occurs is found and the maximum value is determined. Half of the maximum value is the reference level at which the width of the transfer function is estimated, which in turn determines the energy resolution of the detector. The smaller the width, the higher the energy resolution of the detector and the less image distortion caused by a polychromatic X-ray beam.

Afterglow refers to the persistence of the image even after turning off the radiation. To detect fast changes in the transmitted intensity of X-ray beams, crystals are required to exhibit extremely low afterglow, e.g., less than 0.01% after 100 ms, when the radiation is terminated (Kalender 2011).

The response time of the detector is the speed at which the detector can detect an X-ray event and recover itself to detect the successive event. The response time should be as short as possible (typically in microseconds) in order to avoid problems related to afterglow and detector "pile-up" (Seeram 2015).

Dynamic range is the ratio of the largest signal to the smallest signal which can be measured, e.g. if 1 μA and 1 nA are the largest and smallest signals respectively, the dynamic range would be 10^6 to 1 (Parker and Stanley 1981).

Multi-slice CT detectors

Over the years, the major concern of the CT manufacturers has been increasing the number of detection elements simultaneously illuminated by the X-rays, thus improving the scanning speed. Widening the solid angle of the detected X-ray beam enables faster acquisition and improves the thermal efficiency of the X-ray tube, which is due to the larger fraction of the radiation used in imaging. Therefore, increasing the number of detector slices/rows results in competition amongst manufacturers. The evolution of X-ray detectors has been well summarised by Panetta (Panetta 2016).

Scintillation detectors are extendable to multi-row or multi-slice detectors; whereas high-pressure gas detector arrays are not easily extendable to flat area detector systems. Therefore, modern CT systems are equipped with detectors that are mainly based on solid-state scintillation detectors (Buzug 2008). As shown in Fig. 2.18, multi-slice detectors are available in two geometries, either curved or flat resulting from a single slice (curved or linear). Flat-panel detectors (FPD) are mainly used in industrial CT scanners (e.g. μ-CT systems); in contrast, clinical CT systems are normally equipped with curved or cylindrical detectors.

A typical composition of digital FPD based on scintillator detector technology is presented in Fig. 2.19a. Each element consists of a photodiode and a thin-film transistor (TFT) made of amorphous silicon on a single glass substrate. The pixel matrix is coated with an X-ray sensitive layer of caesium iodide (CsI) by a physical deposition process; the final coating with CsI is the required scintillator layer. A thin aluminium cover layer is used as a reflector and a carbon cover protects from mechanical damage. Multi-chip modules are used as read-out amplifiers at the edge of the detector field. A number of physical and chemical processing steps, similar to those used in semiconductor production, are employed to achieve these finely structured detector elements. This results in a high fill factor, which is desirable and cannot be achieved by a simple combination of the basic detector types. Fill factor is the ratio of the X-ray sensitive area of the detector to the total area of the detector, thus

$$\text{fill factor}(f) = \frac{\text{X-ray sensitive area of the detector}}{\text{total area of the detector } (x \times y)} \qquad (2.18)$$

as illustrated in Fig. 2.19b (Buzug 2008). More details about various detector configurations and their effects on image quality are discussed in Chap. 3.

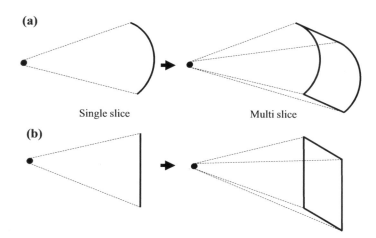

(a)

Single slice   Multi slice

(b)

Fig. 2.18 Detector configurations from single slice to multi-slice: **a** curved and **b** linear

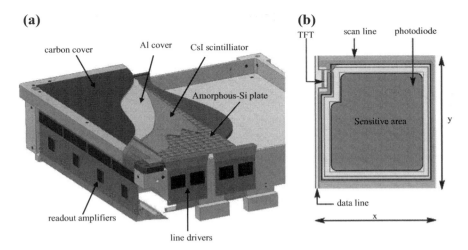

Fig. 2.19 **a** Composition of a digital flat-panel detector (*Courtesy* General Electric CT Systems) and **b** description of the sensitive area for a single element (which determines the fill factor as in Eq. 2.1) (Buzug 2008)

2.2.2 *Image Processing in CT*

In this section, the basics of image digitisation and processing are explained. Reconstruction can also be considered a part of image processing, but the details of reconstruction are covered separately in Sect. 2.3.

As defined by Castleman (1996), digital image processing is "subjecting numerical representations of objects to a series of operations in order to obtain a desired result". Over the years, image processing has emerged as a discipline in itself; therefore, it is not possible to cover the details within this limited scope. Only the concepts and techniques used in the CT field are included here.

Characteristics of a digital image

A digital image is a numerical image arranged in such a manner that the location of each number (pixel) is identifiable using a Cartesian coordinate system; see the (x, y) coordinate system in Fig. 2.20. The x and y axes describe the rows and columns placed on the image respectively. As an example, the spatial location (9, 4) in Fig. 2.20 identifies a pixel located nine pixels to the right of the left-hand side and four pixels down from top of the image. Such an arrangement is said to be in the spatial location domain. In general, images can be represented in two domains based on their acquisition: spatial location domain and spatial frequency domain. Radiography and CT acquire images in the spatial location domain, while magnetic resonance imaging (MRI) acquires images in the spatial frequency domain. The term spatial frequency refers to the number of cycles per unit length, i.e. the number of times a signal changes per unit length. Within an object, small structures produce

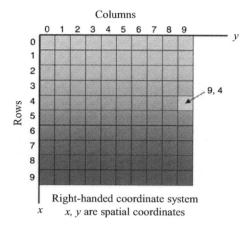

Fig. 2.20 A right-handed coordinate system used to digitally describe an image in spatial location domain. Adapted from (Seeram 2015)

high spatial frequencies whereas large structures produce low spatial frequencies, and they represent detail and contrast information in the image respectively.

The fundamental parameters to describe a digital image are: matrix, pixels, voxels and bit depth (Seeram 2015). A digital image is a made up of a 2D array of numbers termed a matrix. The matrix consists of columns (M) and rows (N) that define small regions called picture elements or pixels. The columns and rows describe the dimensions of the image, and the size (in bits) is given by the following equation:

$$size = M \times N \times 2^k \tag{2.19}$$

where k is the bit depth. When the number of columns and rows are equal, the image is square in shape; otherwise the shape is rectangular. The operator can select the matrix size, which is referred to as the field of view (FOV).

Pixels are the picture elements that make up the matrix and generally are of square shape. Each pixel is characterised with a number (discrete value) representing a brightness value. The numbers represent material characteristics, e.g. in radiography and CT, these numbers are related to the attenuation coefficient of the material (as explained in Sect. 2.1), while in MRI they represent other characteristics of the material, such as proton density and relaxation time. The pixel size can be calculated according to the following relationship:

$$Pixel\ size = FOV/Matrix\ size. \tag{2.20}$$

For digital imaging modalities, the larger the matrix (for the same FOV), the smaller the pixel size and the better the spatial resolution.

Voxel (volumetric pixel) is a 3D pixel that represents the information contained in a volume of material in the imaged object. Voxel information is converted into numerical values contained in the pixels, and these numbers are assigned brightness levels, as shown in Fig. 2.21. The numbers represent signal intensity from the detectors. A high number represents a high intensity and low number represents a low intensity; which are shaded bright and black dark respectively.

The number of bits per pixel is defined as bit depth. As shown in Eq. 2.19, k is the bit depth and is used as an exponent to 2, which is the base of the binary number system. E.g., a bit depth of 8 implies that each pixel has 2^8 (256) grey levels or shades of grey. The effect of bit depth on the image quality can been seen in Fig. 2.22. The results in terms of image quality improve with increasing bit depth. Bit depth affects the number of shades of grey, i.e. the contrast resolution of the image (Seeram 2015).

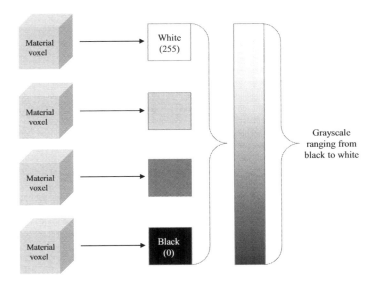

Fig. 2.21 Voxel information converted into numerical values (for a bit depth of 8). Adapted from (Seeram 2015)

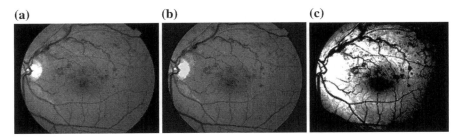

Fig. 2.22 Image quality as a function of bit depth: **a** 8-bit (256 levels), **b** 4-bit (16 levels) and **c** 1-bit (2 levels) (*Image courtesy* http://eye-pix.com/resolution/)

Sinograms

As already discussed, a CT dataset is a set of images acquired over an angular range. Sinogram space is one of the most popular ways to represent the projections. As shown in Fig. 2.23, a sinogram is formed by stacking all of the projections acquired at different angles; in other words, a set of projection data for a given slice through an object is called a sinogram. A sinogram is presented as an image where the number of lines corresponds to the number of projections taken; each line represents a projection at a given angle and the number of columns is equal to the width of the detector. The reconstruction of a slice is typically carried out from the data recorded in a sinogram.

In order to locate the projection of a point in the object specified by its polar coordinate (r, ϕ) in the sinogram space, a rotating coordinate system (x', y') with its y' axis parallel to the X-ray beam is defined (Fig. 2.23). The x' coordinate of the point (and, therefore, its location on the projection) follows the following relationship:

$$x' = r \times \cos(\varphi - \beta) \qquad (2.21)$$

where β is the projection angle formed with the x axis. Equation 2.21 shows that the projection of a single point as a function of the projection angle results in a sinusoidal curve in sinogram space. Any object can be approximated by a collection of points located in space and, for such an object, the sinogram space consists of several overlapped sine or cosine curves. Figure 2.24 shows the cross section of a body phantom and its sinogram. The high-intensity curves near the centre (left to right) of the sinogram correspond to the projections formed by the phantom itself, and the low-intensity curves across the left-right span of the sinogram are formed

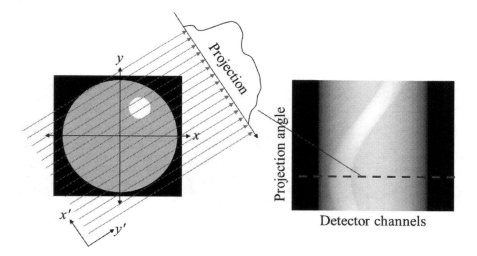

Fig. 2.23 Illustration of mapping between the object space and the sinogram

(a) (b)

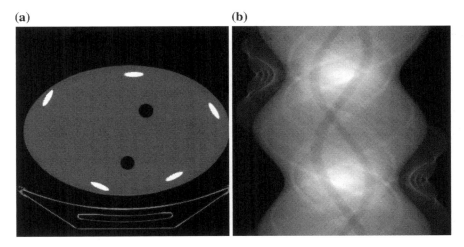

Fig. 2.24 **a** The body phantom and **b** its sinogram (Hsieh 2009)

by the projections of the table underneath the phantom. The two air pockets located near the centre of the phantom are clearly visible in the sinogram as two dark sinusoidal curves near the centre, and the five high-density ribs near the periphery of the phantom are depicted as bright sinusoidal curves.

A sinogram is also useful for analysing the projection data and to detect abnormalities in a CT system. E.g., a defective detector channel manifests itself as a vertical line in a sinogram, since data collected by a single detector channel maintain a fixed distance from the iso-centre (centre of rotation), over all the projection angles. In a similar manner, a temporary malfunction of the X-ray tube produces a disrupted sinogram with horizontal lines, since each horizontal line in the sinogram corresponds to a particular projection view and, therefore, relates to a particular time instance (Hsieh 2009).

Image filtering

Image filtering is a general term, which corresponds to any kind of operation applied to pixels in an image. The common uses of image filtering include suppressing noise, smoothing, edge enhancement and resolution recovery (Lyra and Ploussi 2011; Van Laere et al. 2001).

In the field of CT, filtered back projection (FBP) is the most commonly used reconstruction technique; details of the reconstruction techniques will be explained in the following section. The filtered back projection (FBP) is typically accompanied by a filtering step. The projection profile is filtered to remove typical star-like blurring that is present in simple back projection (Fig. 2.25); more details are given in Sect. 2.3.

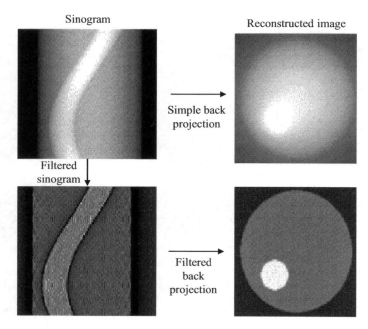

Sinogram   Reconstructed image

Filtered
sinogram

Simple back
projection

Filtered
back
projection

Fig. 2.25 Illustration of filtering step in the reconstruction

Filtering techniques

Filtered back projection is accomplished by combining back projection with a ramp filter. Low spatial frequencies are mainly responsible for the blur caused in unfiltered images. Aliasing of the frequencies that cause the blur is carried out by using a high pass filter. The mathematical function of a ramp filter in the frequency domain is given by

$$H_R\left(K_x, K_y\right) = k = \left(K_x^2 + K_y^2\right)^2 \qquad (2.22)$$

where K_x and K_y are spatial frequencies.

The ramp filter is a compensatory filter as it eliminates the star artefact resulting from simple back projection. A high pass filter increases the edge information of the object, but amplifies the statistical noise present in the image. Therefore, the ramp filter is always combined with a low pass filter in order to remove high spatial frequency noise from a digital image. Low pass filters or smoothing filters block the high frequencies and allow the low frequencies to be remained unaltered. Commonly used low pass filters are: Shepp-Logan, Hanning, Hamming, guided image filtering, median filtering and neighbourhood sliding filtering (Lyra and Ploussi 2011; Ziegler et al. 2007).

The low pass filters are characterised by "cut-off frequency" and "order" which determine the frequency above which noise is eliminated and the slope of the filter respectively. The filter function is defined to be zero for frequencies above the cut-off frequency. A high cut-off frequency results into improved spatial resolution but the image remains noisy. On the other hand, a low cut-off frequency increases smoothing and lowers the image contrast (Lyra and Ploussi 2011).

Shepp-Logan filter

The Shepp-Logan filter is categorised as a low pass filter, which has high resolution and produces the least smoothing (Gui and Liu 2012). The Shepp-Logan filter can be represented mathematically as

$$S(f) = \frac{2f_m}{\left[\pi\left(\sin|f|\frac{\pi}{2f_m}\right)\right]}. \tag{2.23}$$

In Eqs. 2.23–2.25, f and f_m are the spatial frequencies and the cut-off or critical frequency of the image.

Hanning filter

The Hanning filter is also a low pass filter, which is characterised by the cut-off frequency. It is defined by the following equation in the frequency domain

$$H(f) = \begin{cases} 0.50 + 0.50\cos\left(\frac{\pi f}{f_m}\right), & 0 \le |f| \le f_m \\ 0, & \text{otherwise} \end{cases}. \tag{2.24}$$

The Hanning filter is highly effective in image noise reduction as it reaches the zero relatively quickly. However, this filter is not able to effectively preserve edges (Riederer et al. 1978).

Hamming filter

The Hamming filter is another kind of low pass filter, which is able to offer a high degree of smoothing, similar to the Hanning filter. The amplitude at the cut-off frequency is the only difference between Hamming and Hanning filters. The mathematical representation of Hamming filter is

$$H(f) = \begin{cases} 0.54 + 0.46\cos\left(\frac{\pi f}{f_m}\right), & 0 \le |f| \le f_m \\ 0, & \text{otherwise} \end{cases}. \tag{2.25}$$

Guided image filtering

In the guided image filtering approach proposed by He et al. (2013), the filter input or another image is used as a guidance image. The output of the guided filter is locally a linear transform of the guidance image. A general linear translation-variant filtering process is defined, which involves a guidance image I, an input image p, and an output image q. The filtering output at a pixel i is expressed as a weighted average

$$q_i = \sum_j W_{ij}(I)p_j \qquad (2.26)$$

where i and j are pixel indexes. The filter kernel W_{ij} is a function of I and independent of p.

The guided filter has edge preserving and smoothing properties similar to the bilateral filter but does not suffer from unwanted gradient reversal artefacts (Bae et al. 2006; Durand and Dorsey 2002), which are evident when using bilateral filters. The guided filter is used for various applications, which include image smoothing/enhancement, high dynamic range imaging, haze removal and joint upsampling (He et al. 2013).

Median filtering

Median filtering is used to remove the impulse noise (also called salt-and-pepper noise), in which a certain percentage of individual pixels are randomly digitised into two extreme intensities, maximum and minimum. A median filter replaces each pixel value by the median of its neighbours, i.e. the value such that 50% of the values in the neighbourhood are above, and 50% are below. However, median filtering is generally very good at preserving edges. Median filtering is a nonlinear method, which performs better than linear filtering for removing noise in the presence of edges (Arias-Castro and Donoho 2009).

Neighbourhood sliding filtering

A neighbourhood sliding filter is highly useful in noise suppression in the reconstructed image. In this filtering process, images are filtered by sliding neighbourhood processing in which a mask is moved across the image. The mask is a filtering kernel made up of weighting factors (the values of which depend on the design), which can either attenuate or accentuate certain properties of the image that is being processed (Lukac et al. 2005).

2.3   Reconstruction

Image reconstruction is a necessary step for performing tomographic analysis, and in the CT measurement chain, it follows the projections acquisition and image processing step (see Sect. 2.2), as described in Fig. 2.1. Image reconstruction provides the link between the acquired radiographic images and the 3D representation of the object under investigation. In this section, the fundamentals of image reconstruction and the basics of the main reconstruction algorithms are presented.

By intention, a rigorous derivation of the mathematical formulae is not provided; for more details the reader can refer to Hsieh (2009) and Buzug (2008).

2.3.1 Concept of Reconstruction

After the X-ray projections are acquired, a reconstruction process is performed, by means of mathematical algorithms, in order to obtain the 3D volume (or a 2D slice as will be described in the following section) of the object, which describes its internal and external structure. The reconstructed 3D volume consists of voxels, the volumetric representation of pixels.

From the tomographic point of view, the object that has to be reconstructed is a 2D distribution of a function $f(x, y)$, which represents the linear attenuation coefficients of the object. The necessary information to recover this function, the 2D images, are the projection data (see Sect. 2.2.2). In order to understand the principle on which reconstruction from projections is based, the kind of information that is contained in an X-ray projection should first be analysed.

As discussed in Sect. 2.1.3, the X-rays passing through the object are attenuated according to Lambert-Beer law (Eq. 2.15). The X-ray detector records attenuated intensity values after the X-ray beam penetrates the object. The resultant intensity value of each detector pixel is a function of the attenuation coefficient μ and the path followed by the X-rays. The reader should note that in this section, for an easier mathematical treatment, the Lambert-Beer law will be used in its simplified form (Eq. 2.15), valid for the monochromatic case and with the assumption of a homogenous material.

In Fig. 2.26, an object is represented for the sake of simplicity in a 2D view by a circle (section of the object, orthogonal to the rotary axis). Considering an X-ray passing through the object following the path L, the line integral is obtained by writing Eq. 2.14 in the differential form, and then integrating along the path L

$$p_L = \ln(I_0/I) = \int_0^L \mu(s)ds \qquad (2.27)$$

where $\mu(s)$ is the attenuation coefficient at position s along ray L, I_0 is the input X-ray intensity and I is the output X-ray intensity.

Fig. 2.26 Representation of the contribution of each voxel to the total attenuation along the ray path L. According to Lambert-Beer law X-rays are attenuated from the input intensity value I_0 to the output intensity value I

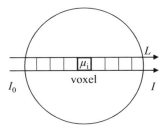

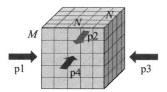

Fig. 2.27 Schematic representation of the problem of tomographic reconstruction. Reconstructing an object means assigning the attenuation coefficient to each voxel of the cubic matrix, by knowing the line integrals recorded by the detector at different angular positions. In this example, p1, p2, p3, p4 represent four projections acquired at different rotation angles

Along path L, each voxel with attenuation coefficient μ_i contributes to the total attenuation. Therefore, each X-ray projection describes the line integral of the distribution of attenuation coefficients, along a particular X-ray path. The purpose of tomographic reconstruction is to assign the correct value of μ to each voxel, knowing only the line integrals p_L; i.e. the attenuated intensity values recorded by the detector, for a different number of orientations of the lines L that are the projections acquired at different angular positions (Fig. 2.27). In other words, considering a detector with $M \times N$ pixels, the reconstruction process assigns the attenuated intensity value to the matrix of $M \times N \times N$ voxels.

The theoretical foundation for reconstructing tomographic images was established in 1917 by Radon, an Austrian mathematician who provided a mathematical solution to uniquely determine the value of a function knowing its line integrals (Radon 1917). The reconstruction of this function is performed by applying an inverse transform, referred to as inverse Radon transform, to recover the function from its set of line integrals (i.e. the Radon transform). In 1963, Cormack proposed a mathematical method for reconstructing images using a finite number of projections (Cormack 1963). Cormack's theory turned into a real application in 1971 when the first CT scanner was developed (see Chap. 1).

There are several techniques for performing image reconstruction, which can be classified into analytical, algebraic and statistical reconstruction algorithms. Analytical techniques model the object as a mathematical function, thus, reconstruct the object by solving a continuous integral equation. Analytical techniques are divided into exact and non-exact algorithms, depending on whether or not the solution of the integral equation is exact. Several analytical techniques are present in the literature (Hsieh et al. 2013; Katsevich 2003; Grangeat 1990). Algebraic reconstruction techniques (ARTs) make use of iterative reconstruction approaches in which several iterations are performed until certain criteria are met (Herman et al. 1991; Buzug 2008). Statistical reconstruction algorithms are also iterative methods, but in this case the unknowns are assigned by means of likelihood principles (Buzug 2008; Herman and Meyer 1993).

Sections 2.3.2–2.3.4 present the mathematical concepts on which analytical techniques are based, and provide a description of filtered backprojection (FBP) and FBP-type algorithms. The most commonly used non-exact algorithm in practice, the Feldkamp Davis Kress (FDK) algorithm, is presented. Section 2.3.5 gives an overview of the principles of algebraic and statistical reconstruction techniques and provides the main advantages and disadvantages compared to conventionally used FBP algorithms. The reader should be aware that the presentation is introductory and, therefore, far from being complete; it is also focused more on FBP-type algorithms. The full derivation of mathematical models can be found elsewhere (Hsieh 2009; Buzug 2008; Kak and Slaney 2001; Feldkamp et al. 1984).

2.3.2 Fourier Slice Theorem

The mathematical basis on which tomographic reconstruction relies is the Fourier slice theorem, also known as the central slice theorem. For introducing the theoretical principles, the case of a 2D image reconstruction and a parallel-beam geometry is presented (Fig. 2.28). Even though the parallel-beam geometry is not the most used geometry on modern industrial CT systems, the results presented here provide the basis for understanding the more complex situations of fan-beam geometry and cone-beam geometry (3D image reconstruction), which are discussed in the following sections.

In the coordinate system defined in Fig. 2.28, the object is represented by the object function $f(x, y)$ and its Radon transform is given by

$$p(\vartheta, \varepsilon) = \int\limits_{-\infty}^{+\infty} f(x,y)\delta(x\cos\theta + y\sin\theta - \varepsilon)dxdy \qquad (2.28)$$

where δ is the Dirac delta function.

The Fourier slice theorem states that the Fourier transform of a parallel projection of an object $f(x, y)$ obtained at angle θ describes a radial line, in the Fourier

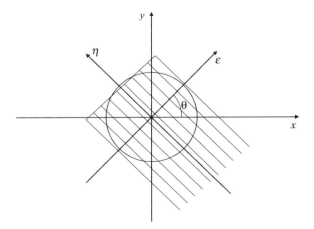

Fig. 2.28 Coordinate system used for the tomographic reconstruction in parallel-beam geometry. The object, which in this case for simplicity is represented as a circle in the 2D view, is described by the function $f(x, y)$

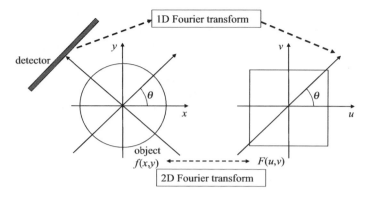

Fig. 2.29 Representation of the Fourier slice theorem

space of the object, taken at the same angle (Fig. 2.29). In other words, if $P_\theta(\omega)$ is the 1D Fourier transform of Eq. 2.28, where ω is the frequency component, $F(\theta, \omega)$ is the 2D Fourier transform of the object

$$P_\vartheta(\omega) = F(\vartheta, \omega). \tag{2.29}$$

The concept expressed in Eq. 2.29 means that, if a sufficient number of projections is collected, the entire Fourier space can be filled by performing the 1D Fourier transform of each parallel projection at angle θ. The object can then be reconstructed by applying the 2D inverse Fourier transform.

Although the Fourier slice theorem provides a straightforward procedure for tomographic reconstruction, there are some difficulties in the practical implementation. After applying the 1D Fourier transform to each projection, the Fourier space is filled on a polar coordinate grid. The fast Fourier transform (FFT), however, requires data on a Cartesian grid. For performing the 2D inverse Radon transform by means of a FFT algorithm, a regridding process is required in which projection data are rearranged from a polar to a Cartesian grid through interpolation. However, interpolation in the frequency domain is not as straightforward as in the spatial domain, and is difficult to implement. To overcome this problem, alternative implementations of the Fourier slice theorem have been proposed (Stark et al. 1981; O'Sullivan 1985; Edholm 1988). The most popular implementation is the FBP.

2.3.3 Filtered Backprojection

The FBP algorithm, is the most widely used reconstruction method, from which is derived the most implemented non-exact algorithm, the FDK algorithm which will be discussed later. The concept of simple backprojection is illustrated in Fig. 2.30. In order to explain the principle of backprojection, an ideal representation, with a simple object consisting of five spheres, and three projections (p1, p2 and p3 in Fig. 2.30) taken at different angles (θ) is considered. Each projection describes the distribution of the attenuation coefficient for the given X-ray path (Fig. 2.30a). According to the backprojection principle, every profile is backprojected along the viewing direction (i.e. the θ angle) from which it was acquired (Fig. 2.30b). In the resulting grid, each contribution is added to each voxel intersecting that ray direction. Of course, in real cases, the situation will be more complex than the simplified one given in Fig. 2.30.

From Fig. 2.30 it is clear how the resultant intensity value of each voxel of the grid is given by the sum of non-negative values corresponding to each projection. Due to the fact that each projection is a non-negative function, positive values are also assigned to voxels that do not contain the object. This leads to a blurred image that is not of sufficient quality. Figure 2.31 represents the result of the simple backprojection of a point object. The superimposed profiles produce a central spike with a broad skirt that falls off as $1/r$.

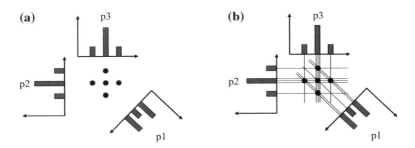

Fig. 2.30 Concept of projection and backprojection. **a** Projection: each radiographic projection describes the distribution of attenuation coefficients along the given ray path. **b** Backprojection: every radiograph is backprojected along the viewing direction from which it was acquired. The *number of lines* depicts the amount of absorption. A grid is generated, in which each contribution is added to the voxel intersecting that ray direction

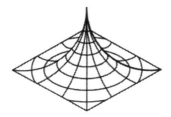

Fig. 2.31 Simple backprojection of a point object. During the reconstruction process, positive values are assigned to each voxel. This leads to a blurred image with a central spike and a broad skirt that falls of as $1/r$. Adapted from (ISO 15708-1, 2002)

Due to resulting image blurring, simple backprojection is not sufficient, as it alters the morphology of the image. The implementation of the FBP reconstruction algorithm solves the problem. FBP involves convolving each object projection with a filtering function, before calculating the inverse Fourier transform to recover the object. Mathematically, the image reconstruction is given by

$$f(x,y) = \int_0^\pi d\vartheta P(\vartheta, \varepsilon) * k(\varepsilon)\big|_{\varepsilon = x\cos\vartheta + y\sin\vartheta}, \qquad (2.30)$$

where $k(\varepsilon)$ is the convolution kernel. The convolution kernel basically consists of a high pass filter that preserves the response of the detector but adds negative tails in order to even out the positive contributions outside the object. In Fig. 2.32, it is shown how, for the same point object as in Fig. 2.31, the convolution with the filtering function removes the blurring caused by the backprojection process. The properties of the filter kernel are very important for image reconstruction. For a detailed description of the influence of different filter kernels on the reconstructed images the reader can refer to Hsieh (2009).

Fan-beam filtered backprojection

In the previous sections, the Fourier slice theorem and the FBP method were presented in the case of a parallel beam geometry. With a divergent beam configuration (i.e. the rays are not parallel lines), such as for fan-beam and cone-beam geometries, the relationship found in Eq. 2.29 becomes more complex.

In practice, there are two possible designs for fan-beam geometries: the equiangular fan-beam and the equal-spaced fan beam (Fig. 2.33).

Figure 2.34 describes the geometrical relationship for equiangular and equal-spaced fan beam geometries. In equiangular fan-beam, any fan-beam ray can be uniquely specified by two parameters: γ which is the angle formed by the ray with the iso-ray, and β which is the angle the iso-ray forms with the y axis. In equal-spaced fan-beam, a ray can be specified uniquely by the parameter set t which

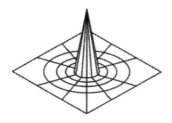

Fig. 2.32 Filtered backprojection of the point object backprojected in Fig. 2.30. The blurring caused by the backprojection process is removed by the convolution with the filtering function. Adapted from (ISO 15708-1 2002)

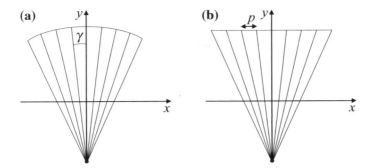

Fig. 2.33 Schematic representation of equiangular and equal-spaced fan-beam geometries. **a** In equiangular geometries, the detector is made of several modules located on an arc. The angular space γ between the modules is constant. **b** In equal-spaced geometries, the detector is flat and the sampling elements are distributed at a constant distance p along the surface

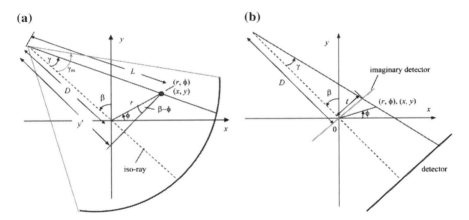

Fig. 2.34 Geometrical relationships for **a** equiangular and **b** equal-spaced fan-beam geometries. The iso-ray is an imaginary ray that connects the X-ray source with the iso-centre. Adapted from (Hsieh 2009)

is the distance between the iso-centre and the intersection of a ray, and β which is defined in the same way as for the equiangular fan-beam.

Starting from Eq. 2.30, derived for the parallel-beam case, the reconstruction formulae for the equiangular and equal-spaced fan beam algorithms are given respectively in Eqs. 2.31 and 2.32 in polar coordinates. For the derivation of the formulas the reader is referred to Hsieh (2009). The terms of the two equations are explained in Fig. 2.34. The equiangular case is given by

$$f(r, \varphi) = \frac{1}{2} \int\limits_{0}^{2\pi} d\beta \int\limits_{-\gamma_m}^{\gamma_m} q(\gamma, \beta) h(L \sin(\gamma' - \gamma)) D \cos \gamma d\gamma, \qquad (2.31)$$

where the equiangular fan-beam projection is denoted by $q(\gamma, \beta)$. The equal-spaced case is given by

$$f(r, \varphi) = \frac{1}{2} \int\limits_{0}^{2\pi} U^{-2}(R, \varphi, \beta) d\beta \int\limits_{-\infty}^{+\infty} \left(\frac{D}{\sqrt{D^2 + t^2}} \right) q(t, \beta) h(t' - t) dt. \qquad (2.32)$$

where

$$U(r, \varphi, \beta) = \frac{D + R \sin(\beta - \varphi)}{D}. \qquad (2.33)$$

and the equal-spaced fan-beam projection is denoted by $q(t, \beta)$. Comparing Eq. 2.31 with 2.30, there are two main differences. The first difference consists of a multiplication of a factor $D\cos\gamma$ prior to performing the filtering operation. The second difference is that, when dealing with fan-beam geometries a weighted backprojection is used, with a weighting factor that depends on the distance from the reconstructed pixel to the X-ray source.

Another way of performing fan-beam reconstruction, is to apply directly the equation derived in the parallel-beam case (Eq. 2.30) after performing a rebinning process. The rebinning process consists of reformatting the fan-beam projections into parallel-beam projections. By means of geometry, a set of fan-beam projections acquired with an equiangular sampling geometry are reformatted into parallel-beam projections $p(\theta, s)$ according to

$$p(\vartheta, s) = g\left(\theta - \arcsin\frac{s}{R}, \arcsin\frac{s}{R} \right) \qquad (2.34)$$

where θ is the angle formed by the parallel-beam ray and the y axis in Fig. 2.34, and s is the length of the perpendicular segment from the origin of the coordinate system to the ray taken into consideration.

In the case of equal-spaced sampling, parallel-beam projections $p(\theta, s)$ are obtained as:

$$p(\vartheta, s) = g\left(\theta - \arcsin\frac{s}{R}, \frac{sR}{\sqrt{R^2 - s^2}} \right). \qquad (2.35)$$

For the rebinning process, interpolation is always necessary. Rebinning negatively affects the spatial resolution of the reconstructed image. On the other hand, the rebinning process improves the noise uniformity in the reconstructed image.

Cone beam filtered backprojection

The methods discussed so far, dealt with 2D image reconstruction. However, for many applications, a 3D representation is necessary. Many of today's industrial and clinical CT systems rely on cone-beam geometries because of the several advantages provided (more details about the cone-beam geometry are given in Chap. 3). For example, when scanning an object of 100 mm height on a system equipped with a fan-beam geometry, if an interslice distance of 0.1 mm is used, then 1000 slices need to be acquired, resulting in a measuring time that might be 1000 times larger than with a cone-beam CT system (in the hypothesis that the rotation time is kept constant for the two cases).

3D tomographic reconstruction is much more complex than 2D image reconstruction, and much of the research effort is currently focussed on improvement of reconstruction algorithms. This section focuses on the widely used Feldkamp-Devis-Kress (FDK) algorithm. For a detailed description of exact reconstruction algorithms the reader can refer to Herman et al. (1991).

The FDK algorithm is the most widely used reconstruction algorithm, and it belongs to the category of non-exact algorithms (or approximate algorithms). As introduced before, this means that, independently from the image quality (e.g. noise), the solution of the integral equation describing the reconstruction problem is not exact.

The FDK algorithm was formulated by Feldkamp, Devis and Kress in 1984 (Feldkamp et al. 1984) and it relies on the FBP method. Initially, the algorithm was provided for a circular trajectory and a planar detector, however, the FDK algorithm can be adapted to other trajectories (e.g. helical trajectories). The basic idea provided by Feldkamp was to treat independently every fan-shaped surface of the cone-beam, defined by the X-ray source and every detector line. This means that the mathematical reconstruction problem can be reduced to the case of 2D fan-beam image reconstruction, which was described in the previous section. Of course in this case, due to the angle between the ray and the *x*-*y* plane (i.e. the cone angle for that particular ray), the distances must be adapted, and a length correction factor that depends on the cone angle is used. The length corrected data are then filtered, as in the case of 2D fan beam FBP, and the backprojection step is subsequently performed in 3D.

2.3.4 Sufficiency Conditions

Despite the several advantages provided, cone-beam geometries using circular scanning trajectories are affected by so-called cone-beam artefacts (more details are given in Chap. 5). In fact, as stated in the Tuy-Smith sufficiency conditions, an exact reconstruction of the object is possible if, on every plane that intersects an object, there exist at least one cone-beam source point (Tuy 1983; Smith 1985). This condition is not satisfied for circular trajectories, where a complete set of

Fig. 2.35 Representation of
the zones in which,
theoretically, Radon date are
available: for circular
scanning trajectories, all
points that can theoretically
be measured in the Radon
space are within a torus

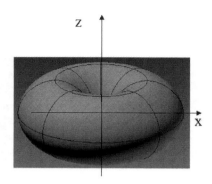

Radon data is not available. Therefore, there will be some shadow zones in which
Radon data are missing. For circular trajectories, as shown in Fig. 2.35, all points
that can theoretically be measured in the Radon space are included in a torus.

The cone-beam artefact, which will be discussed in detail in Sect. 5.2, increases
with cone angle, and with the distance from the mid plane of the detector (where all
Radon data are available). Some solutions to the problem of missing data are
presented in Katsevich (2003) and Zeng (2004).

The use of helical trajectories can eliminate the cone beam artefact, as in this
case the Tuy-Smith sufficiency conditions are met. Helical scanning trajectories are
discussed in Chap. 3.

2.3.5 Algebraic and Statistical Reconstruction Techniques

This section gives a brief overview of the basics of algebraic and statistical
reconstruction algorithms. All the algorithms discussed so far were analytical in
nature and modelled the object as a mathematical function. Reconstructing the
object meant solving a continuous integral equation. Although the computation
effort generally required for analytical algorithms is much less intensive than the
one required for iterative reconstruction techniques, analytical algorithms are based
on some assumptions that are necessary for a less difficult mathematical treatment
of the problem. In FBP-type algorithms, in fact, e.g. the focal spot is assumed to be
a point, and the radiation is considered as monochromatic. In reality, however, the
focal spot size has finite dimensions and the radiation is polychromatic. Due to the
polychromaticity of the radiation, beam hardening artefacts (for more details see
Chap. 5) can affect the quality of reconstruction, due to the lack of FBP algorithms
to deal with polychromatic beams. Moreover, noise-free projections are assumed,
which is far from reality.

One way of overcoming some of the above mentioned problems is the use of
iterative reconstruction (Buzug 2008). The basic idea on which iterative reconstruction algorithms rely is straightforward, and consists of a trial and error

approach. The tomographic image that has to be reconstructed is seen as a discrete array of unknown variables, i.e. the attenuation coefficients. The set of projections, therefore, can be represented as a linear system of equations. Solving the reconstruction problem means solving this linear system. For every viewing angle, a forward projection is calculated by summing the intensities of all pixels along the particular ray-path. The projection is then compared to the actual recorded projection, and the process is iteratively repeated until convergence to the desired result. Statistical reconstruction algorithms are also iterative methods, but in this case the unknowns are assigned by means of likelihood principles. For more details on statistical image reconstruction the reader can refer to Fessler (2000).

Compared to FBP-type algorithms, iterative reconstruction methods can provide several advantages, such as a better quality of results when data are missing (i.e. limited angle tomography), better handling of truncated projections and a decrease of metal artefacts (Hsieh 2009). However, one of the major limitations of iterative algorithms is the computation time, which is normally significantly higher with respect to FDK-based algorithms. This limitation is also the main reason why in industrial practice iterative reconstruction techniques are not yet frequently implemented.

References

Agarwal BK (2013) X-ray spectroscopy: an introduction. Springer, Berlin

Arias-Castro E, Donoho DL (2009) Does median filtering truly preserve edges better than linear filtering? Ann Stat 1172–206

Bae S, Paris S, Durand F (2006) Two-scale tone management for photographic look. ACM Trans Graph (TOG) 25(3):637–645

Banhart J (2008) Advanced tomographic methods in materials research and engineering. Oxford University Press, Oxford

Berger MJ, Hubbell JH, Seltzer SM, Chang J, Coursey JS, Sukumar R, Zucker DS, Olsen K (2010) XCOM: photon cross section database. National Institute of Standards and Technology, Gaithersburg, MD, http://physics.nist.gov/xcom. Accessed 4 Apr 2016

Buzug TM (2008) Computed tomography: from photon statistics to modern cone-beam CT. Springer, Berlin

Carroll QB (2011) Radiography in the digital age. Charles C Thomas, Springfield (IL)

Castleman KR (1996), Digital image processing. 2nd edn. Prentice-Hall, Englewood Cliffs

Cierniak R (2011) X-Ray computed tomography in biomedical engineering. Springer, London

Cormack AM (1963) Representation of a function by its line integrals, with some radiological applications. J Appl Phys 34:2722

Cunningham IA (2000) Computed tomography. In: Bronzino JD (ed) The biomedical engineering handbook, vol I. CRC, Boca Raton, pp 61–62

Curry TS, Dowdey JE, Murry RC (1990) Christensen's physics of diagnostic radiology. Lippincott Williams & Wilkins, Philadelphia (PA)

De Chiffre L, Carmignato S, Kruth J-P, Schmitt R, Weckenmann A (2014) Industrial applications of computed tomography. CIRP Annals—Manufact Technol 63(2):655–677. doi: 10.1016/j.cirp.2014.05.011

Durand F, Dorsey J (2002) Fast bilateral filtering for the display of high-dynamic-range images. InACM transactions on graphics (TOG) 21(3):257–266

Edholm P, Herman G, Roberts D (1988) Image reconstruction from linograms: implementation and evaluation. IEEE Trans Med Imaging 7(3):239–246

Feldkamp LA, Davis LC, Kress JW (1984) Practical cone-beam algorithm. J Opt Soc Am A 1 (6):612–619

Fessler JA (2000) Statistical image reconstruction methods for transmission tomography. Hand book of medical imaging, vol 2. Medical Image Processing and Analysis. SPIE Press, Bellingham

Fuchs T, Kachelrieß M, Kalender WA (2000) Direct comparison of a xenon and a solid-state CT detector system: measurements under working conditions. IEEE Trans Med Imaging 19 (9):941–948

Grangeat P (1990) Mathematical framework of cone beam 3D reconstruction via the first derivative of the radon transform. In: Herman GT, Louis AK, Natterer F (eds) (1991) Mathematical methods in tomography. Springer, Berlin, pp 66–97

Gui ZG, Liu Y (2012) Noise reduction for low-dose X-ray computed tomography with fuzzy filter. Optik-Int J Light Electron Opt 123(13):1207–1211

He K, Sun J, Tang X (2013) Guided image filtering. IEEE Trans Pattern Anal Mach Intell 35 (6):1397–1409

Herman GT, Meyer LM (1993) Algebraic reconstruction techniques can be made computationally efficient. IEEE Trans Med Imaging 12(3):600–609

Herman GT, Louis A K., Natterer F (1991) Mathematical framework of cone beam 3D reconstruction via the first derivative of the radon transform. Springer, Berlin

Hsieh J (2009) Computed tomography: principles, design, artifacts, and recent advances. SPIE, Bellingham

Hsieh J, Nett B, Yu Z, Sauer K, Thibault JB, Bouman CA (2013) Recent advances in CT image reconstruction. Curr Radiol Rep 1(1):39–51

Jackson DF, Hawkes DJ (1981) X-ray attenuation coefficients of elements and mixtures. Phys Rep 70(3):169–233

Jenkins R, Snyder RL (1996) Introduction to x-ray powder diffractometry. Chem Anal 138

Kak AC, Slaney M (2001) Principles of computerized tomographic imaging. Society of Industrial and Applied Mathematics

Kalender WA (2011) Computed tomography: fundamentals, system technology, image quality, applications. Wiley, New York

Katsevich A (2003) A general schedule for constructing inversion algorithm for cone beam CT. Int J Math Math Sci 21:1305–1321

Leroy C, Rancoita PG (2004) Principles of radiation interaction in matter and detection. World Scientific Publishing, Singapore

Lifshin E (1999) X-ray characterization of materials. Wiley, New York

Lukac R, Smolka B, Martin K, Plataniotis KN, Venetsanopoulos AN (2005) Vector filtering for color imaging. Sig Process Mag IEEE 22(1):74–86

Lyra M, Ploussi A (2011) Filtering in SPECT image reconstruction. J Biomed Imaging 2011:10

Mannigfaltigkeiten, Saechsische Akademie der Wissenschaftten, Leipzig, Berichte über die Verhandlungen, 69, 262–277

Müller P, De Chiffre L, Hansen HN, Cantatore A (2013) Coordinate metrology by traceable computed tomography, PhD Thesis. Technical University of Denmark

O'Sullivan J (1985) A fast sinc function gridding algorithm for Fourier inversion in computer tomography. IEEE Trans Med Imaging 4(4):200–207

Panetta D (2016) Advances in X-ray detectors for clinical and preclinical computed tomography. Nucl Instrum Methods Phys Res Sect A 809:2–12

Parker DL, Stanley JH (1981) Glossary. In: Newton TH, Potts DG (eds) Radiology of the skull and brain: technical aspects of computed tomography. Mosby, New York

Pfeiffer F, Weitkamp T, Bunk O, David C (2006) Phase retrieval and differential phase-contrast imaging with low-brilliance X-ray sources. Nature Phys 2(4):258–261

Pfeiffer F, Bech M, Bunk O, Kraft P, Eikenberry EF, Brönnimann C, Grünzweig C, David C (2008) Hard-X-ray dark-field imaging using a grating interferometer. Nature Mater 7(2): 134–137

Pfeiffer F, Bech M, Bunk O, Donath T, Henrich B, Kraft P, David C (2009) X-ray dark-field and phase-contrast imaging using a grating interferometer. J Appl Phys 105(10):102006

Poludniovski G, Evans P, DeBlois F, Landry G, Verhaegen F, SpekCalc (2009) http://spekcalc.weebly.com/. Accessed 4 Apr 2016

Radon J (1917) Über die Bestimmung von Funktionen durch Ihre Intergralwerte Längs Gewisser

Riederer SJ, Pelc NJ, Chesler DA (1978) The noise power spectrum in computed X-ray tomography. Phys Med Biol 23(3):446

Requena G, Cloetens P, Altendorfer W, Poletti C, Tolnai D, Warchomicka F, Degischer HP (2009) Sub-micrometer synchrotron tomography of multiphase metals using Kirkpatrick—Baez Optics. Scripta Mater 61(7):760–763.

Seeram E (2015) Computed tomography: physical principles, clinical applications, and quality control. Elsevier Health Sciences

Smith BD (1985) Image reconstruction from cone-beam projections: necessary and sufficient conditions and reconstruction methods. IEEE Trans Med Imaging 4(1)

Stark H, Woods WJ, Paul I, Hingorani R (1981) Direct Fourier reconstruction in computer tomography. IEEE Trans Acoust Speech Signal Process 29(2):237–245

Stock SR (2008) Microcomputed tomography: methodology and applications. CRC Press, Boca Raton, FL

Tesla Memorial Society of New York. http://www.teslasociety.com/pbust.htm. Accessed 5 Apr 2016

Tuy HK (1983) An inversion formula for cone-beam reconstruction. SIAM J Appl Math 43 (3):546–552

Van Laere K, Koole M, Lemahieu I, Dierckx R (2001) Image filtering in single-photon emission computed tomography: principles and applications. Comput Med Imaging Graph 25(2): 127–133

Wittke JH (2015) Signals. Northern Arizona University, Flagstaff, AZ. http://nau.edu/cefns/labs/electron-microprobe/glg-510-class-notes/signals/. Accessed on 29 April 2016

Zeng K, Chen Z, Zhang L, Wang G (2004) An error-reduction-based algorithm for cone-beam computed tomography. Med Phys 31(12):3206–3212

Ziegler A, Köhler T, Proksa R (2007) Noise and resolution in images reconstructed with FBP and OSC algorithms for CT. Med Phys 34(2):585–598

Chapter 3
X-ray Computed Tomography Devices and Their Components

Evelina Ametova, Gabriel Probst and Wim Dewulf

Abstract This chapter focuses on the hardware of current and emerging X-ray computed tomography (CT) systems. First, the basic configuration of industrial cone-beam X-ray CT systems is recalled. Subsequently, individual hardware components, such as sources, detectors and kinematic systems, are described. The chapter continues with an overview of novel developments that expand the application domain of X-ray CT towards, for example, larger objects or fully integrated in-line systems. The chapter is concluded by a description of key safety elements for operation of X-ray CT systems.

3.1 Basic Configuration of Industrial Cone-Beam X-ray Computed Tomography Systems

Most conventional industrial X-ray computed tomography (CT) systems employ a cone-beam acquisition geometry, and are composed of four main components (Fig. 3.1): (i) an X-ray tube for generating X-ray beams, (ii) an X-ray detector for measuring the extent to which the X-ray signal has been attenuated by the workpiece, (iii) a set of mechanical axes for positioning the object between the source and detector and providing the rotation required for CT, and (iv) a computer for data acquisition, reconstruction and subsequent analysis. Essentially, the characteristics and performance of the first three components directly influence the quality of the acquired data. The measurement volume is limited by the detector size and by the dimensions of the cabinet. Typical flat panel detectors are characterised by 1000×1000 or 2000×2000 pixels, with an area coverage up to 400 mm $\times$ 400 mm. Most industrial CT systems use full-angular (360°) scanning; up to 3600 images

E. Ametova · G. Probst · W. Dewulf (✉)
Department of Mechanical Engineering, KU Leuven, Leuven, Belgium
e-mail: wim.dewulf@kuleuven.be

© Springer International Publishing AG 2018
S. Carmignato et al. (eds.), *Industrial X-Ray Computed Tomography*,
https://doi.org/10.1007/978-3-319-59573-3_3

are commonly acquired for each scan. The rotation speed depends both on the exposure time and on the mechanical stability of the rotary stage. The X-ray source should have enough power to penetrate high-absorption materials. Commercial standard tubes are normally limited by 450 kV, but special tubes up to 800 kV are also available (De Chiffre et al. 2014).

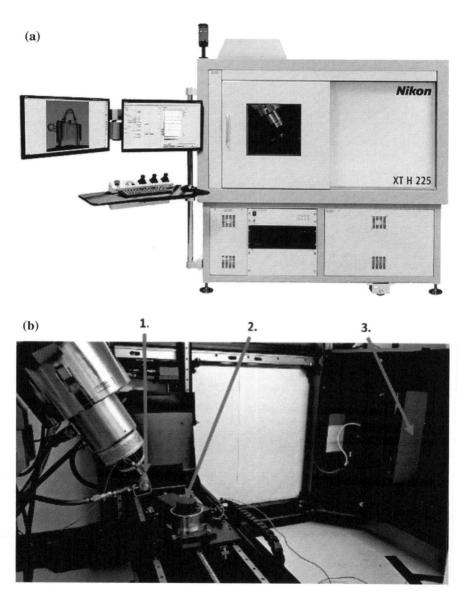

Fig. 3.1 Industrial CT system with micro-focus source and thermally controlled cabinet (*top*) (courtesy of Nikon Metrology), and cabinet of the industrial CT system, showing (*1*) X-ray source, (*2*) rotary stage, and (*3*) detector

3.2 Source

The basic concept of an X-ray source has been introduced in Sect. 2.1.1. This section provides more details on the key elements of an X-ray source: the filament (cathode), the electron optics and the target (anode). All components are enclosed in a vacuum chamber (typically <0.1 μbar) to prevent arcing and oxidation of the electrodes (see Fig. 3.2). At the exit of the X-ray tube, the X-ray beam is "shaped" by passing through a circular aperture or diaphragm (beryllium window). Since periodic replacement of the filament and/or the target are required, most current industrial X-ray tubes are of the open-tube type. Therefore, the vacuum is maintained using vacuum pumps. After target or filament replacement, the vacuum needs to be gradually restored by conditioning the machine.

3.2.1 Filament

The filament (usually a thoriated tungsten wire—see Fig. 3.3; also other materials such as lanthanum hexaboride [LaB$_6$] are available) is heated to a temperature of approximately 2400 K using the filament current and the associated Joule effect.

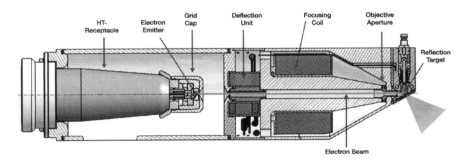

Fig. 3.2 Schema of an X-ray tube with reflection target (courtesy of X-RAY WorX GmbH)

(a)   **(b)**   **(c)**

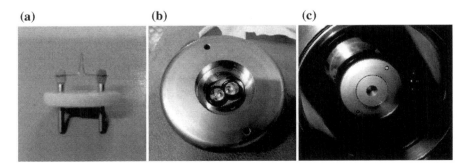

Fig. 3.3 **a** The filament, **b** the focusing cup where the filament is mounted and **c** the focusing cup mounted on the X-ray source

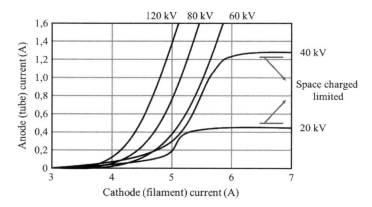

Fig. 3.4 Relation between tube current and filament current respectively, with tube voltage (from Bushberg and Boone 2011)

At this temperature, some electrons can overcome the binding energy of the metal filament through a process called thermionic emission. These electrons will escape the filament and be accelerated in the electric field within the tube towards the target material, thus creating a tube current. The thorium in the tungsten wire increases the efficiency of the thermionic process, thus prolonging the service life of the filament. The service life typically amounts to 500 to 1000 operating hours, after which replacement of the filament is required.

The eventual tube current is a function of both the filament current and the tube voltage (i.e. the voltage between the cathodic filament and the anodic target) (see Fig. 3.4). Due to the non-linearity of the thermoelectric effect, a small increase in filament current, hence temperature, results in a large increase in tube current. However, when a high filament current is combined with a low acceleration voltage between cathodic filament and anodic target, the space charge cloud around the filament is dense and hinders further electrons to leave the filament. Therefore, a further increase in filament current no longer increases the tube current and the tube is said to be working in space charge limited operation. At higher acceleration voltages in contrast, thermionic electrons are readily attracted by the anode, thus the maximum filament temperature—hence current—limits the tube current. In this region, increasing the tube voltage has only a limited effect on the tube current.

3.2.2 Electron Optics

The electrons liberated from the filament are accelerated towards the target anode and are focused on a small region, called the focal spot. This leads to penumbra regions that are illuminated by the X-rays emitted only from part of focal spot, which results in image blur (unsharpness). The size and shape of the focal spot exerts a large influence on the final quality of the radiographs. Small focal spots produce less blurring and better visibility of detail in the radiographs. Large focal spots have the benefit of a greater heat-dissipating capacity (see Sect. 3.2.3), with the drawback of increasing blur and degrading the visibility of details; especially at high magnifications (see Fig. 3.5).

The size of the electron beam arriving from the filament largely determines the size of the focal spot. In order to produce a small focal spot at the envisaged position on the target material, the trajectories of the accelerated electrons are controlled by electron optics.

An electric field near the filament is created, through Wehnelt optics, such that the accelerated electrons are directed to a small spot. Magnetic lenses steer the position of the electron beam to an envisaged location on the target material. Changing the source power, however, shifts the positioning of the electron beam—hence of the focal spot—within the X-ray tube. This leads to poor focus, edge unsharpness and contrast loss in the radiographs. Therefore, proper electron beam alignment needs to be performed in order to maintain high CT quality. A correctly aligned electron beam leads to increased image sharpness, a better distinction of material peaks in the gray-value histogram of the reconstructed volume, reduced noise and an overall improved surface quality of the thresholded volume. Figure 3.6 shows the effect of a misaligned electron beam on the overall quality of a reconstructed 3D CT model.

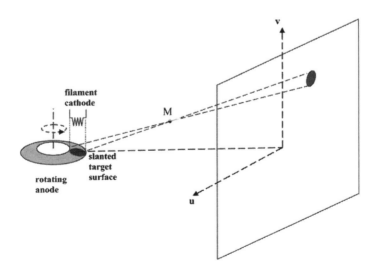

Fig. 3.5 An illustration of projection of focal spot blurring (Chen et al. 2008)

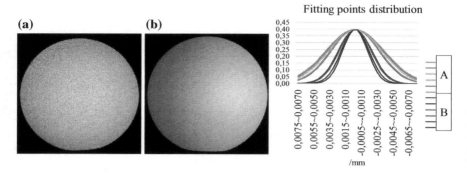

Fig. 3.6 Influence of proper electron beam alignment in an X-ray CT system at high magnification. **a** Illustrates an object measured with poor electron beam alignment; **b** shows the same object measured with a proper electron beam alignment. The graph, at the right hand side, depicts the fit point deviations of spheres that were fit to both objects. As can be observed, the deviations are higher when the electron beam is not properly aligned

3.2.3 Target

Tungsten is commonly used as a target material, due to its high melting point and its X-ray generation efficiency. However, other materials, such as molybdenum, copper and silver are available. Different target materials have characteristic radiation at different photon energies. For instance, characteristic radiation happens just below 70 kV for tungsten, yet just below 20 kV for molybdenum. Figure 3.7 shows the resulting spectra for four major target materials with their respective continuous (bremsstrahlung) radiation superimposed with their characteristic radiation. Some manufacturers offer multi-metal targets that are fitted on an indexing head in order

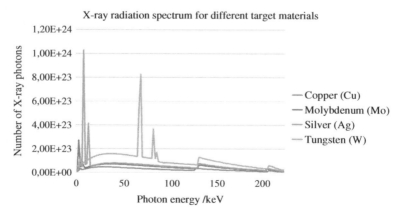

Fig. 3.7 X-ray radiation spectrum for different target material. Copper (Cu), molybdenum (Mo), silver (Ag) and tungsten (W). Graph generated via the simulation software BAM aRTist

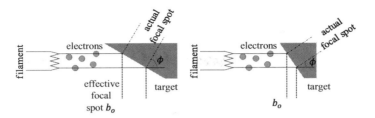

Fig. 3.8 Effective focal spot in relation to angle of which the electron beam hits the target material

to facilitate the selection of an appropriate target material as a function of the workpiece to be scanned.

Most X-ray sources are equipped with a reflection target. The size of the effective focal spot is determined by the orientation of the surface of the target normal relative to the electron beam (see Fig. 3.8). However, this implies that the X-ray intensity distribution also varies in a directionally dependent way, which can necessitate additional considerations in special applications. Moreover, large angles between the incident electron beam and the target material are not desired, due to the probability that the electrons are elastically reflected from the surface of the target material, not contributing to the generation of X-ray radiation.

The kinetic energy of the electrons hitting the focal spot is only partially converted into X-rays (about 1%), whereas the remainder is converted into heat (about 99%). X-ray tubes are, therefore, actively cooled. Whereas in some cases oil is circulated inside the anode, the use of a high thermal conductivity, high mass copper block as carrier for the actual target material is also commonly applied in order to conduct the heat towards a cooling medium.

Despite cooling, focal spot drift might occur during the acquisition of radiographs, hence decreasing the quality thereof. Figure 3.9a illustrates the temperature change of an X-ray tube at the X-ray window for three power levels; whereas the corresponding X and Y components of focal spot drift are depicted in Fig. 3.9b, c

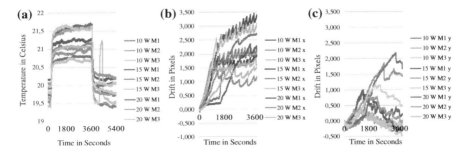

Fig. 3.9 a The temperature of an X-ray tube at the X-ray window during one hour of scanning time; **b** and **c** depict how the focal spot drifts in X and Y directions due to thermal expansion of the target material, respectively (from Probst et al. 2016)

respectively. The shifting in positioning of the focal spot, throughout a CT scan, adds blur to the radiographs, and affects the quality of the reconstructed CT model particularly for high magnification measurements.

Heat generation makes it difficult to conciliate sharper radiography images with the market demand for higher X-ray power, either for fast scanning (shorter scanning times) or increased penetration lengths (higher voltage). An energy density of the focused electron beam that is too high may lead to pitting of the target due to localised melting at the focal spot. This might, amongst other things, lead to self-absorption of the X-rays by the target material around the pit, hence affecting the spectrum of the X-ray beam, which becomes angle dependent. This effect (the Heel effect) increases with the lifetime of the target.

To avoid pitting, focal spot sizes are correlated to the tube power. Typical focal spot sizes for industrially relevant micro-focus sources are in the range of 1 μm to 20 μm, but nano-focus sources with focal spot sizes below 1 μm are commercially available for lower power applications. When increasing the power settings for a given tube, the electron beam is often slightly defocused by software. Within the industrially relevant power ranges, a correlation of 1 μm per 1 W can be used as a rule of thumb, as illustrated by Table 3.1. Indexing of the target can be allowed in order to avoid pitted regions.

Other solutions offered commercially to reconcile the requirement for small focal spot diameters with higher power demands, concern the use of rotating targets

Table 3.1 Correlation between focal spot sizes and tube power for a selection of Nikon Metrology—X-TEK tubes

Micro-focus source	Max. voltage (kV)	Max. power (W)	Focal spot size	
160 kV Xi, reflection target	160	60	3 μm up to 7 W	60 μm at 60 W
160 kV reflection target	160	225	3 μm up to 7 W	225 μm at 225 W
180 kV transmission target	180	20	1 μm up to 3 W	10 μm at 10 W
225 kV reflection target	225	225	3 μm up to 7 W	225 μm at 225 W
225 kV rotating target option	225	450	10 μm up to 30 W	160 μm at 450 W
320 kV reflection target	320	320	30 μm up to 30 W	320 μm at 320 W
450 kV reflection target	450	450	80 μm up to 50 W	320 μm at 450 W
450 kV high brilliance source	450	450	80 μm up to 100 W	113 μm at 450 W
750 kV with integrated generator	750	750	30 μm up to 70 W	190 μm at 750 W

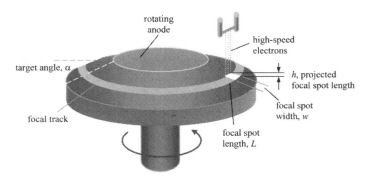

Fig. 3.10 X-ray tube with rotating target anode (from Hsieh 2003)

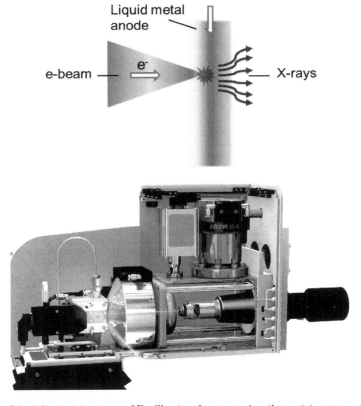

Fig. 3.11 Principle (*top*) (courtesy of Excillum) and cross section (*bottom*) (courtesy of Bruker) of an X-ray tube with liquid metal jet anode

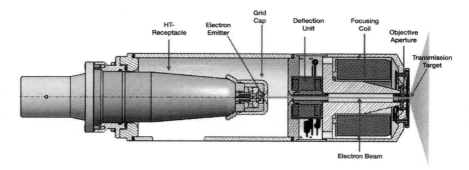

Fig. 3.12 X-ray source with transmission target. Actual versus effective focal spot for beveled reflection targets (courtesy of X-RAY WorX GmbH)

(see Fig. 3.10) or the use of metal jet liquid targets (see Fig. 3.11). In both cases, the heat is dissipated over a larger, moving area while continuously renewing the material exposed to the high-energy electron beam. Rotating targets are commonly used in the medical field, and have been introduced in the industrial CT field primarily for the higher power systems, whereas liquid metal jet sources are more suited for lower power systems. Current generation liquid metal jet targets use gallium or indium based metal alloys that melt close to room temperature, and have approximate emission characteristics of some regular solid anodes. The potential for increasing power over focal spot size by approximately three orders of magnitude has been reported (Otendal et al. 2008).

For smaller power CT systems, X-ray tubes can be equipped with a transmission target that generates X-rays collinearly with the electron beam within the vacuum chamber (see Fig. 3.12). In this case, the target is a thin material component that allows the electrons to pass through its entire body. The thin layer of target material cannot resist high temperatures, thus limiting the allowable power range of the X-ray tube.

Figure 3.13 shows positioning of the different X-ray tube technologies in terms of tube power and focal spot sizes, thus emphasising the complementarity of the approaches. CT practitioners should, however, always carefully interpret the values, since measurement of focal spot size is nontrivial and little non-commercial comparative information on actual performance is available. Moreover, the focal spot size can vary over time.

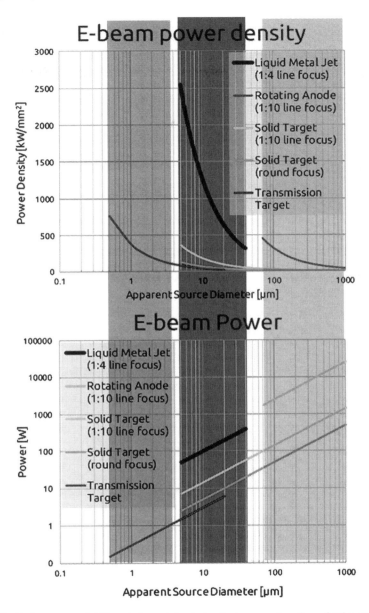

Fig. 3.13 Performance of different X-ray tube technologies as a function of tube power and apparent source diameter (courtesy of Excillum)

3.3 Detectors

3.3.1 Flat Panel Detectors Versus Linear Diode Array Detectors

As discussed in Chap. 2, most industrial CT systems use flat panel detectors (FPDs) based on a matrix array (for example, 2000 × 2000 pixels) in combination with a cone beam of X-rays. An alternative set-up to flat panel detectors are linear diode array detectors (LDAs), which rely on a single line of pixels for the capture of X-ray images, and are hence used in combination with X-ray fan beams. A variant of the LDA, called curved linear diode array (CLDA) detectors, are constructed along a circular curve, such that every detector pixel has the same source to detector distance. When a component is scanned, in a similar way as with a traditional FPD, only one single slice of the component can be registered. Figure 3.14 illustrates the basic principle of both FPD and LDA detectors.

The CLDA detector has some advantages and disadvantages when compared to the FPD. Some of the advantages are: LDA features inter-pixel shielding, resulting in less pixel interaction from adjacent pixels. LDA optimises the collection of X-rays by eliminating scatter phenomena, which results in a reconstructed volume of higher quality due to less noise. Moreover, an LDA's scintillator thickness is greater, resulting in an overall better signal-to-noise ratio. For CLDA detectors, their arc centre curvature is coincident with the X-ray focal spot, which reduces the image deformation when moving away from the central pixel. The disadvantage is that data acquisition with LDA is much more time consuming than with FPD. The object will be scanned slice by slice, which requires one full rotation of the object per slice. During the acquisition phase, the object also has to move in the vertical direction. This extra vertical movement leads to extra inaccuracy of the reconstructed object (Welkenhuyzen 2015).

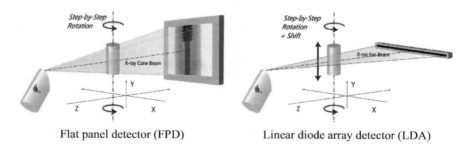

Flat panel detector (FPD)   Linear diode array detector (LDA)

Fig. 3.14 Cone beam geometry (flat panel detector) versus fan beam (linear diode array detector) (courtesy of General Electrics)

3.3.2   *Energy Integrating Detectors Versus Photon Counting Detectors*

Scintillator-based detectors (Chap. 2) do not allow discrimination between different photon energies, hence the polychromatic nature of bremsstrahlung spectra leads to beam hardening artefacts, in particular hindering inspection of multi-material components. Recently, photon counting detectors (PCDs) have received increasing attention in CT (Taguchi and Iwanczyk 2013). PCDs are based on direct conversion of X-ray photons into an electric charge proportional to the photon's energy (see Fig. 3.15). Typically, direct conversion materials, such as cadmium telluride (CdTe) or cadmium-zinc-telluride (CZT), are used. The charge produced in direct conversion is approximately ten times higher than in scintillator-based detectors, consequently the number of photons can be counted and photon energies can be resolved (Suetens 2009).

PCD technology allows discrimination between different photon energies reaching the detector during a single projection, typically discretising the X-ray spectrum into two to eight energy bins. Advantages when scanning multi-material components not only relate to eliminating beam hardening artefacts, but also to increasing the image contrast for the low attenuating components, by giving more weight to the lower photon energy bins. Moreover, PCDs allow information from the lowest energy bins to be ignored, hence eliminating noise (Panetta 2016). Unfortunately, the count-rate capability of PCDs is still limited, hence they are not yet capable of coping with the high tube currents (hence photon fluxes) of current industrial CT applications.

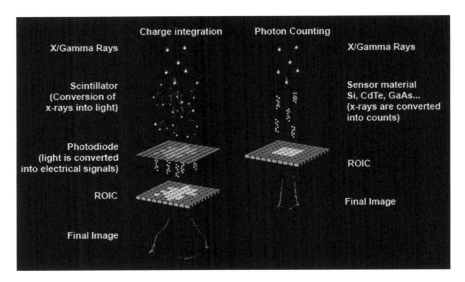

Fig. 3.15 Energy integrating detectors versus photon counting detectors (courtesy of Direct Conversion)

3.4 Frame, Kinematic System and Cabinet

As previously indicated, stationary tube-detector systems are used for industrial CT (see Fig. 3.16). Three axes span a Cartesian coordinate system. The magnification axis (Z-axis) is defined along the line from the X-ray source to the detector. The Y-axis is parallel to the rotation axis. The X-axis is orthogonal to both the Y- and Z-axes. The lateral opening angle of the X-ray beam is called the fan angle and the longitudinal opening angle is called the cone angle. Due to the high requirements for measurement performance of industrial CT systems, the geometrical alignment and stability of the kinematic system are crucial for overall system performance.

A traditional cone-beam CT system is considered aligned when it satisfies several conditions (see Fig. 3.17) (Ferrucci et al. 2015):

- the intersection of the magnification axis with the detector is coincident with the centre of the detector,
- the magnification axis is normal to the detector,
- the magnification axis intersects the axis of rotation at a 90° angle, and
- the projection of the axis of rotation is parallel to the detector columns.

For a square detector, the first two conditions are equivalent to equality of cone and fan angles.

Relative movement of CT components during a scan must also satisfy the following requirements:

- the relative distance between all components is constant and the position of the rotation axis is fixed, and
- the plane of the rotation is perpendicular to the rotation axis.

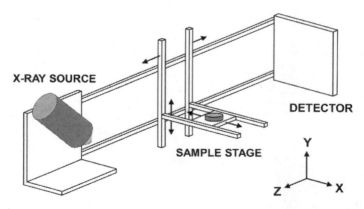

Fig. 3.16 The linear path from the X-ray source to the detector defines the magnification axis. The Y-axis is parallel to the rotation axis. The X-axis is orthogonal to both the Y- and Z-axis, forming a Cartesian coordinate system. A flat panel detector, which corresponds to cone-beam CT systems, is shown. This diagram shows one type of construction of CT systems; however, other architectures are possible (Ferrucci et al. 2015)

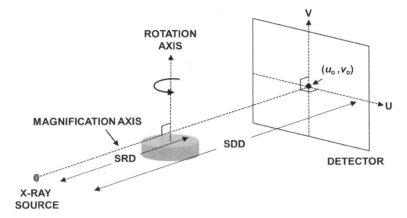

Fig. 3.17 An aligned cone-beam XCT system satisfies a series of conditions shown in this diagram (Ferrucci et al. 2015)

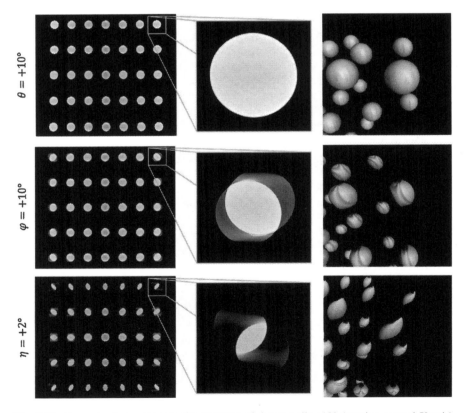

Fig. 3.18 Reconstructed volume in the presence of detector tilt +10° (rotation around X-axis), detector slant +10° (rotation around Y-axis), and detector skew +2° (rotation around Z-axis). *Left* Grey-value slice along XY-plane before grey-value thresholding. *Centre* Magnified portion of grey-value image. *Right* Three-dimensional view of the reconstructed sphere objects after applying grey-value thresholding (Ferrucci et al. 2016 © Crown Copyright, reproduced with permission by NPL)

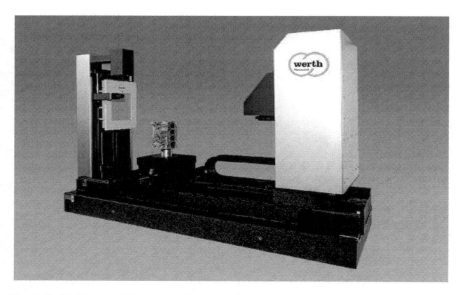

Fig. 3.19 Multi-sensor CT system with granite base (courtesy of Werth)

Misalignments in the kinematic structure lead to image artefacts, as illustrated in Fig. 3.18. The frame supporting all the kinematic components, therefore, requires high dimensional stability, damping capacity and stiffness. Following traditional good practice in coordinate measuring machine (CMM) construction, a number of manufacturers rely on granite as a base material (see Fig. 3.19) (Hocken and Pereira 2011). Also, temperature control of the cabinet and active cooling of the source is required. Dimensional metrology should be carried out at 20 °C (this is the internationally-agreed standard temperature). A different temperature will result in expanding or shrinking of the object and possible thermal distortion of the structure of the CT scanner. Not only the temperature but also temperature gradients in time and space will affect thermal distortions and hence the measurements.

The accuracy and stability of kinematic system depend on the hardware components, such as the drive system, bearing system and displacement transducers. The following subsections deal with the technical characteristics of these components.

3.4.1 Drive Systems

The purpose of a drive system, in an industrial X-ray CT system, is to position the object within the field of view of the detector at a desired magnification. Positional information, where the object is placed, is not provided by the drive system. It is the function of displacement sensors or scales to accurately define where the manipulator is within the measurement volume of the system. The most common used drive systems are: leadscrew, rack-and-pinion and belt drives.

Leadscrew drives convert the rotational movement of a motor attached to it, into linear translational motion to a carriage attached to the nut system. Leadscrew drives possess high dynamic stiffness. The conversion of rotational movement into linear movement happens at the nut, usually made of a polymer composite or bronze, that slides through the helical threads of the leadscrew. This process demands high energy, as friction at the mating interface between the leadscrew and nut is high. To overcome this problem, ball bearings are mostly used. The screw and nut on a ball screw have matching helical grooves that allow the ball bearings to re-circulate in those races, which decreases the total energy needed to convert rotation motion into linear translational motion when compared to a leadscrew.

Rack-and-pinion drives are comprised of a pair of gears that are used to convert rotational movement into linear movement. The pinion gear that drives the system engages onto the linear rack which is then driven. Form errors and backlash are a problem with these drives, as they limit the positioning accuracy. However, problems with positioning accuracy can be minimised with software.

Belt drives are composed of a motor, a belt and a multistage reducer. Belt drives are used when axial aligned shafts are not requited, as they have high tolerance for misalignment. Belt drives provide transmission of power to the moving axes, helping to protect the machinery from overload and jam. Besides that, they damp and isolate noise and vibration coming from the driving system, as they act as a low-pass filter, preventing high-frequency motor oscillations from entering the measuring structure.

3.4.2 Bearing Systems

Bearings are elements that allow relative movement between mechanical components of a mechanical system. For most X-ray CT systems, bearings are used to move the manipulator in respect to the source and detector, as both are usually stationary in industrial CT machines. Different manufacturers opt for different types of bearing systems depending on the chosen design of their CT system. CT systems comprising granite tables usually rely on non-contact bearing systems, such as air bearings, whereas more traditional CT systems with metallic frames make use of contact bearing systems.

Contact bearing systems have contact elements that may include sliding, rolling and flexural bearings. They are made of an inner and outer metal surface which the balls roll against. The balls (rollers) bear the load so that the device functions more efficiently. These bearings usually have very high mechanical stiffness normal to the direction of motion, when compared to non-contact bearing systems. When designing a machine, criteria such as wear, dynamic behaviour and total error motion should be given first priority, as the manipulator positioning accuracy is intrinsically correlated to the measurement accuracy of the CT system.

Non-contact bearings, such as air bearings, make use of a thin layer of air under pressure, usually between 1 μm and 10 μm due to the low viscosity of air, that provides support to the structure in motion. The greatest advantage of air bearings, when compared to mechanical bearings, is that, since there is no physical contact between the moving parts, no abrasion is present during their operation. This makes the durability of air bearings practically unlimited if properly dimensioned and aligned.

3.4.3 Displacement Transducers

In X-ray CT, the placement of the object to be measured within the measurement volume of the machine is highly importance as it will define voxel size and, therefore, measurement resolution. Misplacement of the manipulator, even if by a couple of micrometres, will yield errors in the measured volume due to over or under-estimation of voxel size. Therefore, it is of paramount importance that positioning is done as accurately as possible. A large variety of displacement transducers are commercially available to guarantee that relative movement between two components is carried out appropriately. X-ray CT systems commonly make use of linear scales to guarantee positioning, where a scale is fixed on to the frame of the machine and an electro-optical head fixed to the manipulator.

Transmission scales work with the principle of light passing through a grating scale (also called optical encoders) (Hocken and Pereira 2011). Movement between the scale (photo-receptor) and photo-emitter generates an alternating light signal, which after conversion to an electrical signal, gives the relative position of the moving component. Modern transmission scale encoders, due to the generation of a quadrature signal, are able to identify the direction of movement. Commercially available transmission scales can have resolutions down up to 5 μm, and even better.

Reflection scales are similar to transmission scales, as they use diffraction patterns to identify the position between two moving components. The diffraction pattern is established by alternating reflective lines and diffusely reflective gaps.

Interferential scales likewise use grating scales with line pitches of 100 lines or more per millimetre. The output signal for these sensors is electronically subdivided for finer resolution and then digitized for a position readout.

Laser interferometer scales are based on the principle of light interference. A laser beam is split into two beams by means of a beam-splitting mirror; one beam is then used as a reference and one used to create interference with the reference beam. By moving the mirror, there is a phase change between the two beams. Upon recombination of the two beams, an interference (or fringe) pattern is produced that can then be counted to account for displacement between the moving components.

As is the case for traditional coordinate measurement systems, software compensation is often used to correct for systematic deviations of the frame and kinematic system.

3.5   Integrated and Special Purpose CT Systems

3.5.1   Multi-sensor Systems

As discussed in Chap. 7, establishing traceability of X-ray CT systems is still challenging, and tactile coordinate measuring systems (CMSs) can still reach higher accuracies, particularly when inspecting complex multi-material parts. Therefore, multi-sensor CMSs have been developed that comprise not only the equipment for X-ray CT metrology, but also tactile and optical probes (see Fig. 3.19).

3.5.2   Four-Dimensional X-ray CT Systems

Four-dimensional (4D) X-ray CT systems, also referred to as in-situ CT systems, are used to investigate the behaviour of workpieces or specimens over time, including the effects of, for example, creep, fatigue and phase changes. Dedicated

Fig. 3.20 Micro-CT system with in-situ load cell

4D CT set-ups have been developed to allow, for example, tensile tests inside a CT system (Fig. 3.20). Due to the considerable measurement time required for a full CT scan, it is sometimes necessary to temporarily interrupt the load in order to acquire stable images. The consequent effects of, for example stress relaxations, need to be accounted for in the interpretation of the CT scan results.

3.5.3 Dual Energy CT Systems

Dual energy CT (DECT) employs two energies in the data acquisition. Strictly speaking DECT utilises a certain geometry twice at two different energy levels. The concept of DECT was coincident with the early development of CT. However, it is only recently that DECT has been widely applied, which is due to the evolution of scanning geometries. DECT can be carried out by scanning twice at different peak tube voltages (kVps) with a third or fourth generation CT system. However, DECT can also be acquired with hardware systems which are designed exclusively for DECT. These include a rapid kVp switching system, layer detector, and dual-source/detector system, as shown in Fig. 3.21. DECT is often used for material decomposition in medical CT. For industrial CT, DECT has superior performance in terms of beam hardening artefact minimisation and low contrast detectability.

3.5.4 At-Line and In-Line CT Systems

Within an Industry 4.0 context, there will be a move towards in-line X-ray CT systems that allow fast yet robust quality control of workpieces (Hermann et al. 2016). A first solution offered today by industrial suppliers is robot-operated CT systems, where industrial robots autonomously load and unload CT systems (see Fig. 3.22).

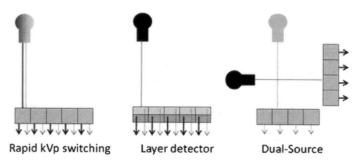

Rapid kVp switching   Layer detector   Dual-Source

Fig. 3.21 Three hardware approaches for DECT. The X-ray tube of rapid kVp switching system produces low and high energy spectra by operating at rapidly modulating X-ray tube voltages. Layer detector system makes use of energy sensitive detector to exploit difference between spectra (from Fornaro 2011)

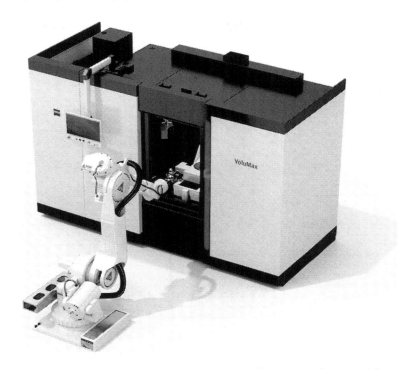

Fig. 3.22 Zeiss Volumax Systems for integrated at-line CT based quality control (courtesy of Zeiss)

An alternative in-line system has a conveyor belt through a gantry system realising a helical CT scanning trajectory (Fig. 3.23). Helical scanning trajectories can be obtained by perfoming a simultaneous rotation of the object and translational movement along the rotation axis. The translational movement can be achieved by transalting the rotary stage, or with a combined movement of source and detector (Aloisi et al. 2016). In this case helical CT acquires data while the X-ray tube and detector pair rotates at a constant angular velocity, and the table moves along the axis perpendicularly penetrating the tube-detector circle at a constant speed. Hence, the object by definition moves in a helical pattern around the object being scanned (see Fig. 3.23). Helical scanning has the advantage of the high speed of a cone-beam CT system, yet does not suffer from the cone-beam effects. Therefore, helical CT has a great potential in industrial applications (Aloisi et al. 2016). Inspection speeds of one to five minutes per part are targeted.

3.5.5 SEM CT for Small Samples

Scanning electron microscopes (SEMs) allow inspection of the internal structure of specimens at nanometre to micrometre resolution (Goldstein et al. 2012). A traditional SEM utilises a focused beam of high-energy electrons to irradiate the

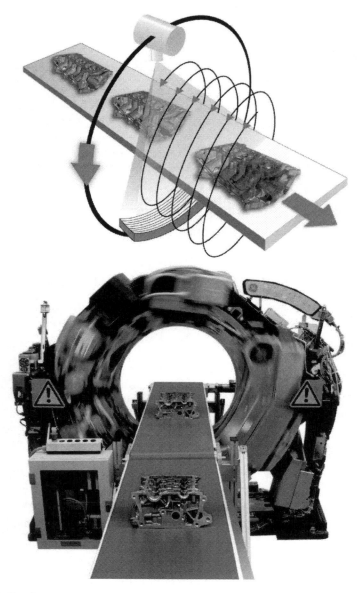

Fig. 3.23 Concept sketch (*top*) and in-line system (cover removed) for integrated in-line CT based quality control (courtesy of GE)

volume to be analysed. Interactions between the sample and the electron probe produce several types of signal including secondary electrons, backscattered electrons and characteristic X-rays. These signals are obtained from specific emission volumes within the sample and can be used to examine characteristics of the

(a)   **(b)**

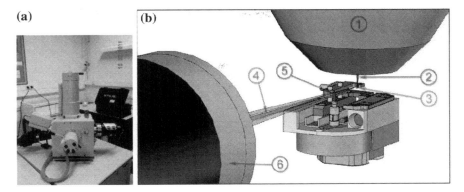

Fig. 3.24 Schematic overview of **a** SEM CT system (Pyka et al. 2011) and **b** the set-up to add micro-tomography capabilities to a SEM: (*1*) objective lens of the SEM, (*2*) electron beam, (*3*) metal target for X-ray generation, (*4*) X-ray cone-beam, (*5*) sample for micro-CT positioned on rotation stage, (*6*) camera assembly (courtesy of Bruker MicroCT)

sample, such as surface topography, crystallography and composition. With its high magnification power (>100 000×), SEM can image samples on the centimetre-scale at nanometre resolution, and even sub-nanometre resolution is now quoted for some commercial instruments (Sunaoshi et al. 2007).

Today, SEM systems can be adapted into fully fledged CT systems for micro-scale samples (see Figs. 3.24 and 3.25). SEM takes advantage of the focused electron beam available within the standard SEM system, which is complemented by an X-ray CCD camera and an assembly to be mounted in place of standard specimen holders. The holder comprises both a target and an object manipulation system (magnification axis and rotation stage).

Fig. 3.25 Comparison of SEM image and SEM-CT based images of a used filter from a vacuum cleaner filled with dust particles. From left to right: (*1*) SEM image of the filter surface; (*2*) front view of SEM-CT reconstruction where the filter material is shown in silver, and dense particles in *red*; (*3*) back view of the same SEM-CT reconstruction; and (*4*) side view with semitransparent filter material. It can be noticed that most particles are absorbed within the front surface of the filter and cannot pass through the filter material (courtesy of Bruker Micro-CT)

3.5.6 X-ray CT of Large and Highly Attenuating Objects

The achievable penetration capability of CT systems is directly related to the X-ray source power. With up to 450 kV (or recently 800 kV) acceleration voltages, conventional industrial CT systems remain limited by maximum allowable specimen dimensions as a function of the material composition. Second, the scanning volume is limited by the detector size and the dimensions of the cabinet. However, examination of specimens, such as complete vehicles, engine blocks and aircraft structural parts, is in high demand. Recently, several high-energy CT systems have been developed, using linear accelerators (LINACs) (see Fig. 3.26).

Typically in a LINAC, electrons are accelerated, in a hollow pipe vacuum chamber through a high frequency modulated series of cylinder electrodes, towards a metal target (see Fig. 3.27). The length of the drift tubes increases from the electron source towards the target. The driving frequency and the spacing of the electrodes are tailored to maximise the voltage differential at the moment when the electron crosses the gap between the electrodes, hence accelerating the electrons. Due to the low mass of electrons, they can be rapidly accelerated, hence the total length of a LINAC for high-power CT can remain much smaller compared to similar devices used for proton acceleration.

Scanning large parts also requires the availability of large detectors, or of methods that allow stitching of multiple radiographs or reconstructed volumes (Maass et al. 2011). Due to the fundamental physics of LINAC systems, pulsed X-ray are generated, which requires careful consideration when selecting the read-out time of the detector (Salamon et al. 2013).

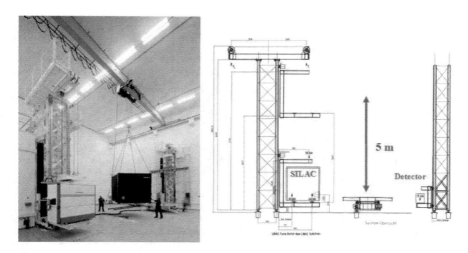

Fig. 3.26 Photograph of the CT scanner in Fraunhofer EZRT (*left*). Drawing of the setup with a 9 meV Siemens Linear Accelerator (SILAC), a 3 m diameter turntable with a carrying capacity of 10 t, and a 4 m wide line detector with a pixel pitch of 400 μm (Salamon et al. 2013). The system allows scanning of cars (for example, before and after crash tests), containers, etc

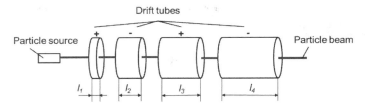

Fig. 3.27 Schema of a LINAC

The X-ray attenuation at energies above 1 MeV is increasingly based on pair production, i.e. the formation of an electron-positron pair (see Sect. 2.1.3). Whereas higher penetration thicknesses for more attenuating materials can be achieved (see Fig. 3.28), a drawback concerns the increased number of scattered photons, contributing to higher noise levels (see Fig. 3.29).

Another new development concerns robot-CT systems, in which the X-ray source and detector are both mounted on cooperative robots (see Fig. 3.30), hence allowing the generation of flexible trajectories and the scanning of large parts. The positioning accuracy of the robots is the main challenge of this novel and promising development.

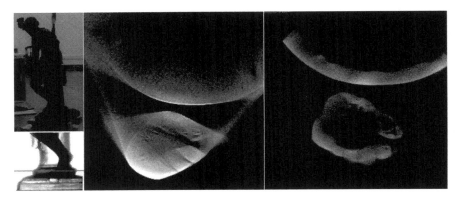

Fig. 3.28 Illustration of the advantages of a 9 MV LINAC X-ray system (*right*) over a 450 kV system with insufficient penetration (*left*). The images on the left show a bronze statue positioned on the rotation stage and a radioscopy of the region of interest marking the cross-section (*blue line*). The inner structures of the statue parts are only visible in the 9 MV scan (Salamon et al. 2012)

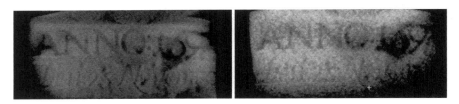

Fig. 3.29 Illustration of the disadvantages of a 9 MV LINAC X-ray system (*right*) over a 450 kV system with sufficient penetration (*left*). The increased noise level due to additional photon scattering is obvious (Salamon et al. 2012)

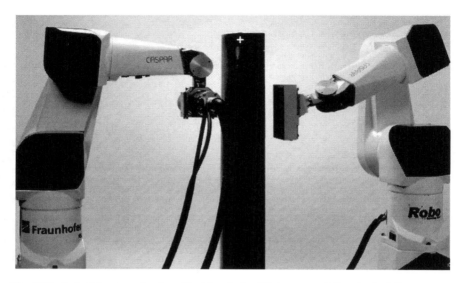

Fig. 3.30 RoboCT system developed by Fraunhofer IIS (courtesy of Fraunhofer IIS)

3.5.7 Synchrotron Industrial Tomography

Since the 1980s, large synchrotron radiation facilities have also been used to generate X-rays for CT applications (see Fig. 3.31). Synchrotron radiation offers unique X-ray beam properties, such as: a monochromatic beam, a high photon flux, coherence, collimation and high spatial resolution. The source itself offers a broad band of X-ray energies with a very high flux and high brilliance. Such a quasi-parallel beam eliminates common problems associated to cone or fan beams and image magnification. Synchrotron radiation is produced when a moving

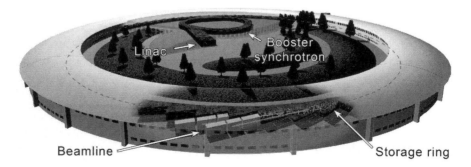

Fig. 3.31 Diagram of the ESRF synchrotron Grenoble, France (courtesy of ESRF)

electron changes its direction, resulting in the emission of a photon. When the electron is moving fast enough, the emitted photon is at X-ray wavelengths. The X-ray beams are guided through a set of lenses and instruments (beamlines), where they illuminate and interact with the objects being studied. Synchrotron X-ray beamlines are often provided with monochromators producing a beam with a very narrow energy distribution and a tunable energy. Monochromatic X-ray beams are a remedy to the problems of beam hardening typical to wide beam spectra. Even when monochromatised, synchrotron X-ray beams offer a higher intensity compared to classical X-ray tubes.

3.5.8   Laminography for Flat Parts

For a number of applications, it is unfeasible to realise radiographs from a full 360° perspective due to, for example, a high penetration length in at least one direction, yet digital laminography can be utilised. In laminography systems, the source and detector move synchronously (typically by orbiting around a common rotation axis) such that points on one plane (the so-called focal plane) are always projected onto the same position, whereas points on all parallel planes are projected onto different positions. Averaging all images hence leads to sharp images for the focal plane, whereas the information from all other planes is blurred (see Fig. 3.32).

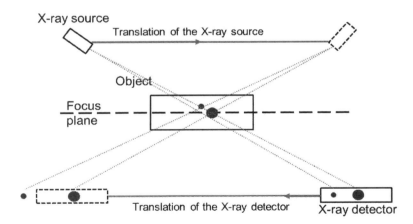

Fig. 3.32 Basic principle of laminography. X-ray source and detector move simultaneously. Features in the focus plane are always projected at the same position of the detector. Features out of the focus plane are projected at different positions of the detector therefore producing blurry reconstructed image

3.6   Safety Aspects of X-ray CT Systems

To reinforce the application safety of the industrial X-ray CT system, full attention should be paid to the radiological protection from the manufacturer to the customer. The safety system for an industrial CT system is normally composed of a shielded cabinet fitted with a warning system (light and/or sound), safety interlock equipment, radiation warning sign, and other related measures.

1. Shielded cabinet fitted with a warning system

The majority of the X-ray beam should be well attenuated within the cabinet. The purpose of the warning system is to evacuate the personnel before an X-ray leak. It consists of sound and light alarms, and broadcasting equipment. All CT systems should be equipped with external warning systems. Larger machines should also have internal warning systems.

2. Safety interlock equipment

Safety interlock equipment includes the emergency stop switch, interlock hatches and interlock key. External emergency stop switches should be equipped at convenient locations. In the case of larger machines, internal switches are also required. When the stop switch is pressed, the radiation and the motion of the scanning gantry stop. The machine can only be manually reset. Interlocked hatches are equipped at the shielded cabinet door. X-rays can only be generated with the hatch closed. In any emergency, the personnel should press at least one emergency stop button, unplug the machine if possible, and remove the interlock key from the switch.

3. Radiation warning sign

Radiation warning signs incorporating the international radiation symbol and a legend such as 'Caution X-ray radiation' (see Fig. 3.33) should be attached to the machine.

The following precautions should be kept in mind by all the personnel who could potentially be exposed to an industrial CT system.

1. Radiation safety training programs must be provided to all the personnel before commencing work with CT equipment.

Fig. 3.33 Radiation warning sign

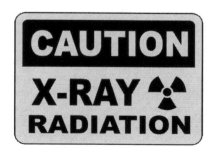

Basic radiation safety training by means of a short presentation should be given to all the personnel who have the chance to be exposed to the industrial CT system in their daily work. The training should provide adequate information about the fundamental concepts, measurements and principles of radiation protection. In addition to the minimum level of radiation safety training provided to everyone, the personnel who work closely with CT systems should have further training so they can perform their specific role responsibly. If necessary, refresher training should be given to ensure the knowledge and skills of the personnel are up-to-date.

2. The equipment should be tested and certificated before use. The certificate of the equipment should be regularly renewed.
3. Unauthorised personnel should never be allowed to operate the system, unless they are closely supervised by trained personnel.
4. Any failure of the security system should be reported to the manufacturer as soon as possible.
5. Always switch off the X-rays before opening the door of the machine. Do not rely on the interlock hatch to switch the X-rays off.
6. Before entering the interior of the cabinet, the interlock key should be removed and always kept with the personnel.
7. Always be aware of the signs and warnings.

References

Aloisi V, Carmignato S, Schlecht J, Ferley E (2016) Investigation on metrological performances in CT helical scanning for dimensional quality control. In: 6th conference on industrial computed tomography (iCT), 9–12 Feb, Wels, Austria
Bushberg JT, Boone JM (2011) The essential physics of medical imaging. Lippincott Williams & Wilkins
Chen L, Shaw CC, Altunbas MC, Lai C-J, Liu X (2008) Spatial resolution properties in cone beam CT: a simulation study. Med Phys 35:724–734
De Chiffre L, Carmignato S, Kruth JP, Schmitt R, Weckenmann A (2014) Industrial applications of computed tomography. Ann CIRP 63:655–677. doi:10.1016/j.cirp.2014.05.011
Ferrucci M, Leach RK, Giusca C, Dewulf W (2015) Towards geometrical calibration of X-ray computed tomography systems—a review. Meas Sci Technol 26:092003
Ferrucci M, Ametova E, Carmignato S, Dewulf W (2016) Evaluating the effects of detector angular misalignments on simulated computed tomography data. Precis Eng 45:230–241. doi:10.1016/j.precisioneng.2016.03.001
Fornaro J, Leschka S, Hibbeln D, Butler A, Anderson N, Pache G, Scheffel H, Wildermuth S, Alkadhi H Stolzmann P (2011) Dual- and multi-energy CT: approach to functional imaging. Insights Imaging 2:149–159
Goldstein J, Newbury DE, Echlin P, Joy DC, Romig AD Jr, Lyman CE, Fiori C, Lifshin E (2012) Scanning electron microscopy and X-ray microanalysis: a text for biologists, materials scientists, and geologists. Springer Science & Business Media
Hermann M, Pentek T, Otto B (2016). Design principles for industrie 4.0 scenarios. In: System Sciences (HICSS), 2016 49th Hawaii international conference, pp 3928–3937
Hocken RJ, Pereira PH (2011) Coordinate measuring machines and systems. CRC Press

Hsieh J (2003) Computed tomography: principles, design, artifacts, and recent advances, vol. 114. SPIE Press

Maass C, Knaup M, Kachelriess M (2011) New approaches to region of interest computed tomography. Med Phys 38(6):2868–2878

Otendal M, Tuohimaa T, Vogt U, Hertz HM (2008) A 9 keV electron-impact liquid-gallium-jet X-ray source. Rev Sci Instrum 79(1):016102

Panetta D (2016) Advances in X-ray detectors for clinical and preclinical computed tomography. Nucl Instrum Methods Phys Res Sect A 809:2–12

Probst G, Kruth J-P, Dewulf W (2016) Compensation of drift in an industrial computed tomography system. In: 6th conference on industrial computed tomography, Wels, Austria (iCT 2016)

Pyka G, Kerckhofs G, Pauwels B, Schrooten J, Wevers M (2011) X-rays nanotomography in a scanning electron microscope–capabilities and limitations for materials research. SkyScan User Meeting, Leuven, pp 52–62

Suetens P (2009) Fundamentals of medical imaging. Cambridge University Press

Salamon M, Errmann G, Reims N, Uhlmann N (2012) High energy X-ray Imaging for application in aerospace industry. In: 4th international symposium on NDT in Aerospace, pp 1–8

Salamon M, Boehnel M, Reims N, Ermann G, Voland V, Schmitt M, Hanke R (2013) Applications and methods with high energy CT systems. In: 5th international symposium on NDT in Aerospace, vol 300

Sunaoshi T, Orai Y, Ito H, Ogashiwa T (2007) 30 kV STEM imaging with lattice resolution using a high resolution cold FE-SEM. In: Proceedings of the 15th european microscopy congress, Manchester, United Kingdom

Taguchi K, Iwanczyk JS (2013) Vision 20/20: single photon counting X-ray detectors in medical imaging. Med Phys 40(10)

Welkenhuyzen F (2015) Investigation of the accuracy of an X-ray CT scanner for dimensional metrology with the aid of simulations and calibrated artifacts. Ph.D. thesis. KU Leuven

Chapter 4
Processing, Analysis and Visualization of CT Data

Christoph Heinzl, Alexander Amirkhanov and Johann Kastner

Abstract In an almost inexhaustible multitude of possibilities, CT allows to inspect highly complex systems and materials. Compared to other testing techniques CT provides results in a quick way: It is nondestructive and does not interfere with the specimen, it allows non-touching characterizations and what is most important CT allows to characterize hidden or internal features. However, CT would not have reached its current status in engineering without the achievements and possibilities in data processing. Only through processing, analysis and visualization of CT data, detailed insights into previously unachievable analyses are facilitated. Novel means of data analysis and visualization illustrate highly complex problems by means of clear and easy to understand renderings. In this chapter, we explore various aspects starting from the generalized data analysis pipeline, aspects of processing, analysis and visualization for metrology, nondestructive testing as well as specialized analyses.

4.1 Generalized Data Analysis Pipeline

In the following section, the generalized data analysis pipeline is outlined, which covers the following steps: (1) preprocessing and data enhancement, (2) segmentation, feature extraction and quantification, and (3) rendering of the results (see Fig. 4.1). We base this chapter on the assumption that the data is available as 3D volume, i.e. our discussion of the analysis pipeline starts from reconstructed volumetric images (reconstruction itself is covered in Sect. 2.3). The focus of this chapter will be put on preprocessing and data enhancement, feature extraction and quantification as well as aspects of visual analysis. It should be further noted that this chapter can only provide a coarse overview of the different areas with respect to industrial CT and that the discussion of the individual techniques is non exhaustive.

C. Heinzl (✉) · A. Amirkhanov · J. Kastner
University of Applied Sciences—Upper Austria, Wels Campus, Wels, Austria
e-mail: Christoph.Heinzl@fh-wels.at

© Springer International Publishing AG 2018   99
S. Carmignato et al. (eds.), *Industrial X-Ray Computed Tomography*,
https://doi.org/10.1007/978-3-319-59573-3_4

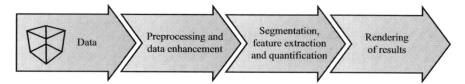

Fig. 4.1 Generalized data analysis pipeline

4.2   Preprocessing and Data Enhancement

This part covers different aspects of processing and enhancement of the generated CT data, which are required for quantitative characterization. We start our considerations with smoothing filters, and continue with segmentation and feature extraction techniques. Then we move on to the characterization of features, and end with surface determination methods.

4.2.1   Smoothing

Smoothing techniques may be classified with regard to their sensitivity range in global and local smoothing techniques. Both areas show advantages and drawbacks to be considered for data evaluation which we address in the following sections.

4.2.1.1   Global Smoothing

The overall goal of global smoothing techniques is found in the reduction of (additive) noise as well as other irregularities of the input data (see Fig. 4.2). Global smoothing techniques are typically based on convoluting images with constant filter kernels which are pre-set in accordance with the targeted smoothing technique. To compute the filter output the filter kernel is virtually moved voxel by voxel over the input data. For each position of the kernel the output data value is calculated as the weighted sum of the weights as given in the filter kernel voxels multiplied by the input data voxel values which are currently covered by the filter kernel voxels. A potentially critical drawback of global smoothing techniques is that they tend to blur edges and, in the result, to generate unsharp images.

One of the simplest implementations of global smoothing is the mean filter. This filter considers each voxel within the kernel with the same weight in order to compute the output voxel value as the weighted sum of all voxels. The weight per voxels is specified as $\frac{1}{N}$, where N is the number of voxels of the filter kernel. The weights allow the algorithm to avoid general drifts in grey values. For global smoothing, other kernels may be used as well. One very prominent smoothing filter

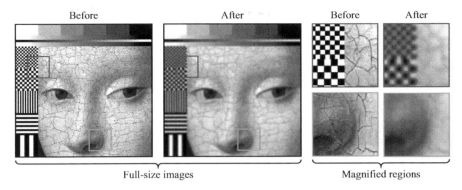

Before   After   Before   After

Full-size images   Magnified regions

Fig. 4.2 Example of Gaussian smoothing is performed using following parameters: variance = (15; 15), maximum error = (0.01; 0.01), maximum kernel width = 32

of this kind is found in Gaussian smoothing. The Gaussian kernel is defined in 3D-space by following equation:

$$G_\sigma(x, y, z) = \frac{1}{\left(\sqrt{2\pi}\sigma\right)^3} e^{-\frac{x^2 + y^2 + z^2}{2\sigma^2}} \qquad (4.1)$$

where σ is the standard deviation of the Gaussian distribution, and x, y, z are distances from the origin along x-axis, y-axis, and z-axis respectively. In general, the Gaussian kernel guarantees that voxels around the centre of the kernel are considered with the highest weights, while voxels with higher distance to the centre have lower influence on the result. As a side note, convolution based filters may also be computed in the frequency domain, where a simple multiplication replaces the convolution operation. Especially for larger filter kernels convolutions in the frequency domain show better performance in terms of faster computation times but implicitly require transformations of the data to the frequency domain and back.

Another filter type for global smoothing using kernels is found in median filtering (see Fig. 4.3). In contrast to mean and Gaussian smoothing, the median filter is not using weights but returns as output voxel the median of the voxels in the specified kernel neighbourhood. So basically, the kernel here defines a neighbourhood of considered voxels. Those considered voxels of the kernel are then sorted in an ascending manner. Of these sorted values the filter returns the median value as output. To compute the median filter output for the complete dataset, the filter kernel is again virtually moved voxel by voxel over the input data in order to compute the output data value. The median filter does not compute new values but uses existing grey values. It is especially useful for removing single outliers within the data values such as salt and pepper noise, e.g., generated by defective pixels.

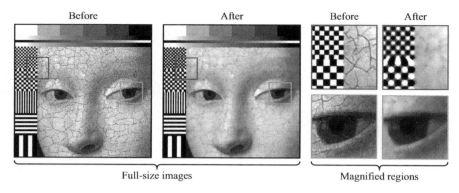

Fig. 4.3 Example of median smoothing performed using following parameters: radius = (4; 4)

4.2.1.2 Local Smoothing

In contrast to global smoothing, local (non-linear) smoothing techniques try to adapt the smoothing process in accordance to regional features of the input datasets. Local smoothing may be accomplished using a variety of filters. Aside their main advantage of adapting the smoothing process on the input data, these filters tend to be computationally more complex, which results in higher computation times and higher memory requirements. In the following paragraphs, we focus on two of them: bilateral filtering, and anisotropic diffusion.

Bilateral filtering as presented by Tomasi and Manduchi (1998) is a frequently used local smoothing technique (see Fig. 4.4). Bilateral filtering also computes for each voxel a weighted average of its neighbourhood. However, by considering the differences in the value ranges of the input voxels in the respective neighbourhood, the bilateral filter preserves edges while smoothing homogeneous regions. The key idea of the bilateral filter is therefore that voxels influencing other voxels should not

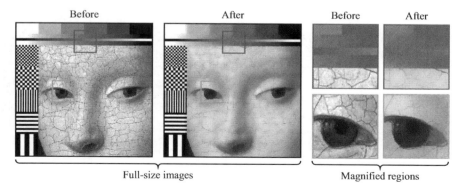

Fig. 4.4 Example of bilateral filtering. Following parameters are used: domain sigma = (10; 10), domain mu = 2.5, range sigma = 50, radius = (1; 1), number of range Gaussian samples = 100

only be nearby spatially, but they should also show a similar value. In this sense, the equation for the bilateral filter is given as follows:

$$BF[I_p] = \frac{1}{W_p}\sum_{q\in S}G_{\sigma_s}(\|p - q\|)G_{\sigma_r}(|I_p - I_q|)I_q \qquad (4.2)$$

G_{σ_s} and G_{σ_r} mainly determine the bilateral filtering process. G_{σ_s} is a Gaussian weight, which decreases the influence of the filtering spatially while G_{σ_r} is a Gaussian weight, which controls the influence of the filtering within the voxel value ranges. p and q are the considered voxel locations, while I_p and I_q are the intensities at the corresponding voxel locations. The weighting factor W_p ensures that the voxels sum to 1 and is specified as follows:

$$W_p = \sum_{q\in S}G_{\sigma_s}(\|p - q\|)G_{\sigma_r}(|I_p - I_q|) \qquad (4.3)$$

Anisotropic diffusion as introduced by Perona and Malik 1990 may be used similarly to smooth homogenous interior and exterior regions of the data while preventing from degrading or moving edges (see Fig. 4.5). To achieve this behaviour, anisotropic diffusion is based on a multi-scale description of images. Anisotropic diffusion embeds an image $U(\boldsymbol{x})$ in a higher dimensional function of derived images $U(\boldsymbol{x},t)$, which are considered to represent the solution of a heat diffusion equation. A variable conductance term C is applied to control the diffusion in the filter. Given the initial condition of $U(\boldsymbol{x},0) = U(\boldsymbol{x})$ (which represents the original image), the solution of the heat equation is outlined as follows:

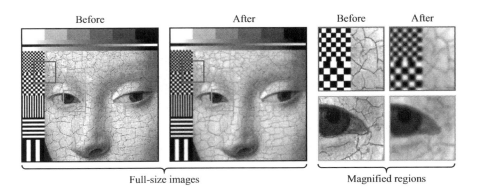

Fig. 4.5 Example of anisotropic diffusion filtering. Following parameters are used: time step = 0.01, conductance = 1, conductance scaling update interval = 1, conductance scaling = 1, fixed average gradient magnitude = 1000, number or iterations = 250, maximum Root-mean-square deviation error = 10, Root-mean-square deviation change = 50

$$\frac{dU(x,t)}{dt} = C(x)\Delta U(x,t) + \nabla C(x)\nabla U(x,t) \qquad (4.4)$$

The variable conductance term C is of core importance: the conductance term allows selectively preserving or removing features by anisotropically varying the diffusion strength. To ensure that the diffusion mainly takes place in homogeneous regions without affecting boundary regions, C is typically specified as a nonnegative and monotonically decreasing function as indicated in the following equation:

$$C(x) = e^{-\left(\frac{\|\nabla U(x)\|}{K}\right)^2}, \quad K = const \qquad (4.5)$$

Variants of the anisotropic diffusion filter for example take the local curvature into account to control the diffusion process, which may produce reasonable results in certain applications but again elevates the computational costs.

4.3 Segmentation, Feature Extraction and Quantification

One of the main steps in the generalized data analysis pipeline is found in the extraction and quantification of features within the data. To this end, it is assumed that the datasets have been preprocessed and enhanced in order to get an optimal result in the feature extraction step. We consider in this section segmentation and feature extraction techniques and end with regimes to quantify the features of interest. As for dimensional measurement applications interfaces between different materials are of interest, this section also gives an overview of surface extraction techniques.

4.3.1 Segmentation

Segmentation, i.e., the partitioning of an image or a volume into two or more regions, is of utmost importance for processing CT images into interior or exterior regions or to mask interesting regions of a scanned specimen. When it comes to quantification of features of interest such as voids, pores, inclusions or fibres the segmented result is the basis for any succeeding calculation such as the computation of lengths, diameters, or shape factors of the features of interest. As a wide range of applications implement tailored algorithms, many approaches are found in the literature. Most of those approaches share the problem of over- and under-segmentation, i.e. a too large or too small region of the feature is covered by the segmentation result, related to suboptimal parametrization of the algorithm. Aside from conventional segmentation techniques returning a single label for a voxel or region, more recently also fuzzy segmentation techniques are gaining importance.

4.3.1.1 Threshold-Based Techniques

Threshold-based segmentation algorithms are typically used when it is possible to distinguish individual objects in an image by their grey values. Threshold-based techniques select a threshold to differentiate between two objects. Depending on the segmentation technique this threshold can be global or local. The difference between global and local thresholds is that a global threshold is used for the whole image while a local threshold is adjusted for each voxel based on its neighbourhood. Threshold-based techniques also differ by how they select the used threshold. For example, Otsu's method (Otsu 1979) and ISO50 segmentation compute the corresponding thresholds between two peaks in the dataset's histogram (see Fig. 4.6c): While Otsu's method maximizes the separability of grey value classes in bimodal histograms, the ISO50 threshold assumes the correct threshold to be at the exact middle of two material peaks of interest. Furthermore, there are thresholding methods based on clustering methods such as k-means to segment an image (i.e., clustering-based thresholding). Regarding further threshold based techniques the reader is referred to the survey published by Sezgin and Sankur 2004, who provide a detailed overview of threshold based segmentation techniques.

4.3.1.2 Boundary-Based Methods

Boundary-based methods as well as threshold-based methods are used when it is possible to differentiate components by the grey value. The key difference between these two types of methods is that only boundary-based methods use (relative) differences of grey values of voxels within a neighbourhood to detect the borders between components. The main advantage of boundary-based methods is that they find local borders between components, which might be crucial for some application scenarios, e.g., where a grey value of a component is varying depending on location. In contrast to segmentation methods, edge detection filters only return regions where edges (i.e., material interfaces) are assumed. A well-established filter

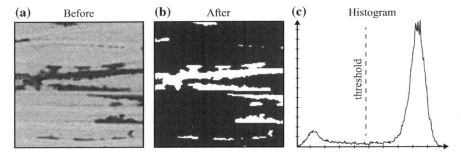

Fig. 4.6 Otsu thresholding converts a grey-scale image **a** into a binary mask, **b** the algorithm calculates a histogram, **c** and then automatically finds a proper threshold

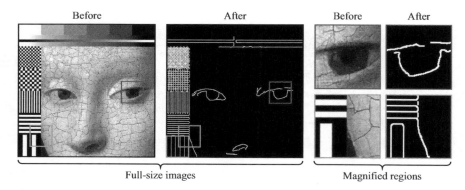

Fig. 4.7 Canny edge detection filter finds sharp edges in a grey-scale image (*left*). The result is a binary image, (*middle*) where pixels with high intensity (*white colour*) indicate sharp edges. Parameters: variance = (10; 10), maximum error = (0.2; 0.2), upper threshold = 14, lower threshold = 0

of edge detection is found in the Canny edge detector (Canny 1986). A result of the Canny filter is shown in Fig. 4.7.

4.3.1.3 Region-Growing

Region-growing methods start from one or multiple regions and grow these regions as long as a homogeneity criterion is valid (Adams et al. 1994, Zhigeng and Jianfeng 2007). The start regions, i.e., seeds, are either manually specified by the user or automatically set by an algorithm. A region-based segmentation approach is also found in the watershed transform, where the image is considered as a height field. (Beucher and Lantuejoul 1979, Beucher 1991) first applied this principle to image segmentation. Homogenous regions, also called catchment basins, are constructed by a flooding process. This process typically leads to strong oversegmentations. Felkel et al. 2001 therefore proposed to resolve this problem by placing

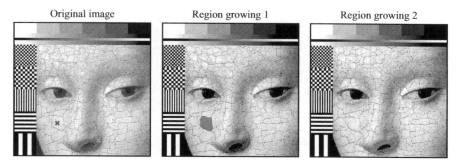

Fig. 4.8 A threshold interval based 2D region growing algorithm enlarges a region from a seed in the input image (*left*). The region is represented by *masks of the same color* superposing the input image (*middle and right*)

markers manually. Al-Rauush and Papadopoulos (2010) applied the watershed transform to segment touching or overlapping particles and to calculate the distribution of the local void ratio from 3D images. Figure 4.8 shows an example growing a region based on the red marked seed.

4.3.2 Feature Extraction

Features of interest are highly application and user specific and therefore lack a general description, apart from typically being regionally connected and corresponding to real structures of the specimen. In the medical context features of interest may be for example tumours, vessels, or bones; in the biological field these might be cells or agglomerations. In the materials science domain features of interest cover pores, cracks, inclusions, etc. Therefore, various feature extraction methods exist depending on their intended application case. In this section, we will address the most common practices to extract features.

4.3.2.1 Connected Component Analysis

Connected component analysis (Ballard and Brown 1982) was initially developed in graph theory, in which subsets of connected components get uniquely labelled using a given heuristic. For image processing purposes, this method separates an image into regions, which are connected based by their voxels. In this context connected component analysis is often confused with image segmentation, which are completely different domains: when features can be detected by using a segmentation algorithm, connected component analysis is used in addition to extract all separate connected regions of the resulting binary segmented image, which ideally correspond to the features of interest. Following this idea, connected component analysis is typically performed on binary images, where voxels with grey value equal to 0 represent the background, and voxels with grey value 1 represent the foreground, which is targeted to be to be separated. The main aspects in the classification of a voxel to a new connected component are different types of voxel connectivity for binary images. The most common neighbourhood types are listed below (see Fig. 4.9):

- 6-connected (voxels sharing a common face are neighbours);
- 18-connected (voxels sharing a common face or edge are neighbours);
- 26-connected (voxels sharing a common face, edge or corner are neighbours).

For every voxel of the binary segmented image the corresponding neighbourhood is checked. Depending on whether the considered voxel fulfils the neighbourhood criterion, the considered voxel is classified as part of the connected component. If the neighbourhood criterion is not fulfilled, a new region value, i.e.

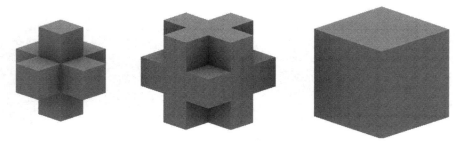

Fig. 4.9 Three types of voxel-connectivity: 6-connected (*left*), 18-connected (*middle*), 26-connected (*right*)

(a)   **(b)**

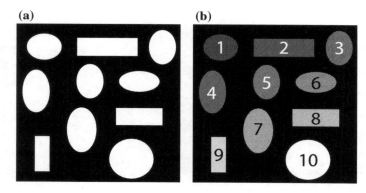

Fig. 4.10 Connected component analysis assigns a unique label to each component in a binary mask **a**. The result labelled image **b** maps pixels/voxels into unconnected objects

label, is assigned, incrementing the region label counter. Figure 4.10 shows the principle of connected component analysis. Connected component analysis may also be performed on greyscale or colour images. In this case however, more complex metrics of voxel connectivity incorporating information about grey value or colour are used.

4.3.2.2 Principal Component Analysis

Principal component analysis (PCA) is a mathematical method which is primarily used to discover and understand patters in the data (Wold et al. 1987, Jolliffe 2002) by reducing the dimensionality of the analysed data. Prominent application areas of PCA are found in face recognition applications or the analysis of satellite images but also in other high dimensional data domains. PCA calculates the orthogonal transformation where new axes (i.e., the principal components) are linear combinations of variables. Principal components are calculated in such a way that the first

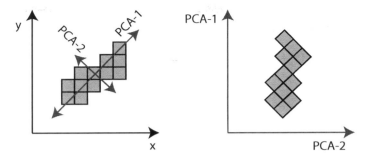

Fig. 4.11 PCA may be used to calculate dimensions of objects in volumetric images. The first Eigen vector (PCA-1) of the covariance matrix indicates the main object orientation (*length*), the second Eigen vector indicates the width direction

component features the largest possible variance and each succeeding component shows again the highest possible variance under the constraint, that it has to be orthogonal to all preceding components.

Since the first principal component features the largest variance, it is used as the main or the primary direction of the considered object. Accordingly, the variance of the first component returns the length of the object. The variances for the second and third components are width and depth correspondingly. An example is shown in Fig. 4.11.

4.3.2.3   Hessian Analysis

Another method to extract orientations of a feature or region of interest is found in the Hessian analysis which allows the user to extract the main directions of features by computing the Hessian matrix (Frangi et al. 1998). Mathematically, the Hessian matrix describes the local curvature of a function of many variables, i.e., in case of CT data applied to a scalar field (Gradshteyn and Ryzhik 2000). The Hessian matrix itself is a square matrix of second-order partial derivatives of a scalar field. For an n-dimensional image the Hessian matrix is calculated by using the following formula:

$$H = \begin{bmatrix} \frac{\partial^2 f}{\partial x_1^2} & \frac{\partial^2 f}{\partial x_1 \partial x_2} & \cdots & \frac{\partial^2 f}{\partial x_1 \partial x_n} \\ \frac{\partial^2 f}{\partial x_2 \partial x_1} & \frac{\partial^2 f}{\partial x_2^2} & \cdots & \frac{\partial^2 f}{\partial x_2 \partial x_n} \\ \vdots & \vdots & \ddots & \vdots \\ \frac{\partial^2 f}{\partial x_n \partial x_1} & \frac{\partial^2 f}{\partial x_n \partial x_2} & \cdots & \frac{\partial^2 f}{\partial x_n^2} \end{bmatrix} \tag{4.6}$$

The Hessian matrix is used in many algorithms to find and detect edges, ridges, critical points, and to characterize the structure's orientation. In the case of fibre

Table 4.1 Correlation between material structures in the data and eigenvalues (L low, H+ high positive, H− high negative)

λ_1	λ_2	λ_3	Structural orientation
L	L	L	Noise (no structure with preferred direction)
L	L	H−	Brighter structure in the form of a sheet
L	L	H+	Darker structure in the form of a sheet
L	H−	H−	Brighter structure in the form of a tube
L	H+	H+	Darker structure in the form of a tube
H-	H−	H−	Brighter structure in the form of a blob
H+	H+	H+	Darker structure in the form of a blob

reinforced composites, it is also used for extracting the main orientation of individual fibres. Furthermore, the analysis of eigenvalues of the Hessian matrix returns important information about image structure orientation. Table 4.1 gives an overview about the correlation between material structures in the data and eigenvalues.

4.3.2.4 Template Matching

Template matching is a technique for discovering areas in an image, which are similar to a given template/patch image (see Fig. 4.12). The idea of template matching approaches is similar to the idea of image convolution: the patch image is moved voxel-by-voxel over the source image and for each position of the patch a similarity is calculated based on a predefined heuristic (Cole et al. 2004). A modification of the algorithm also considers changing orientations of the template to account for similar structures in different orientations. In template matching, areas where the similarity is high lead to the conclusion, that searched feature is

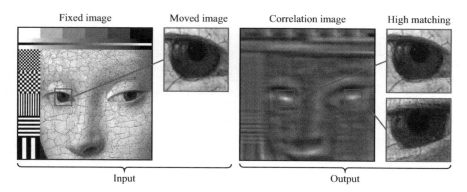

Fig. 4.12 Normalized correlation image filter is a template matching technique using an input image (*left*) and a patch to calculate normalized cross correlation image (*right*). Pixels with high intensity in the normalized cross correlation image indicate high similarity of corresponding regions in the input image to the patch image

located in the investigated area. Various modifications of the template matching approach are used to extract blood vessels (Frangi et al. 1998) in the medical field. In the industrial field, an application area of template matching is found in extracting fibres (Bhattacharya et al. 2015).

4.3.2.5 Clustering

Clustering is a process of grouping objects in such a way that similar objects are united in the same cluster and thus assigned the same cluster ID. Clustering plays an important role in the quantification of features. For example, features may be clustered by size, shape, or location in order to derive statistical information. When the number of expected clusters is known (e.g., through the number of the different features), the user presets of the number of clusters can be used as input for the clustering algorithm. The clustering algorithm iteratively divides a set of objects into clusters to maximize predefined criteria of object similarity. One of the most popular clustering algorithms is the K-means (Selim and Ismail 1984) which separates an input set of objects into k clusters, where k is a user specified number. K-means clustering determines that an object only belongs to a specific cluster if the distance between this object and the cluster is shorter than the distance to any other cluster.

Another popular group of clustering algorithms is hierarchical clustering (Kohonen 1990), which are helpful when the number of clusters is not known before the clustering step. There are two approaches to build the hierarchy of clusters: bottom-up and top-down approaches. Algorithms that use the bottom-up approach start from placing each observation into a separate cluster. Then, on each cycle of the algorithm, similar clusters are merged until there is only one cluster containing all observations. Algorithms that use the top-down approaches, in contrast, initially place all observations into one cluster. Then in each iteration clusters are split until each cluster contains only one observation. Other clustering algorithms use other preconditions of an object belonging to a cluster, e.g., connectivity-based clustering, proximity-based clustering, or distribution-based clustering.

4.3.2.6 Feature and Cluster Quantification

When all features are separated, they may be quantified in terms of the characteristics of interest. Aside simple general characteristics (e.g., lengths, diameters, surface, volume) and application specific characteristics (e.g., main directions, shape factors, start and endpoint, aspect ratio). Also derived characteristics (e.g., the total volume or surface area of all features) are computed directly using the results of connected component analysis.

Often the extracted features are clustered into classes of similar features. For example, pores in glass fibre reinforced polymer composites might be clustered/filtered by volume, size, shape or direction. For many applications, also statistical analyses are of interest. A statistical evaluation, e.g., of classes of extracted features of interest in order to compute mean or median characteristics (Weissenböck et al. 2014) as well as shapes (Reh et al. 2013), is highly relevant for subsequent analyses such as finite element simulations. In addition, further information may be computed using the underlying grey values covered by the individual features.

4.3.3 Surface Determination

Surface extraction as well as the determination of the surface positions for dimensional measurement features is crucial in dimensional metrology. In contrast to segmentation, surface extraction techniques find a precise border between components in the image (e.g. material and air). Often, surface extraction techniques involve segmentation as the first step to roughly estimate regions of components and the borders between them for subsequent refinement in the surface determination step. In the following section an overview of surface extraction techniques is presented followed by a consideration of various subpixel approaches for edge detection.

4.3.3.1 Surface Extraction

Regarding surface extraction for industrial metrological applications, surfaces are typically first estimated by specifying a single threshold, i.e. an isovalue, which is supposed to subdivide the data into interior and exterior regions of the materials or specimens to be investigated. Besides empirical determination of this threshold, the so called ISO50 threshold is widely used (VolumeGraphics 2016). ISO50 in essence is calculated from the dataset's grey value histogram. For this purpose, the grey values at the position of the histogram peaks of the material and the air regions are extracted, in order to compute the ISO50 value as the mean of both peaks' grey values. Depending on the analysis tasks either a surface model is extracted for further processing (e.g. variance comparison between surface model and CAD model), or the dimensional measurement features of interest are directly extracted from the volumetric dataset.

In case of surface model based data evaluation, a polygonal mesh is extracted using the empirically determined isovalue or using the ISO50 value by means of a surface creation algorithm. The algorithm which is up to now mostly used is the marching cubes algorithm as originally proposed by Lorensen and Cline (1987). Marching cubes creates triangle models for surfaces of constant (grey) value from scalar fields. The algorithm goes through the data taking 8 neighbourhood locations into account, which form a small cube. For this cube the algorithm determines for all the 8 neighbours whether they are inside or outside of the targeted surface based on the specified grey value (e.g., ISO50). If the grey value of a considered neighbour is below the threshold it is supposed to be outside, if it is higher, it should be inside. Based on the inside/outside signature of the neighbours in the cube, which is stored in an 8-bit value, one out of 256 unique cases (2 states to the power of 8 cases of how the triangles or polygons may pass through the cube) of the surface passing through a cube may be assigned. To match the underlying data, the triangles passing through the cube as well as their vertices have to be adapted by linear interpolation with regard to the grey values of the cube. In the last step, all individual polygons are fused into the final surface model.

There are a lot of algorithms improving on marching cube. For instance, Flying Edges as presented by Schroeder et al. (2015) is a recent algorithm for extracting surfaces which performs according to the authors up to 10 times faster than the original marching cubes algorithm. Flying Edges is a high-performance iso-contouring algorithm, which is subdivided in 4 passes. Only on the first pass the data is traversed entirely. All subsequent passes operate on local metadata or locally on rows to generate intersection points and/or gradients. At the end, the algorithm produces the output triangles into the preallocated arrays.

4.3.3.2 Subpixel Approaches to Edge Detection

Under ideal conditions global surface extraction techniques would extract the optimal surface for homogeneous single material components. However, due to artefacts in the CT data, irregularities of the scan process, the fact that specimens tend to be heterogeneous, as well as numerous other effects, erroneous surface models are created. For example, artefacts may modify the grey values of the generated CT datasets in regions of complex geometry or high penetration lengths. Using global surface extraciton techniques air regions may be added to the surface model, while in other areas material regions may be thinned and even structural changes such as holes may be introduced (Heinzl et al. 2006). Therefore, more sophisticated methods are required for surface extraction. We distinguish here the following classes of subpixel techniques for edge detection: moment-based methods, reconstructive methods, and curve-fitting methods.

Moment-based methods are using either intensity moments of the image, which are based on the voxel grey values, or also spatial moments, using spatial information about the voxel neighbourhood, in order to determine the position of the edge. As an example of this class we indicate the subpixel edge detection method of Tabatabai and Mitchell 1984, which aims to fit three image intensity moments to the ideal step edge. In their approach the moments are not considering any spatial information but define the moments as a sum of pixel intensity powers.

Reconstructive methods target to retrieve the continuous information of edges in the discrete CT image. Using various interpolation and approximation techniques the continuous image function is reconstructed based on discrete sample values. Reconstructive methods for edge detection may focus on image intensity (e.g., Xu 2009), on the first image derivative (e.g., Fabijańska 2015) or on the second image derivative (e.g., Jin 1990). Following this idea in the domain of point sets and surface models, Steinbeiss (2005) locally adapt initial surface models extracted from the volumetric CT datasets. Using an initial surface model of the specimen, Steinbeiss does not calculate the full gradient fields but for performance reasons only profiles. These grey value profiles are calculated in the direction of the surface normal of each point in the surface model. The location of each point is subsequently adjusted to correspond to the position at maximal gradient magnitude of the grey value profile. A modified version of this algorithm was used by Heinzl et al. (2007) for extracting surfaces of CT datasets from multi-material components.

Curve-fitting methods for edge detection try to extract the edges by fitting curves into points which are determined by conventional edge detection operators in pixel accuracy. In this sense edge points as determined by the Canny operator were for example used by Yao and Ju (2009), fitting cubic splines into the extracted spatial edge points using Canny edge detection. Following this principle Gaussian curve fitting, as presented by Fabijańska 2015, is used for precise determination of edge location. For Gaussian curve fitting a profile for the point is calculated by sampling gradient along the gradient's direction. Then the Gaussian curve is fitted to the samples. The location of the fitted Gaussian peak is then considered as the edge location. Aside cubic splines and Gaussian curves also 2nd, 3rd, or higher order polynomial curves are used for the fitting using the same principle as before (e.g. as proposed by Lifton (2015)).

4.4   Visual Analysis

The last and one of the most important steps in the generalized data analysis pipeline is the visualization mapping and rendering of the final results for visualization and visual analysis of the underlying data (see Fig. 4.1). In this step data is visualized in order to be interpreted by the domain experts. The main task of visualization is therefore to represent and transport important information by means of visual metaphors. The way how a dataset is visualized strongly depends on the data characteristics as well as on the application the data is used for. In this section, we first indicate non-exhaustive general concepts of visual analysis tools (e.g. colour mapping, juxtaposition, superposition), and subsequently explore various visual metaphors (e.g. scatter plots, glyphs or heat maps) they are using.

4.4.1   Common Concepts

In the following sections, we present common visualization concepts related with color theory, and positioning of objects in the scene. These concepts are basic and they are used in many visualization tools.

4.4.1.1 Colours and Colour Maps

The colour model describes the mathematical model of how colours are represented. Colours are typically used as supportive concept to stress features of interest. There are many different colour models designed for various applications. However, the most popular are RGB, CMYK, and HSL and HSV colour models. The RGB colour model represents a colour as a mixture of three basic colours: red, green, and blue. Any colour can be obtained by mixing these basic colours, e.g., yellow is obtained by mixing red and green. In contrast the CMYK colour model uses four basic colours: cyan, magenta, yellow and key (black). The HSL and HSV colour models are based on hue, saturation, lightness, and value. Hue is an important colour property which is used to denote "pure" colours such as red, yellow, green, blue etc. Often hue is represented in a colour wheel (see Fig. 4.13).

Typically, a colour wheel is drawn in such a way that similar colours are located closely to each other and the most dissimilar colours are located in opposite sides of the wheel. For example, in Fig. 4.13 opposite colours are: red and aqua, yellow and blue, lime and fuchsia. Often, the hue property is used to highlight dissimilarity of features in a volumetric image. For example, the combination of yellow and blue colour can be used to highlight two features. The combination of red, lime, and blue colour can be used to highlight three features.

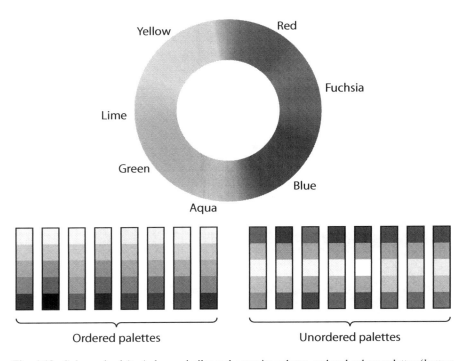

Fig. 4.13 Colour wheel (*top*) shows similar and opposite colours, ordered colour palettes (*bottom left*), and unordered colour palettes (*bottom right*)

A colour map sets the correlation between a data value and a colour (Munzner 2014). For example, a colour map can be designed to distinguish the size or orientation of features in the data or to highlight the variance in a certain feature. Typically, palettes are used to define colormaps. A palette is a set of colours logically related with each other. Palettes can be either ordered or unordered. Ordered palettes are typically used to highlight transactions between two or more objects in an image (see Fig. 4.13). Unordered palettes are used to differentiate objects from themselves without focusing on their transactions (see Fig. 4.13).

4.4.1.2   Juxtaposition

Juxtaposition visualization shows two or more images together (see Fig. 4.14). The images can be shown in rows, in columns or in arbitrary order. Juxtaposition is also sometimes referred as dual-viewing, showing only two images as side-by-side views with the images ordered in a row. Typically, juxtaposition visualization is used to show different aspects of an object or objects of the same class.

4.4.1.3   Superposition

Superposition visualization overlaps two or more images in the same viewport (see Fig. 4.15). It allows the user to perceive how the images are connected in space, however this visualization hides some parts of the original images. Often, the overlapping mode may be adjusted. For example, it is possible to change the

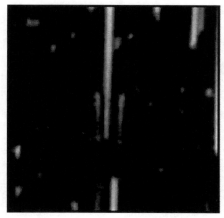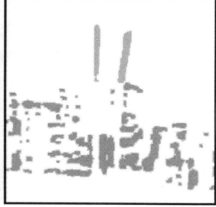

Fig. 4.14 Juxtaposition visualization shows an original CT image of fibre reinforced composite with defects inside (*left*) and the corresponding defect map showing defect regions and their types (*right*)

Fig. 4.15 Example of
superposition visualization
combining an original CT
image of fibre reinforced
composite and a defect map

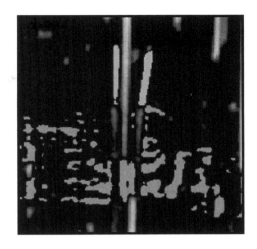

transparency of one or another image or set the overlapping order. Superposition is
used in many visualizations to emphasize some certain features or regions.

4.4.1.4   Ben Shneiderman's Visual Information-Seeking Mantra

Shneiderman (1996) proposed a set of guidelines called visual information-seeking
mantra to design advanced graphical user interfaces. The visual information-
seeking mantra includes three subsequent steps of an analysis: overview first, zoom
and filter, and details-on-demand. In the first step the user explores the entire
dataset. The way, how the dataset should be presented depends on data dimensionality and type. For example, for a 3D volumetric dataset a zoomed-out volume
rendering can be used to provide an overview. Typically, the user can find interesting regions or features or features. In the second step the user emphasizes these
regions by performing a zooming operation or applying a filter to hide uninteresting
features or features. In the third step, the user browses and selects individual items
to get additional information. This information can be displayed in context
using pop-up windows or in a specific place of the user interface.

4.4.2   Visual Metaphors

In following sections, we present common visual metaphors concerning volume
rendering, visualization of features, and comparative visualization. The presented
set of visual metaphors is not exhaustive.

4.4.2.1 Volume Rendering

Volume rendering is a type of visualization techniques designed to display 3D volumetric images. Typically, volume rendering techniques use colour and opacity transfer functions to map a volumetric image to a data structure consisting of elements with opacity and colour. Then this structure is visualized by means of a volume rendering algorithm. Aside the specification of transfer functions, the main challenge related with volume rendering is that rendering 3D images is a computationally-costly process. The size of a CT image can go up to a several hundreds of gigabytes which requires a powerful hardware for real-time rendering. There is a large variety of volume rendering techniques balancing between speed and quality of the final image. For example, volume ray casting techniques allow the user to achieve a high quality of the rendering while the spatting technique was especially designed to minimize rendering time.

Colour and opacity mapping to the features of interest is a very important aspect related with volume rendering. A colour transfer function establishes the correspondence between grey values in the image and colours. An opacity transfer function establishes the correspondence between the grey values and opacity. Figure 4.16 shows an example of different setting of a colour transfer function.

4.4.2.2 Slicing

A simple but efficient visual metaphor is slicing. Slicing refers to visualization techniques that allow the user to virtually cut a volumetric dataset, i.e., to extract and render an arbitrary 2D slice from a volumetric image. Slicing techniques are very important and widely-used in the field of visual analysis of CT data. These techniques provide an intuitive representation of CT data. However, the most

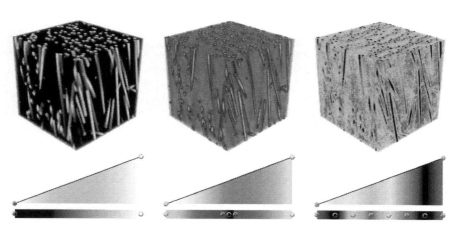

Fig. 4.16 Example of volume rendering with different colour transfer functions

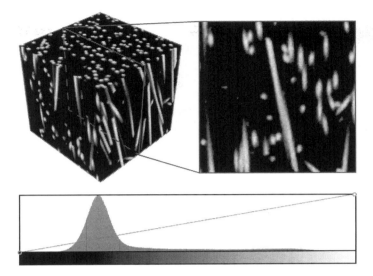

Fig. 4.17 Example of slicing: the slicer (*right*) shows a slice from the CT image (*left*)

important advantage is that slicing allows the user to compare also different modalities such as microscopy at the same position in the dataset. Typically, slicing techniques are interactive and they allow the user to set the slice position and orientation. An example of a slicing technique is shown on Fig. 4.17.

4.4.2.3 Heat Maps

A heat map graphically represents the density of the occurrence of a feature of interest in a dataset by using colour mapping. Typically, regions with different densities are therefore differentiated by colours (see Fig. 4.18). When the feature is represented as a point cloud, a proper density function has to be calculated beforehand. Density functions may be calculated in different ways: The most common ways to compute density functions are indicated as follows: (1) dividing

Fig. 4.18 Example of the heat map visualization of voids in a CT dataset. The colour transfer function (*right*) sets the correspondence between a feature density and colour (e.g., high porosity-density regions have red colour while low porosity-density regions have *blue colour*)

the image into sectors and calculating the average value for each sector; (2) using a kernel based approach. Heat maps are also an important means of data abstraction, e.g. features in a 3D dataset are projected onto a 2D plane as shown in Fig. 4.18.

4.4.2.4 Scatter Plots and Scatter Plot Matrices

A scatter plot is a visual metaphor, which is typically used to show the correlation of two variables in a set of data points (Weissenböck et al. 2014; Reh et al. 2015). Abscissa and ordinate describe two characteristics of interest. All data points are integrated in the scatter plot at the position of their magnitude with respect to the selected characteristics. As each point in the scatter plot represent values for two user-defined variables (see Fig. 4.19), clusters may be identified in this view. If more than two variables of a dataset are analysed, each characteristic may be plotted over the other one. To analyse two and more scatter plots, they may be organized in a matrix., i.e., the scatter plot matrix. Due to perception and screen space constraints, typically scatter plot matrices feature up to 10 plots. For higher dimensional data, the variables to be shown in the scatter plot matrix may be pre-selected.

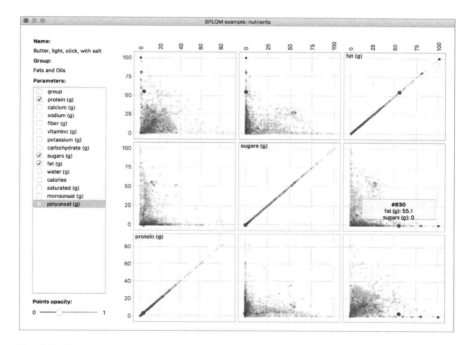

Fig. 4.19 Example of scatter matrices showing correlations between protein, sugars and fat in 7637 different food products. The selected spot (*blue dot*) is butter

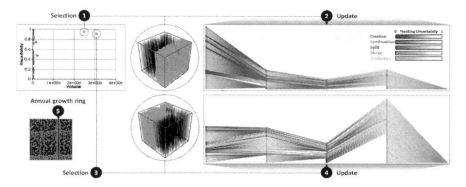

Fig. 4.20 Example of a tracking graph of pores in wood during the drying process. Two points, showing big agglomeration of pores, are consequently selected in a scatter plot. These agglomerations show similar behavior to each other.

4.4.2.5   Graphs

A graph is a graphical representation of a structure consisting of vertices and edges connecting the vertices. Graphs are a very convenient way to represent the connectivity of features in data. For example, graphs are widely used in the field of visualization of time-varying data since they can graphically represent which features are connected and how they are connected in the different time-steps (Reh et al. 2015). An example of such a graph is shown in Fig. 4.20. One of the biggest challenges related to graph visualization is calculating the best graph layout.

4.4.2.6   Parallel Coordinates

Parallel coordinates is a method of visualizing n-dimensional data. Variables are represented by n parallel bars, i.e. the parallel coordinates. Each data point is represented by a polyline connecting the bars. The place where the polyline crosses a bar (i.e., the characteristic) shows the variable's value (see Fig. 4.21). The main

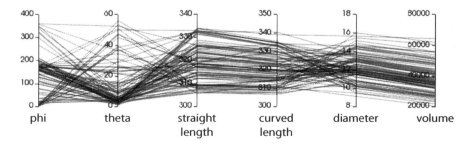

Fig. 4.21 Example of parallel coordinates showing fibre characteristics (phi and theta angles showing fibre orientation, straight and curved fibre lengths, diameter, and volume) extracted from a CT scan of fibre reinforced composites

advantage of parallel coordinates is that they represent the entire set of data in one 2D diagram and show hidden correlations in data. Often, parallel coordinates are implemented as an interactive widget. The user can select ranges of interests on bars for each characteristic to filter points. It therefore provides a detailed overview of how variables are correlated. Linked to 2D slicers and 3D views parallel coordinates may be used to explore features as well as characteristics of interest extracted from CT data.

4.4.2.7 Glyphs

Glyphs in visualization are symbols and markers such as ellipsoids or arrows (see Fig. 4.22). Glyphs can be used to encode vector based information, tensor based information to mark specific higher order behaviour in the spatial context of an object. For example, glyphs are used to show wind speeds and directions in meteorology, fibre orientations when analysing fibre reinforced composites (see Fig. 4.22), or regions of high uncertainty. Glyphs may have simple (e.g. arrows or cylinders for encoding directions of distances—Heinzl et al. 2006) or complex shape encodings (e.g. superquadrics for encoding tensor glyphs—Kindlmann 2004), in accordance with the corresponding application. Despite being simple and easy to understand glyphs are also used to encode higher dimensional data using more complex shapes such as superquadrics (Barr 1981).

Fig. 4.22 Glyphs show the main orientation of fibres in a CT scan. Glyphs are represented as superquadrics, the longest edges of them point towards to the local fibre orientation

Original   Comparison   Reproduction

Fig. 4.23 Example of a checkerboard comparative visualization approach showing the difference between the original portrait of Mona Lisa and a copy made by unknown artist

4.4.3   Comparative Visualization

Comparative visualization refers to methods that are used to compare two or more images in order to find similarities and dissimilarities. Comparative visualization plays an important role in understanding dynamic processes or the difference between modalities used to produce CT images. The simplest comparative visualization technique is a juxtaposition visualization (side-by-side rendering of two images). Another approach is found so-called checkerboard visualization which divides the viewport into alternating cells (i.e., the black and white cells of a checkerboards) where each cell renders one out of two modalities. So, cells connected by an edge render different modalities of the same specimen which allows the user to directly compare them (see Fig. 4.23).

4.4.4   Uncertainty Visualization Concepts

Most visualization techniques assume that the data to be visualized is precise and has no uncertainty. This assumption might lead in some cases to wrong results of a data analysis. Uncertainty visualization techniques are overcoming this issue by incorporating information about the underlying uncertainty in the visualization process. In the field of CT, uncertainty visualization is mostly used to visualize the uncertainty induced by applying data processing techniques such as segmentation or surface extraction. The most typical approaches are mapping uncertainty values on the surface (Rhodes et al. 2003) or using point-based probabilistic surfaces (Grigoryan and Rheingans 2004). In the first case, the uncertainty is encoded using a colour transfer function, e.g., green when uncertainty is low and red when uncertainty is high. If uncertainty for some point is high, the colour of this point will be shifted toward red and vice versa (see Fig. 4.24). In the second case, a surface is represented as a cloud of points laying on the surface. Points can be shifted in the surface normal direction according the following formula:

$$P' = P + Uncertainty \times \vec{v}_n X \sim U([-1, 1]) \qquad (4.7)$$

where $X \sim U([-1, 1])$ is a random variable uniformly distributed on $[-1, 1]$, *Uncertainty* is uncertainty value, $\vec{v}_n$ is the normal vector of the surface, P is the previous position of the point.

4.5   Processing, Analysis and Visualization for Metrology

Dimensional measurements as well as geometric tolerancing are crucial instruments of metrology in the area of quality control for industrial components. To perform dimensional and geometric measurements, typically tactile or optical coordinate measurement machines (CMMs) are widely used in industry. However, these conventional techniques share a major drawback: both techniques are not able to measure hidden or internal features but only surface based and accessible areas.

Due to its intrinsic principle, CT allows capturing both internal and external structures and features within one scan. Due to the development of systems with

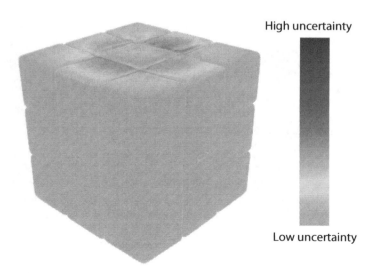

High uncertainty

Low uncertainty

Fig. 4.24 Example of uncertainty visualization: uncertainty values are mapped on the surface

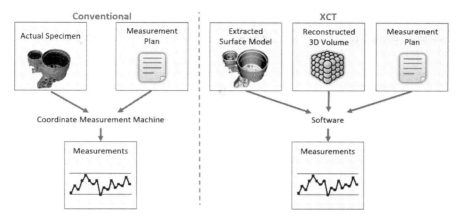

Fig. 4.25 General data processing workflow for dimensional metrology, conventional workflow versus 3DXCT

high accuracy, CT has been increasingly adopted and applied for metrology over the last decades. Nevertheless, CT also faces critical issues which may hinder precise measurements: For CT, the location of the surface has to be extracted from the volumetric dataset of the specimen, as discussed before, as the surface is only implicitly covered in the discrete grid of the CT dataset. Depending on artefacts, additive noise and other irregularities in the scanned CT data, the used preprocessing (e.g., reconstruction, prefiltering), as well as the surface extraction algorithm, this implies a certain positional uncertainty which should not be neglected. However, at the time of writing this chapter, most commercial CT metrology software tools do not account for uncertainty within CT data. Figure 4.25 outlines simplified data processing workflows of conventional metrology as well as CT based dimensional metrology using surface or volumetric data as input.

Below we present several approaches targeting to improve data by fusion concepts and accounting for uncertainty prevalent in CT data and solving the following tasks:

- Account for uncertainty arising in the CT metrology data flow by estimating material (Heinzl et al. 2008) or material interface probabilities (Amirkhanov et al. 2013) and encode the prevalent uncertainty information on the surface or in direct volume renderings (Heinzl et al. 2008).
- Enhance conventional visualization tools such as measurement plot, tolerance tags with information on uncertainty (Amirkhanov et al. 2013).
- Fuse data to enhance dimensional measurement precision (Heinzl et al. 2007; Schmitt et al. 2008; Heinzl et al. 2010; Amirkhanov et al. 2011).
- Reduce artefacts by data fusion (Schmitt et al. 2008; Heinzl et al. 2010; Oehler and Buzug 2007, Wang et al. 2013, Amirkhanov et al. 2011).

4.5.1 Data Fusion of CT Data for Metrology

One of the most critical problems of metrology using computed tomography CT is found in artefacts and their effects on the data as well as the derived measurements thereof. A general definition of artefacts in the domain of CT describes artefacts as artificial structures in the CT datasets, which do not correspond to real structures of the specimen. Industrial computed tomography systems based on cone beam geometry are prone to artefacts such as noise-induced streaks, aliasing, beam hardening, partial volume and scattered radiation effects (Hsieh 2003) (see Sect. 5.2). Artefacts are especially pronounced in terms of multi material components with high density material such as steel or copper within low density material such as polymers or reinforced polymers, which may be critical for dimensional measurements and even prevent from reliable metrology. In literature, e.g. (Schmitt et al. 2008), artefacts of this kind are also subsumed using the term "metal artefacts". Unfortunately, a large number of industrial components features this kind of material, e.g., using copper for wiring and polymer composites for the housing.

Data fusion as a means of combining image information take complementary strengths and weaknesses into account. A simple and straight forward approach of data fusion for CT data to reduce metal artefacts is found in the concept of dual viewing (artefact reduction techniques are closely related to the preprocessing and data enhancement step which are covered in Sect. 5.3). For applying this concept, a specimen is scanned multiple times (at least twice) using different positions, orientations or placements of the specimen on the rotary plate of the CT system (Schmitt et al. 2008; Heinzl et al. 2010). As the characteristic of artefacts changes spatially due to the specimen's geometry in accordance with the used position, orientation or placement, the differences in the directional artefacts may be detected by suitable data processing pipelines and corrected in the fused dataset. These pipelines first require a registration step in order to guarantee that the structures of both datasets are coinciding. The next step focuses on changes in the data of each scan (i.e. edges, interfaces) which are detected by calculating the gradient magnitude information. By a voxelwise detection of regions with strongly deviating gradient magnitude information artefacts are identified. This information of the regional distribution of artefacts is the basis to calculate the weights of the subsequent arithmetic fusion step. For local arithmetic fusion, the contribution, i.e. the weight, of a considered voxel in the individual scans for the fused dataset is computed by linearly weighting differences in the corresponding gradient magnitude images of the individual scans. With this method, a large part of artefacts may be reduced. However, orientation invariant artefacts or artefacts which are affecting in both datasets the same region cannot be eliminated.

Another strategy to overcome aforementioned artefacts exploits the concept of dual energy computed tomography (DECT) or multi energy CT. Applying this DECT strategy, the previously discussed concept of dual viewing is not necessarily a prerequisite, but may be used in addition for providing improved metrology results. Using DECT, the specimen is scanned twice (or multiple times for multi energy CT), with different X-ray energy setups. Preferably each setup uses optimal CT parameters for the different materials, generating datasets with complementary strengths. Using multiple exposures different features in the corresponding attenuation images are targeted to be pronounced or revealed. However, due to the material specific X-ray energy optimization also artefacts are introduced. Data fusion is used in several approaches either in the domain of projection images or in the domain of reconstructed CT images to integrate the complementary characteristics of DECT datasets in a resulting fused dataset, which features the strengths of each dataset but suppresses the weaknesses of the individual datasets. In order to consider spatial features most of those algorithms are operating in the domain of reconstructed data. In contrast, Oehler and Buzug (2007) describe a method for artefact suppression in domain of projection images which is used as basis for related data fusion techniques. The main idea of this approach is to identify artefact affected regions in the sinogram of the corresponding CT dataset. To achieve this identification, strongly absorbing parts (i.e. metal) are segmented. The segmented metal parts are then forward projected onto the sinogram. Using interpolation techniques all metal parts are removed. Wang et al. (2013) are building their fusion prior-based metal artefact reduction approach on a similar concept, extending it by fusing the metal-removed uncorrected image and a precorrected image, respectively. Amirkhanov et al. (2011) similarly segment metal regions and forward project the metal parts onto the sinogram. Using interpolation, the metal regions are replaced by surrogate data computed from close by regions in order to reconstruct a nearly artefact free dataset. In the final step data fusion is applied to reintegrate the metal parts into the artefact reduced dataset.

A method directly operating in the reconstructed image domain was proposed by Heinzl et al. (2007). This approach makes use of the different setups of a dual source CT system: A high energy setup using a minifocus X-ray source generates nearly artefact free but blurry data, while a low energy setup using a high precision microfocus X-ray source generates artefact affected but highly precise data. The declared goal of this method is use the structure of the specimen from the low precision, blurry, high energy dataset while the sharp edges are adopted and fused into the resulting image from the high precision, crisp, low energy dataset. In their data fusion scheme, the authors mainly focus on fusing regions around edges or interfaces. In these areas, the contribution of each dataset is linearly weighted according to the absolute-value difference between the input datasets. The drawback of edge based fusion is that the method only works if the datasets feature similar grey values for similar materials, which might require a further preprocessing step for data mapping. In a succeeding work the authors proposed a concept to combine complete datasets instead of only fusing edge regions (Heinzl et al. 2008) and thus avoids irregularities in the histogram. For both of their approaches reliable surface models are extracted for dimensional metrology from the fused dataset using local adaptive surface extraction techniques.

4.5.2   Visual Analysis of Dimensional Measurement Features

Compared to conventional dimensional metrology using tactile or optical coordinated measurement machines, the surfaces as well as the interfaces are not explicitly given in CT data, which required the extraction of surfaces. As the extracted surface is used as ground truth for further geometric inspection, the used technique for extracting surface and interface positions has a major influence on the quality of the extracted measurement results and therefore constitutes a challenge in the area of metrology using CT. Various surface extraction techniques and subpixel approaches to edge detection techniques have been discussed before. However, we target to raise the awareness of the reader on a slightly shifted topic answering the questions: how precise are the extracted edges or dimensional measurement features, what is the uncertainty prevalent in the CT datasets, and how can we visualize this information?

Conventional dimensional measurement software for CT data does not account for uncertainty prevalent in the data. Tools are pushing hard to extract dimensional measurement features as precise as possible and render them for example in annotations to the measurement features in 2D or 3D views. Unfortunately, at this point all information on the regional uncertainty in dataset due to noise, artefacts etc. is not considered any further and no visual representation on the quality of a

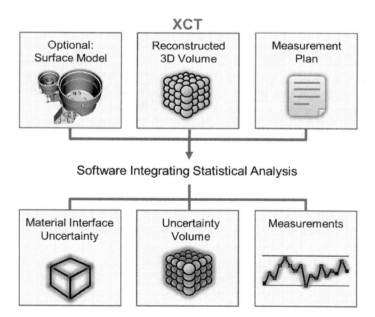

Fig. 4.26 Data processing workflow for dimensional metrology of CT data using statistical analysis

measurement is given. In addition, conventional dimensional measurement software tools are black boxes and users have to qualify each new release on their measurement tasks. Accounting for uncertainty arising in the CT metrology data and enhancing conventional visualization tools such as measurement plots, tolerance tags or measurement strategies with information on uncertainty are therefore considered as a main challenge for future metrology. In the following paragraphs two approaches are explained in more detail, which account for uncertainty in the CT data (Amirkhanov et al. 2013; Heinzl et al. 2008). Similarly, to the conventional workflow, the extended workflow in these approaches integrates statistical analysis on the CT data to compute uncertainty information. Figure 4.26 outlines data processing workflows for dimensional metrology using of CT data using statistical analysis.

By means of statistical analysis, in the work by Heinzl et al. (2008), the grey values of the CT datasets of multi-material components are transformed to probabilities of each individual material. Based on Bayesian decision theory information on uncertainty is introduced, which makes detailed characterizations of single materials as well as material interfaces possible. Using a local histogram analysis technique, probability volumes are computed in the statistical analysis step for each individual material. The extracted probabilities are either rendered directly using suitable colour and opacity transfer functions for direct volume rendering, or used to evaluate extracted surface models by using colour coding, which maps the interpolated uncertainty value on the surface.

Fuzzy CT metrology (Amirkhanov et al. 2013) provides users with various techniques encoding uncertainty in several linked views, which extend the general CT metrology workflow. In their approach the authors determine a material interface probability based on Bayes decision theory. Dimensional measurement features as well as their measurement strategies may be specified on the structures of interest, similarly to conventional tools. Using linked 3D and measurement plot views, and 3D labels the data is explored. An overview of the tool is given in Fig. 4.27. The Fuzzy CT metrology system visualizes the uncertainty of measurements on various levels-of-detail. In Smart Tolerance Tags, which are linked to the corresponding measurement feature in the 3D View common geometric tolerance labels are provided and enhanced with box plots encoding the probability of the material interface throughout the measurement feature. This way the user gets an intuitive overview whether the measurement feature is subject to irregularities as noise or artefacts. The measurement strategy is rendered in the 3D view using reference shapes such as circles for evaluating diameters or meander shapes for inspecting the evenness of a plane. The underlying uncertainty information is mapped to these reference shapes, adapting the thickness of the employed reference shapes as well as color-coding the shape accordingly. In the most detailed setting, uncertainty prevalent in the data is provided as overlay to commonly used measurement plots. Using these techniques an augmented insight into dimensional measurement tasks may be provided without interfering with the conventional dimensional measurement workflow.

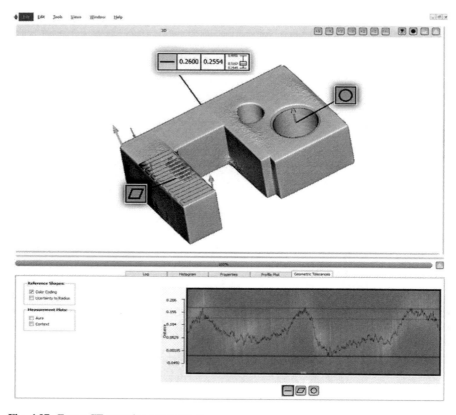

Fig. 4.27 Fuzzy CT metrology user interface

4.6 Processing, Analysis and Visualization for Non-destructive Testing

Aside CT being on the edge of advancing to a fully standardized means of dimensional measurement for every day industrial use, a core application area of X-ray computed tomography is nondestructive testing (NDT). NDT was the first and is still one of the largest application areas using CT (Huang et al. 2003) in the following fields:

- Assembly analysis: When an object is assembled from many components and features a lot of hidden elements, CT is used to investigate the assembly quality. In this application, the main advantage of CT is that it allows the domain experts to get insight into the object without disassembling it.
- Material analysis: While manufacturing, different types of defects might occur in the resulting product, such as micro-cracks or inclusions. CT allows the domain

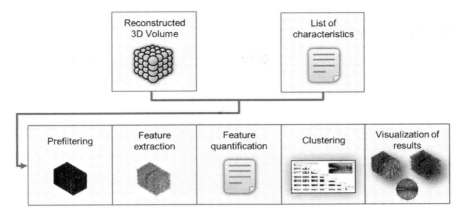

Fig. 4.28 Generalized data processing workflow for non-destructive testing

experts to detect and quantify these defects, as well as internal features (Redenbach et al. 2012; Ohser et al. 2009; Straumit et al. 2015).

- Reverse engineering: CT is widely used in reverse engineering when it is not possible to disassemble an object to understand how it works. For example, CT is used to reverse engineer musical instruments which were produced centuries ago.

Due to the large number of different applications and requirements numerous workflows in the domain of NDT are in use. A generalized workflow for NDT using CT data is shown in Fig. 4.28.

A currently highly active field of research concerns the characterization of novel composite materials. Therefore, we focus in the next paragraphs on this highly challenging area and more specifically on the domain of characterizing fibre reinforced polymers. Fibre reinforced polymers have emerged as new candidate materials to fulfil the highly demanding requirements of aeronautics (Weissenböck et al. 2014), automotive (Bhattacharya et al. 2015), and construction (Fritz et al. 2009) industries. All applications share the need to characterize the individual components, especially the reinforcement component (i.e. fibres, fibre bundles) as well as potential defects (i.e. pores/voids, delamination, inclusions, cracks). In the following paragraphs, we introduce several systems providing methods for performing the following tasks:

- Define fibre classes and visualize fibre regions as well as the relationships between them (Weissenböck et al. 2014). Extract, analyse and visualize fibre lengths and orientations as well as their distributions (Weissenböck et al. 2014; Fritz et al. 2009).
- Extract, analyse and visualize fibre bundles as well as their weaving pattern (Bhattacharya et al. 2015).
- Visualize the individual properties of voids and pores and render the spatial porosity as well as its homogeneity (Reh et al. 2013).
- Visual analysis of dynamic processes (Amirkhanov et al. 2014).

4.6.1   Visual Analysis of Voids

Porosity Maps (Reh et al. 2012) encode a visual metaphor for the characterization
of porosity in carbon fibre reinforced polymers (CFRP). Besides a quick and easy
evaluation regarding quantitative porosity and the calculation of pore properties
(i.e., volume, surface, dimensions, shape factors, etc.) as well as regarding the
homogeneity of the pores within the specimen, porosity maps allow for comparisons to ultrasonic testing or active thermography data. For this purpose, the system
allows the user to first specify a region of interest to be evaluated. In an
exchangeable segmentation step the individual pores are extracted. A connected
component filter identifies and labels the individual objects and subsequently each
object/pore is characterized in terms of its properties. The extracted pore data is then
rendered in a 3D view as well as in a parallel coordinates widget for interaction. The
parallel coordinates widget shows individual pores as connected polylines between
the parallel coordinates' axes and allows the user to focus on specific ranges of the
characteristics of interest. The porosity maps themselves are typically calculated for
the three axis-aligned directions but may also be calculated in arbitrary directions.
To compute the data of the porosity maps, the segmented pores are aggregated in
the targeted directions by summing all segmented pore voxels along a ray in slice
direction. Finally, the computed porosity map data is mapped to colours (see
Fig. 4.18).

4.6.2   Visual Analysis of Fibres

FiberScout (Weissenböck et al. 2014) is an interactive tool used for exploring and
analysing fibre reinforced polymers (see Fig. 4.29). The tool takes labelled data of
fibres as input to calculate the characteristics of interest. The fibres themselves are
extracted by the algorithm as proposed by Salaberger et al. (2011). Using parallel
coordinates and a scatter plot matrix linked to the 3D renderer, classes for features
of interest may be defined containing voids, pores, fibres, inclusions etc. with
similar characteristics. For explanation purposes, we focus here on fibres, but the
tool similarly operates on pores, voids, inclusions, or other features of interest.
While individual fibres are indicated as dots in the individual scatter plots of the
scatter plot matrix, plotting each characteristic over the other, the parallel coordinates indicate an individual fibre as a polyline connecting the parallel coordinate
axes of the different fibre characteristics at their respective magnitude. Selections in
both widgets update the fibres of interest in the 3D renderer. These concepts allow
an in-depth analysis of hidden relationships between individual fibre characteristic.
In addition, an intuitive rendering of the fibre orientation is provided using a polar
plot. Once the different fibre classes are set up, the system also allows for computing hulls around fibres of each class or mean objects (Reh et al. 2013) as
representative for the considered class. As an overlay for 2D and 3D views the

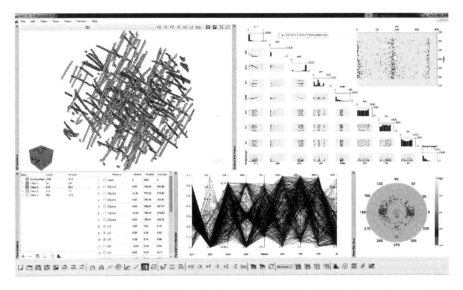

Fig. 4.29 FiberScout interface (Weissenböck et al. 2014). The 3D Renderer shows a GFRP specimen linked to a set of visualizations methods to explore the underlying data: a scatter plot matrix (*top left*), a polar plot (*bottom right*), parallel coordinates (*middle*), and a class explorer for saving and retrieving classifications (*bottom left*)

system calculates heat maps or density maps in order to provide regional information of particular material characteristics. The insights gained using FiberScout can be used in the design office to improve on the component design as well as in material development and simulation providing highly detailed input data for further processing and material optimization.

4.6.3 Visual Analysis of Fibre Bundles

MetaTracts (Bhattacharya et al. 2015) robustly extracts and visualizes fibre bundles in CT datasets of fibre reinforced composites. The MetaTracts tool is typically used for carbon fibre reinforced polymers (CFRP) with endless fibres. CFRPs are manufactured of individual carbon fibres, woven into sheets of carbon fibre cloth which are then layered in the corresponding mould. The matrix component (e.g. epoxy resin) combines the composite similarly as a glue and gets finally cured. The weaving pattern in the cloth and thus the directions of the fibre bundles mainly create the required strength and stiffness properties of the resulting component. MetaTracts was designed to handle CT datasets with large view ports to analyse larger areas of the dataset showing the recurring weaving pattern (unit cell). As larger areas of the dataset are directly related to lower resolution due to the magnification geometry of the used CT device, the challenge of the MetaTracts

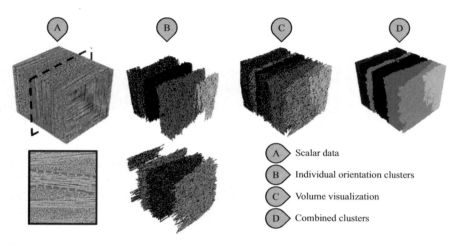

Fig. 4.30 Data set with flat thin and compact bundles. **a** Shows the volume rendering and a 2D slice with one of the boundaries marked in *green*, **b** shows the clusters according to individual orientation, **c** shows the complete result, **d** shows the voxelization of (**c**)

approach is found in the fact that individual fibres are barely visible in the input CT datasets. To solve this challenge, methods originally developed for diffusion tensor imaging (DTI) were applied and extended in order to extract the individual fibre bundles: A coarse version of integral curves extracts subsections of the individual fibre bundles. For this purpose, a fibre bundle is approximated by several series of connected cylinders of a fixed diameter (meta tracts). Then the extracted meta tracts are clustered in a 2-stage approach to the final fibre bundles. As subsections of the meta tracts should show a similar direction, the first step of the clustering is required to be done according to the subsections' main direction. In the second step the subsections are clustered according to proximity. For further processing the surface models or voxelized models may be computed, e.g., for material simulation purposes. An example of an analysis using MetaTracts is shown in Fig. 4.30.

4.6.4 Visual Analysis of Dynamic Processes

Recent developments in visualization also focus on the analysis of dynamic processes, analysing 3D datasets in time. These dataset series of 3D plus time are also referred to as 4DCT. In the following paragraphs, we present two tools: Fuzzy feature tracking for visual analysis of 4DCT data (Reh et al. 2015) and an approach for visual analysis of defects in glass fibre reinforced polymers during interrupted in situ tests (Amirkhanov et al. 2016). Fuzzy Feature Tracking (Reh et al. 2015) targets to analyse and visualize CT dataset series of dynamic processes and more specifically in situ tests of specimens. During these tests a series of CT scans of an object is performed under different conditions, e.g., changing temperature, increasing/decreasing loading. For these tests, it is of highest priority to find corresponding features in the individual time-steps, i.e. the individual CT scans of the dataset series, and to understand, how the features are changing over time. As the test specimen, might change significantly during the process, a tracking uncertainty is introduced, which is calculated by Eq. (4.8):

$$U(a_i, b_j) = 1 - (O(a_i, b_j) \times w_0 + R(a_i, b_j) \times w_r), \qquad (4.8)$$

where $O(a_i, b_j)$ denotes the volume overlap and $R(a_i, b_j)$ is the ratio of volumes between the features a_i and b_j. w_0 and w_r specify user-defined weights, which sum to 1 ($w_0 + w_r = 1$). To analyse 4DCT data, the tool provides a visualization toolbox consisting of four methods: the volume blender, a 3D data view, an event explorer, and a tracking graph. The volume blender provides a 3D overview of all selected time-steps by blending two adjacent volumes over each other in order to provide an impression of how the features of interest evolve over time. The 3D data view arranges and renders all time-steps in a line using the same viewport. The event explorer depicts creation events, continuation, merge events, split events and dissipation in scatter plots (one plot per time-step) and shows the correlation between two feature characteristics of interest. One characteristic is chosen for the X-axis, another characteristic for the Y-axes. The user is free to choose parameters for the features in accordance with the targeted analysis task. Using different colours, the event type is distinguished. In addition, features may be selected in the event explorer and are consequently highlighted in all time-steps. Once a selection is done in any scatter plot of the event explorer, the tracking graph widget shows the corresponding graph of the selected features as well as their evolution over all time-steps. To avoid visual clutter, the tracking graph employs an algorithm for minimizing the number of crossing edges.

Amirkhanov et al. (2016) have presented a related tool, which was designed for the visual analysis of defects in 4DCT datasets acquired using interrupted in situ tensile tests. The tool allows the user to extract individual defects from each time-step as well as to classify them into four possible types: fibre pull-outs, fibre fracture, fibre/matrix debonding, matrix fracture. To visualize the defects and the fracture evolution the tool includes several visualization techniques: the defect viewer shows data in slices and allows the user to highlight the different defects as well as their type using colours. Each defect type is assigned a unique colour to distinguish defects by type and to observe patterns in the material. The defect density maps accumulate quantitative information about defects in maps revealing areas of high defect occurrence. Furthermore, the current fracture surface may be estimated from each time-step based on the identified defects. Finally, a 3D magic lens allows the user to reveal internal features by combining different visualization techniques in one viewport. While the 3D rendering shows the computed crack surface, the magic lens reveals the underlying CT data. Figure 4.31 shows defect density maps of glass fibre reinforced composites for different loadings.

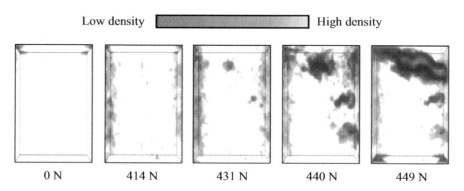

Low density ▨▨▨▨▨▨▨▨▨ High density

0 N   414 N   431 N   440 N   449 N

Fig. 4.31 Defects in fibre reinforced composites using 4DCT datasets of interrupted in situ tensile tests. *Yellow* indicates high defect density, *blue* low defect density

4.7 Specialized Analyses

Most state of the art data processing and visualizations tools are focusing on a single imaging modality using single scalar data such as CT. Approaches such as Dual Energy Computed Tomography allow extracting material information from the data. However, especially in the domain of material analysis, elemental decomposition using X-ray computed tomography is reaching its limits and may end up in ambiguous results. To overcome these issues recent advances are combining CT data with data from additional nondestructive testing modalities (Amirkhanov et al. 2014). A second topic which gets increasing attention concerns the exploration of parameter spaces of data analysis pipelines. Typically, papers provide suitable parameter ranges for the used datasets but setting these parameters empirically. For scalar volumetric datasets, this challenge is difficult to solve, but it is even more challenging for multi-modal data (Froehler et al. 2016). In the following paragraphs two approaches are introduced, which present their solutions, methods and techniques to overcome the previously mentioned challenges.

4.7.1 InSpectr

The InSpectr tool (Amirkhanov et al. 2014) addresses the increasing demand to combine scalar data with spectral data for multi-modal, multi-scalar data analysis. The volumetric datasets used as input are considered as multi-modal as they are imaged using different techniques such as CT and X-ray fluorescence (XRF), which provide scalar volumetric data in the case of CT and spectral volumetric data in the case of XRF. They are multi-scalar in a sense that they show different resolutions, different scales of the same scanning volume. InSpectr also allows the user to integrate element maps from additional modalities (e.g. K-edge imaging, Dual

Energy CT) as well as reference spectra for calibration. Regarding its interface, InSpectr features various data domain specific views to provide insight into a specimen's spatial characteristics as well as its elemental decomposition and thus transports spatial and non-spatial information. 3D views are used to visualize the global structure of the data and the global material composition. In the latter case volume renderings, may be used to explore composite and individual 3D element maps. The composite 3D view integrates all materials of interest in a single view, while the individual 3D element maps are rendered separately. To facilitate a combined exploration, all 3D views are linked. This means interactions such as zoom, pan, or rotate in one view triggers all other views to update in the same manner. In addition, the 2D slicers are extended with overlays. Typically, the slicers show the high-resolution CT dataset. Overlays for the XRF data may be faded in on demand. Furthermore, pie chart glyphs showing the local material decomposition may be activated and used as another visual metaphor. To solve the task of spectral data analysis, InSpectr integrates a spectral view for aggregating and exploring the spectral lines within the XRF data. The aggregate spectral lines allow for a cumulative representation of all materials prevalent in the dataset. Based on the aggregated spectral lines, the spectral view is also used to specify transfer functions for rendering the spectral data in the 3D view. For local analysis of the material decomposition InSpectr provides spectral magic lenses to reveal the materials present at a specific region, local spectrum probing by hovering with the mouse over 2D slicer, which linked to the corresponding spectral lines, as well as and elemental composition probing of points of interest using pie-chart and a periodic table view, which indicates the local material decomposition as bars mapped to a periodic table. An overview of the InSpectr interface is given in Fig. 4.32.

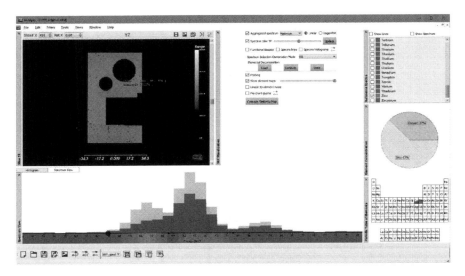

Fig. 4.32 Main interface of InSpectr: slice view (*left-top*), spectrum view (*left-bottom*), periodic table view (*right-bottom*), pie-chart view (*right-middle*), and visualization settings (*right-top*)

4.7.2   GEMSe

GEMSe (Froehler et al. 2016) facilitates the exploration of the space of possible parameter combinations for a segmentation framework and its ensemble of results (see Fig. 4.33). The tool therefore samples the parameter space defined by the parameters of the segmentation framework and computes the corresponding segmentation masks. The generated results are subsequently hierarchically clustered in an image tree, which provides an overview of the resulting segmentation masks. For the selected clusters of the tree exemplary samples are chosen by the system and rendered. Furthermore, for the selected clusters the used parameters are shown in histograms for the parameters of interest as well as the derived outputs of interest. Using scatter plots the correlation between derived output and the parameters of the segmentation framework is analysed and the set of suitable candidate results may be further refined.

4.8   Summary and Outlook

In this chapter, we have presented an overview of the data analysis pipeline of industrial CT data. We focused on preprocessing and data enhancement, feature extraction and quantification as well as aspects and approaches of visual analysis in metrology and nondestructive testing. We provided an overview of various

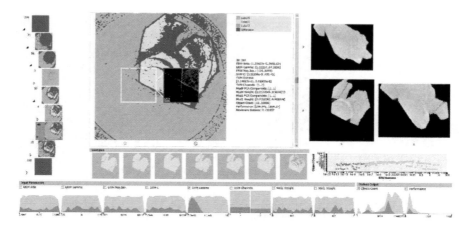

Fig. 4.33 Main interface of GEMSe: (*left*) hierarchically clustered image tree; (*middle*) detail view with magic lens widget and used segmentation parameters, as well as exemplary samples below; (*right*) camera view for selecting slice views and slice number, with scatter plots the correlation between derived output and the parameters of the segmentation framework below. (*bottom*) histogram view showing the used parameters as well as the corresponding derived outputs

approaches which succeed in their data evaluation tasks by tailored visual analysis concepts. However, despite all achievements and advancements in CT hardware only in combination with thorough analysis integrating preprocessing and data enhancement, feature extraction and quantification, visualization mapping and rendering of the results, novel insights are facilitated.

As an outlook, data analysis and visualization is currently in a strongly changing environment. Novel and upcoming approaches will face the challenges of increasingly complex, multimodal and multi-channel data, which need to be provided to the user in a convenient and intuitive way. A strong focus of research will also be found in visual analysis of dynamic processes such as tensile or loading tests. 3D plus time or any other additional dimension will be the basis for data evaluations of the future. Furthermore, due to the ever-changing environments of future analysis a research priority will be found in the domain of parameter space analysis used for tweaking and optimizing input parameters of the data analysis pipelines in order to generate optimal results. And finally, comparative and uncertainty visualization techniques need to be further explored to analyse complex data.

References

Adams R, Bischof L (1994) Seeded region growing. IEEE Trans Pattern Anal Mach Intell 16:641–647. doi:10.1109/34.295913.3

Al-Rauush R, Papadopoulos A (2010) Representative elementary volume analysis of porous media using X-ray computed tomography. Powder Technol 200(1–2):69–77. doi:10.1016/j.powtec.2010.02.011.3

Amirkhanov A, Heinzl C, Reiter M, Kastner J, Gröller E (2011) Projection-based metal-artifact reduction for industrial 3D X-ray computed tomography. IEEE Trans Vis Comput Graph 17(12):2193–2202

Amirkhanov A, Heinzl C, Kastner J, Gröller E Fuzzy CT (2013) Metrology: dimensional measurements on uncertain data. In: SCCG proceedings (digital library) Smolenice castle, Slovakia, 2013, pp 8

Amirkhanov A, Fröhler B, Kastner J, Gröller E, Heinzl C (2014) InSpectr: Multi-modal exploration, visualization, and analysis of spectral data. Comput Graph Forum 33(3):91–100

Amirkhanov A, Amirkhanov A, Salaberger D, Kastner J, Gröller E, Heinzl C (2016) Visual analysis of defects in glass fiber reinforced polymers for 4DCT interrupted in situ tests. Comput Graph Forum 35(3):201–210

Ballard D, Brown C (1982) Computer vision, Chap. 2. Prentice-Hall

Barr AH (1981) Superquadrics and angle-preserving transformations. IEEE Comput Graph Appl 1 (1):11–23. doi:10.1109/MCG.1981.1673799. URL http://ieeexplore.ieee.org/stamp/stamp.jsp?tp=&arnumber=1673799&isnumber=35130

Beucher S (1991) The watershed transformation applied to image segmentation. Scanning Microsc Int 3:299–314

Beucher S, Lantuejoul C (1979) Use of watersheds in contour detection. In: International workshop on image processing: real-time edge and motion detection/estimation, Rennes, France, Sept 1979

Bhattacharya A, Heinzl C, Amirkhanov A, Kastner J, Wenger R (2015) MetaTracts—a method for robust extraction and visualization of carbon fiber bundles in fiber reinforced composites. In:

Proceedings of IEEE pacific visualization symposium (PacificVis) 2015, Hangzhou, China, 2015, pp 191–198

Canny J (1986) A computational approach to edge detection. IEEE Trans Pattern Anal Mach Intell 8:679–698. doi:10.1109/TPAMI.1986.4767851.3

Cole L, Austin D, Cole L (2004) Visual object recognition using template matching. In: Australian conference on robotics and automation, 2004

Fabijańska A (2015) Subpixel edge detection in blurry and noisy images. Int J Comput Sci Appl 12(2):1–19

Felkel P, Bruckschwaiger M, Wegenkittl R (2001) Implementation and complexity of the watershed-from-markers algorithm computed as a minimal cost forest. Comput Graph Forum 20(3):26–35

Frangi AF, Niessen WJ, Vincken KL, Viergever MA (1998) Multiscale vessel enhancement filtering. In: Proceedings of the 1st international conference of medical image computing and computer-assisted intervention (MICCAI 1998), pp 130–137

Fritz L, Hadwiger M, Geier G, Pittino G, Gröller E (2009) A visual approach to efficient analysis and quantification of ductile iron and reinforced sprayed concrete. IEEE Trans Vis Comput Graph 15(6):1343–1350

Froehler B, Möller T, Heinzl C (2016) GEMSe: visualization-guided exploration of multi-channel segmentation algorithms. Comput Graph Forum 35(3):8

Gradshteyn IS, Ryzhik IM (2000) Hessian determinants. §14.314 in tables of integrals, series, and products, 6th edn. Academic Press, San Diego, p 1069

Grigoryan G, Rheingans P (2004) Point-based probabilistic surfaces to show surface uncertainty. IEEE Trans Vis Comput Graph 10(5):564–573

Heinzl C, Klingesberger R, Kastner J, Gröller E (2006) Robust surface detection for variance comparison. In: Proceedings of eurographics/IEEE VGTC symposium on visualisation, 2006, pp 75–82

Heinzl C, Kastner J, Gröller E (2007) Surface extraction from multi-material components for metrology using dual energy CT. IEEE TVCG 13:1520–1527

Heinzl C, Kastner J, Möller T, Gröller E (2008) Statistical analysis of multi-material components using dual energy CT. In: Proceedings of vision, modeling, and visualization, Konstanz, Germany, 2008, pp 179–188

Heinzl C, Reiter M, Allerstorfer M, Kastner J, Gröller E (2010) Artefaktreduktion mittels Dual Viewing für Röntgen computer tomographie von Multimaterial bauteilen, DGZFP-Jahrestagung, p 8

Hsieh J (2003) Computed tomography: principles, design, artifacts and recent advances. SPIE-The International Society for Optical Engineering, 2003

Huang R, Ma K-L, Mccormick P, Ward W (2003) Visualizing industrial CT volume data for nondestructive testing applications. In: VIS '03: Proceedings of the 14th IEEE visualization 2003 (VIS'03), pp 547–554

Jin JS (1990) An adaptive algorithm for edge detection with subpixel accuracy in noisy images. In: Proceedings of IAPR workshop on machine vision applications, 1990, pp 249–252

Jolliffe I (2002) Principal component analysis. Wiley, New York

Kindlmann GL (2004) Superquadric tensor glyphs. In: Deussen O, Hansen CD, Keim DA, Saupe D (eds) VisSym. Eurographics Association, pp 147–154

Kohonen T (1990) The self-organizing map. In: Proceedings of the IEEE, 78(9):1464–1480. URL http://ieeexplore.ieee.org/stamp/stamp.jsp?tp=&arnumber=58325&isnumber=2115. doi:10.1109/5.58325

Lifton JJ (2015) The influence of scatter and beam hardening in X-ray computed tomography for dimensional metrology. Ph.D. thesis, University of Southampton, Faculty of Engineering and the Environment Electro-Mechanical Engineering Research Group, Apr 2015

Lorensen W, Cline H (1987) Marching cubes: A high resolution 3D surface construction algorithm. ACM SIGGRAPH Comput Graph 21:163–169

Munzner T (2014) Visualization analysis and design. CRC Press

Oehler M, Buzug TM (2007) A sinogram based metal artifact suppression strategy for transmission computed tomography. In: Geometriebestimmung mit industrieller Computer tomographie, PTB-Bericht PTB-F-54, Braunschweig, 2007, pp 255–262

Ohser J, Schladitz K (2009) 3D images of materials structures: processing and analysis. Wiley, New York

Otsu N (1979) A threshold selection method from gray-level histograms. IEEE Trans Syst Man Cybern 9(1):62–66. URL http://ieeexplore.ieee.org/stamp/stamp.jsp?tp=&arnumber=4310076&isnumber=4310064. doi:10.1109/TSMC.1979.4310076

Perona P, Malik J (1990) Scale-space and edge detection using anisotropic diffusion. IEEE Trans Pattern Anal Mach Intell 12:629–639

Redenbach C, Rack A, Schladitz K, Wirjadi O, Godehardt M (2012) Beyond imaging: on the quantitative analysis of tomographic volume data. Int J Mater Res 103(2):217–227

Reh A, Plank B, Kastner J, Gröller E, Heinzl C (2012) Porosity maps: interactive exploration and visual analysis of porosity in carbon fiber reinforced polymers using X-ray computed tomography. Comput Graph Forum 31(3):1185–1194

Reh A, Gusenbauer C, Kastner J, Gröller E, Heinzl C (2013) MObjects—a novel method for the visualization and interactive exploration of defects in industrial XCT data. IEEE Trans Vis Comput Graph (TVCG) 19(12):2906–2915

Reh A, Amirkhanov A, Kastner J, Gröller E, Heinzl C (2015) Fuzzy feature tracking: visual analysis of industrial 4D-XCT data. Comput Graph 53:177–184

Rhodes PJ et al (2003) Uncertainty visualization methods in isosurface rendering. Eurographics, vol 2003

Salaberger D, Kannappan KA, Kastner J, Reussner J, Auinger T (2011) Evaluation of computed tomography data from fibre reinforced polymers to determine fibre length distribution. Int Polym Proc 26:283–291

Schmitt R, Hafner P, Pollmanns S (2008) Kompensation von Metallartefakten in tomographischen Aufnahmen mittels Bilddatenfusion. In: Proceedings of Industrielle Computertomographie, Fachtagung, 2008, pp 117–122

Schroeder W, Maynard R, Geveci B (2015) Flying edges: a high-performance scalable isocontouring algorithm. In: IEEE 5th symposium on large data analysis and visualization (LDAV), 2015, Chicago, IL, pp. 33–40

Selim SZ, Ismail MA (1984) K-means-type algorithms: a generalized convergence theorem and characterization of local optimality. IEEE Trans Pattern Anal Mach Intell PAMI-6(1):81–87. URL http://ieeexplore.ieee.org/stamp/stamp.jsp?tp=&arnumber=4767478&isnumber=4767466 . doi:10.1109/TPAMI.1984.4767478

Sezgin M, Sankur B (2004) Survey over image thresholding techniques and quantitative performance evaluation. J. Electron Imaging 13(1):146–168

Shneiderman B (1996) The eyes have it: a task by data type taxonomy for information visualizations. In: Proceedings of the IEEE symposium on visual languages, Washington. IEEE Computer Society Press, http://citeseer.ist.psu.edu/409647.html, pp 336–343

Steinbeiss H (2005) Dimensionelles Messen mit Mikro-Computertomographie. Ph.D. thesis, Technische Universität München

Straumit Ilya, Lomov Stepan V, Wevers Martine (2015) Quantification of the internal structure and automatic generation of voxel models of textile composites from X-ray computed tomography data. Compos A Appl Sci Manuf 69:150–158

Tabatabai AJ, Mitchell OR (1984) Edge location to sub-pixel values in digital imagery. IEEE Trans Pattern Anal Mach Intell 6(2):188–201

Tomasi C, Manduchi R (1998) Bilateral filtering for gray and colour images. In: Proceedings of the IEEE international conference on computer vision, pp 839–846

Volumegraphics: VG Studio Max 2.2—User's Manual. http://www.volumegraphics.com/fileadmin/user_upload/flyer/VGStudioMAX_22_en.pdf. Last accessed Mar 2016

Wang J, Wang S, Chen Y, Wu J, Coatrieux J-L, Luo L (2013) Metal artifact reduction in CT using fusion based prior image. Med Phys 40:081903

Weissenböck J, Amirkhanov A, Li W, Reh A, Amirkhanov A, Gröller E, Kastner J, Heinzl C (2014) FiberScout: an interactive tool for exploring and analyzing fiber reinforced polymers. In: Proceedings of IEEE pacific visualization symposium (PacificVis), 2014, Yokohama, Japan, pp 153–160

Wold Svante, Esbensen Kim, Geladi Paul (1987) Principal component analysis. Chemometr Intell Lab Syst 2(1-3):37–52

Xu GS (2009) Sub-pixel edge detection based on curve fitting. In: Proceedings of 2nd International conference on information and computing science, pp 373–375

Yao Y, Ju H (2009) A sub-pixel edge detection method based on canny operator. In: Proceedings of 6th international conference on fuzzy systems and knowledge discovery, pp 97–100

Zhigeng P, Jianfeng L (2007) A bayes-based region-growing algorithm for medical image segmentation. Comput Sci Eng 9:32–38. doi:10.1109/MCSE.2007.67.3

Chapter 5
Error Sources

Alessandro Stolfi, Leonardo De Chiffre and Stefan Kasperl

Abstract In this chapter, an identification and classification of influence factors in X-ray computed tomography metrology is given. A description of image artefacts commonly encountered in industrial X-ray computed tomography is presented together with their quantification. A survey of hardware and software methods developed for correcting image artefacts is presented.

5.1 An Overview of Influence Factors in X-ray Computed Tomography Metrology

There is a large variety of factors influencing the performance of X-ray computed tomography (CT). The German guideline VDI/VDE 2630-1.2 (2008) provides a thorough overview of the factors impacting the measurement workflow. The influence factors can be split into five groups of parameters: system, workpiece, data processing, environment and operator. Table 5.1 shows the factors that are responsible for influencing the performance of an X-ray CT system. The presence of each influence factor can be significant or negligible depending on the type of measurement task.

A. Stolfi (✉) · L. De Chiffre
Technical University of Denmark (DTU), Produktionstorvet Building 425,
2800 Kgs, Lyngby, Denmark
e-mail: alesto@mek.dtu.dk

S. Kasperl
Fraunhofer Development Center for X-ray Technology (EZRT),
Flugplatzstr. 75, Fürth 90768, Germany

© Springer International Publishing AG 2018
S. Carmignato et al. (eds.), *Industrial X-Ray Computed Tomography*,
https://doi.org/10.1007/978-3-319-59573-3_5

Table 5.1 Influence factors in CT

Group	Influence factors
CT system	X-ray source Detector Positioning system
Data processing	3D reconstruction Threshold determination and surface generation
Workpiece	Material composition Dimension and geometry Surface texture
Environment	Temperature Vibrations Humidity
Operator	Workpiece fixturing and orientation Magnification X-ray source settings Number of projections and image averaging Measurement strategy

5.1.1 CT System

An industrial CT system is composed of three main components: (i) an X-ray source, (ii) an X-ray detector, and (iii) a positioning system. The performance of each of these components impinges upon the measurement.

5.1.1.1 X-ray Source

The accelerating voltage, filament current, focal spot size, target material and transmission window material all have significant effects on the final output of an X-ray source. The X-ray voltage determines the energy of the electrons. Changing the X-ray voltage modifies the spectrum, by increasing the average photon energy. The X-ray voltage influences the contrast between low density materials and the background noise level. The filament current controls the number of electrons bombarding the target material. The more current flowing through the filament, the higher the number of electrons emitted, and there is a saturation point at which the emission is at maximum. Increasing the current through the filament after the saturation point would shorten the life of the filament and can damage it. Both current and voltage influence the focal spot size that represents the smallest diameter of the electron beam generated in the X-ray tube. The spot size lies in the micrometre range and increases with increasing power to prevent the target from melting, evaporating and being subject to plastic deformation. In industry, the focal spot is constant for the power range of 6 W to 9 W and then increases at a rate of $1~\mu m~W^{-1}$ (Hiller et al. 2012). Figure 5.1 shows a typical relationship between X-ray power and X-ray focal spot size for a microfocus X-ray tube used for

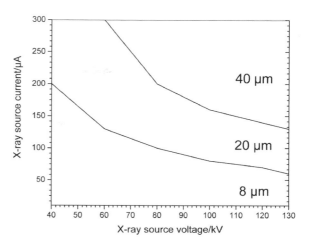

Fig. 5.1 Relationship between X-ray power and X-ray focal spot size for a microfocus X-ray source used for non-destructive inspection. The operational range of the source is only represented (X-ray voltage between 40 kV to 130 kV, and X-ray current between 10 μA to 300 μA)

non-destructive inspection. X-ray power levels below 9 W result in a spot size of 8 μm; X-ray power levels in the range of 8 W to 16 W yield a spot size of 20 μm and above 16 W, the spot size is 40 μm.

The shape of the X-ray spectrum also depends upon the target material and its thickness. A tungsten target is the most widely used in industry because it has a high atomic number which increases the intensity of the X-rays, and because it has a very high melting point (3687 K), low rate of evaporation and mechanical proprieties that are almost independent of operating temperature. Targets made from low atomic number elements, such as copper and molybdenum, are well suited for imaging low absorption workpieces with high contrast. Tan (2015) investigated the extent to which the target material influences dimensional measurements. The investigation was conducted using two different materials. Figure 5.2 shows that the choice of target material has a clear influence on the dimensions of a set of 2 mm diameter spheres, and that the tungsten target yields better imaging at this specific power.

In the case of transmission X-ray sources, the target thickness also influences the intensity of the generated X-rays. If the target thickness is smaller than the average penetration depth of the electrons (Lazurik et al. 1998), electrons can pass through the target without interaction. Consequently, the X-ray emission will also be small. Increasing the target thickness, X-ray attenuation also increases during the penetration of X-rays through the target. Ihsan et al. (2007) found that the X-ray intensity increases with the target thickness until a certain point, at which the intensity abruptly decreases due to the increasing X-ray attenuation occurring during the penetration of X-rays through the target. The surface texture of the targets has also a consequence on the spectrum generated because it locally modifies the amount of material to be penetrated (Mehranian et al. 2010).

The X-ray transmission window is an additional factor impacting on the performance of an X-ray tube. Currently, X-ray windows are made from beryllium due to its distinguishing properties, such as transparency to shorter X-rays, high thermal

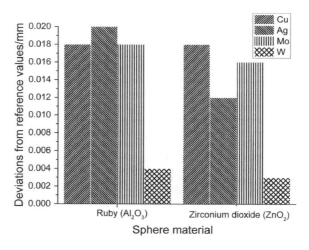

Fig. 5.2 Comparison between contact coordinate measuring machine diameter measurements using two spheres of different materials: ruby sphere (Al_2O_3) and zirconium dioxide (ZnO_2) and four material targets: copper (Cu), molybdenum (Mo), silver (Ag) and tungsten (W). Each bar represents the average value of four CT measurements. The power (18 W) and voxel size (30 μm) were used throughout the course of the investigations (Tan 2015)

conductivity and mechanical strength. Lower Z materials, such as aluminium, lithium and boron, are also used as window materials for X-ray tubes. X-ray windows have thicknesses in the range of 100 μm to 1000 μm, depending on the power of the X-ray source. Simulation studies have shown that the presence of an X-ray transmission window can decrease X-ray intensity by more than 10% compared to the windowless case (Ihsan et al. 2007).

5.1.1.2   CT Detector

There are many parameters affecting image quality in X-ray CT detectors. The most important parameters include pixel pitch size, number of pixels, integration time, linear dynamic range, spectral response and the detector quantum efficiency (DQE). The pixel pitch size has an impact on the spatial resolution and signal-to-noise ratio (SNR). Small pixel pitch sizes improve spatial resolution, making the detector unsharpness, which is of the order of two times the pixel size, negligible.

Large pixel pitch sizes result in better SNR as they present high fill factors, representing the fraction of the pixel area that is sensitive to light. For a fixed sensor frame size, decreasing the pixel size increases the number of pixel elements. This increase yields higher spatial sampling and a potential improvement in the system's modulation transfer function (MTF), provided that the resolution is not limited by other factors, such as the X-ray focus spot size and shape. Reducing the detector pixel size consequently reduces the amount of photons per reconstructed voxel. As a result, longer inspection times and, therefore, scanning times are necessary. A large number of active pixels generates a considerable amount of data, which are often not manageable in common workstations. A possible way around this data size issue is to adopt image binning, which combines adjacent detector pixels into a single effective pixel within the detector (hardware binning) or combines digital values from adjacent pixels in an acquired image (software binning). One major limitation of binning is saturation, which is the point at which a pixel cannot collect further information. Any charge beyond the saturation value will be truncated and the final pixel intensity will reflect the maximum pixel value rather than the actual value. Integration time is the lapse of time during which the detector collects incoming X-rays.

The integration time typically ranges from milliseconds to a few seconds. The linear range represents the range over which the sensitivity of the detector is almost constant. A constant sensitivity ensures the radiation intensity and the grey-scale values are linearly related. Flat panel detectors lose the linearity at pixel intensities of 75% to 90%, depending on noise, of the total dynamic range, representing the maximum extent of signal variation that can be quantified by the detector. At the extreme of the linear dynamic range, detectors may suffer from saturation causing distortion. The sensitivity of a detector depends on the sensitivity of the photodiode array in the vicinity of the peak emission wavelength. For detectors equipped with scintillators made from caesium iodide doped with thallium CsI(Tl), the photodiode should show good sensitivity between 350 nm and 700 nm. The DQE describes how effectively a detector can transfer the energy of the incident X-ray beam out of the detector (Konstantinidis 2011). The DQE can be expressed as:

$$DQE = \frac{SNR_{out}^2}{SNR_{in}^2} \qquad (5.1)$$

where SNR_{in} and SNR_{out} are the input to the detector and the output from the detector, respectively. The DQE is always smaller than unity. Compared to other performance metrics, the DQE allows the simultaneous measurement of the signal and noise performance of an imaging system at different spatial frequencies. An example of DQE for a single detector pixel is shown in Fig. 5.3. It can be seen from the figure that the investigated detector has a DQE in the range of 2% to 50%, depending on the spatial frequency considered. The DQE values decrease as a function of energy, due to the decreased absorption of X-rays from the scintillator at higher energy levels. The geometry and atomic composition of the detector linearly influence the DQE (see Tan 2015). In 2003, the International Electrotechnical Commission (IEC) published a standard method for measurement of the DQE as a function of spatial frequency (IEC 2003). The DQE metric is often not sufficient to describe a detector in a real X-ray CT system in which physical processes can introduce statistical correlations in image signals. More complex metrics have been developed for quantifying the DQE in a way that captures the existing statistical correlations. These metrics are all based on the Fourier transform, and include the noise power spectrum (NPS) and MTF (IEC 2003).

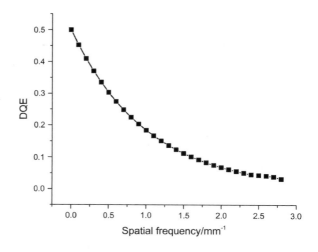

Fig. 5.3 An example of detective quantum efficiency for a single detector element. The element reduces the incoming SNR by about 50% at very low spatial frequency and even more at high spatial frequencies

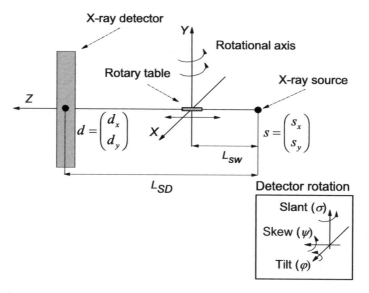

Fig. 5.4 Schema of CT geometry with the nine parameters

5.1.1.3 Positioning System

The geometry of industrial CT systems is defined by the relative position and orientation of the three components: the X-ray source, rotary table and X-ray detector (Ferrucci et al. 2015). Figure 5.4 depicts the experimental geometry of a circular trajectory cone-beam CT set-up. Ideally, the X-ray focal spot, the axes of the rotary table and the centre of the detector should fall in a straight line (Xi et al. 2010).

Additionally, the axis of the rotary table should be parallel to the detector and be projected onto the central column of the detector (Xi et al. 2010), which can lie on a pixel or on the edge between adjacent pixels. Whenever these conditions are not simultaneously satisfied, there will be measurement errors due to geometrical misalignments. In order to quantify these misalignments, nine parameters are necessary, which are the focal spot offsets along the X and Y axes (s_x and s_y), the detector offsets along X and Y axes (d_x and d_y), the perpendicular distance between the X-ray source and the rotation axis L_{SW}, the perpendicular distance from the X-ray source to detector L_{SD}, and finally three orientations of the detector (ψ, φ and σ). The rotation ψ, also known as skew, is an in-plane angle. The rotations, φ and σ, are two out-of-plane angles, known as tilt and slant, respectively. In this description, the misalignments of the rotary table are converted to those of the detector (Bequé et al. 2003). Moreover, this description assumes that all the components of the CT system are rigid bodies, thus their deformation is negligible.

Misalignments associated with the X-ray source are usually due to the difficulties in acquiring its centre coordinates from the projections of a calibration object, including fiducial markers, usually small metal spheres. The position of the markers is subject to uncertainty, the magnitude of which depends on the CT system, scanning parameters, environmental conditions, etc. As a result, the centre coordinates of the source are subject to an uncertainty, which is at least as large as that of the calibration object (by propagation of the uncertainty sources). Kumar et al. (2011) investigated the simulated effect of the source position on the accuracy of measurements by using two ball bars of different size. The results showed that varying the position of the source leads to either overestimating or underestimating distances, depending on the voxel size used for imaging. By increasing the voxel size, the impact of the source position decreases, and vice versa. Source misalignment can also be caused by the X-ray filament not being symmetrically bent into a V-shaped hairpin.

The thermal expansion of the X-ray source causes focal spot drifts, modifying the (x, y) coordinates of the focal spot over the scanning time. Hiller et al. (2012) stated that the drift perpendicular to the detector plane causes a scaling error in the reconstructed volume, whereas a drift parallel to the detector plane causes geometrical form errors in the calculated model. Positioning errors of the rotary table along the z-axis influence the measured dimensions, yielding maximum length measuring errors on the order of 0.15% to 0.20% for size measurements (e.g. a nominal length of 10 mm will result in a measured length in the range of 9.98 mm to 10.02 mm). Figure 5.5 presents an example of deviations from reference values for a set of measurements before and after having corrected the scale error using a reference object based on calibrated sphere-to-sphere distances.

Welkenhuyzen (2016) mapped the geometrical errors of a CT system by moving the rotary table back and forth from the system reference point to the furthest achievable position across the z-axis. Welkenhuyzen noticed that the errors show similar trends for both directions but slightly different magnitudes. The errors are larger when the rotary table is closer to the source than when it is close to the detector. An explanation of this behaviour may be the asymmetrical static response of the guideways on which the rotary table is mounted. Tilting errors of the rotary

Fig. 5.5 A comparison between a corrected data set (*black squares*) and uncorrected data set (*red circles*). Six calibrated lengths of a polymer step gauge were considered: M1 = 2 mm, M2 = 6 mm, M3 = 10 mm, M4 = 2 mm, M5 = 14 mm, M6 = 18 mm, and M6 = 21 mm. The lengths were based on plane-to-plane distances

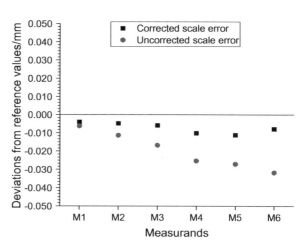

Fig. 5.6 Deviations from reference values of five sphere-to-sphere distances, evaluated at three y-positions, Position 1 = + 174.5 mm, Position 2 = + 204.6 mm, and Position 3 = + 214.7 mm. The positions were defined with respect to the centre of X-ray beam. Deviations are normalised with respect to the ball bar length in order to have a fair comparison

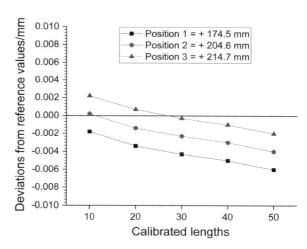

table are known as wobble and eccentricity. Eccentricity describes the displacement of the geometric centre of a rotation stage from the rotation axis in the plane defined by the bearings. Wobble is the tilt of the rotation axis during rotation and is quantified with respect to a reference surface. Wobble causes measurement results that depend on the rotary table height, as shown in Fig. 5.6. The error deviations appear to have similar trends for all three positions but different signs. The selected measurands are strongly underestimated when the workpiece is placed at position 1, while the measurands appear to be both underestimated and overestimated for the other positions. From Fig. 5.6, it can also be seen that the measurements performed in the central X-ray beam, in which X-ray beam is perpendicular to the detector, show the highest accuracy. Wobble and eccentricity similarly impair the structural resolution (See Chap. 6) as they can be deemed fluctuations of the focus spot across the detector.

Wenig and Kasperl (2006) investigated the effects of rotary stage shift and tilt on the accuracy of seven measurands of a workpiece. A lateral shift of the rotation axis along the x-axis of 700 μm, equal to a shift of two pixels on the detector, led to an average relative measurement error of about 2.5%. A tilt of the rotation axis about the y-axis of 1° yielded a relative measurement error of about 0.5% for all features of the workpiece. A further source of error influencing the rotary table performance is the flatness of the plane on which the rotary table is mounted, which should be below a few micrometres.

Aloisi et al. (2017) experimentally investigated the effects of detector tilt φ by purposefully introducing physical misalignments up to 1.5° on a flat-panel detector. The experimental results show that the investigated detector tilt φ up to 1.5° have significant effects when measuring sphere center-to-center errors and diameter errors of a calibrated ball bar positioned in the vertical direction. Moreover it is discussed how for the experimentally tested conditions the sphere center-to-center measurements and diameter measurements exhibit a clear trend due to the detector misalignments. Kumar et al. (2011) investigated the influence of detector misalignments by using simulations. The results showed that out-of-plane tilts are more important than the in-plane tilt ψ. The effect of the out-of-plane tilt φ is to magnify the projections in the vertical direction, while σ magnifies the projection in the horizontal direction. The out-of-plane tilts are somewhat difficult to remove, even after physical adjustments. Errors in the detector position are not a major problem regardless of the size of the workpiece. A detector can be misaligned by more than one rotation angle simultaneously. Geometrical errors of the detector can also be caused by manufacturing errors. Weiß et al. (2012) observed that if the deposited CsI scintillator is not perfectly perpendicular on the photodiode plate, local offsets within the micrometre range may occur. The offsets resemble those caused by position errors along the z-axis. Weiß et al. found the local offsets in their detector to be of approximately 6 μm. The presence of detector distortions can be verified by measuring a reference artefact at different regions across the detector, typically at three different heights (see Chap. 6).

5.1.2   Workpiece

Any workpiece can be measured, provided that it fits inside the detector and that X-rays are able to penetrate it with sufficient contrast. Table 5.2 shows the typical maximum penetrable thicknesses for a typical CT system at different voltages.

Table 5.2 Typical maximum penetrable material thicknesses for common industrial materials

X-ray voltage	130 kV (mm)	150 kV (mm)	225 kV (mm)
Steel	<5	<8	<25
Aluminium	<30	<50	<90
Plastic	<90	<130	<200

All values in the above table ensure a minimum transmission of around 14% (De Chiffre et al. 2014)

Higher penetration lengths are achievable by accepting lower transmission values. The overall size and mass of the workpiece also introduce some limitations in the use of CT. Industrial CT can withstand sample masses of up to 100 kg. The part mass also has a role in the rotary table performance as it modifies the response of the rotary table with respect to the performance stated by the manufacturers. The effect of the workpiece mass on the rotary table performance is difficult to predict as it depends on several factors. In the cases in which CT is used for metrology, the workpiece mass should not exceed 5 kg ($\approx$50 N) in order to maintain mechanical specifications while moving.

A further workpiece feature influencing the CT performance, especially for metrology, is surface texture. Figure 5.7 shows the influence of surface texture on CT measurements compared to those from a tactile coordinate measuring machine (CMM). The CMM is less affected by the workpiece surface due to the intrinsic filtering of the probing sphere. The larger the probing sphere, the less the impact of surface texture. Carmignato et al. (2017) found that surface roughness causes systematic offsets of about $2Rp$ for CT least-squares diameter measurements on periodic roughness profiles (the Rp parameter is the maximum peak height of the roughness profile within the sampling length). The authors report that the result can be generalized also for other types of measurements (e.g. position of flat surfaces) by considering an offset of Rp of the CT least-squares surface from the reference surface measured on the peaks. Tan (2015) observed that surface texture gives surface offsets equal to Rp for aluminium turned parts. The effects of the distribution of the material across the profile, and the voxel size were also investigated in literature. Carmignato et al. (2017) performed experimental and simulated analyses on cylindrical parts to investigate the influence of surface roughness on CT dimensional measurements considering the combined effect of surface morphology and surface filtering characteristics of CT measurements. The results of the analyses report that, in the case of surfaces with periodic roughness profiles, the mean deviations between least-squares diameters and reference diameters take place independently from the profile roughness characteristics (bearing curves) and are confirmed for different voxel sizes. The bearing area curve, also known as the Abbott-Firestone curve, is the cumulative probability density function of the surface profile's height and can be quantified by integrating the profile measurement (Stachowiak and Batchelor 1993). In many references, the effect of surface texture

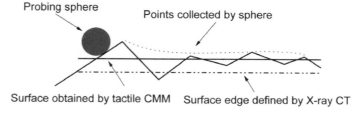

Fig. 5.7 Schema of the influence of surface texture on the difference between CT and tactile CMM measurements

is considered either as a systematic error and/or as an uncertainty contribution. In cases in which the surface texture is assumed to be an uncertainty contribution, it should be modelled using a rectangular distribution (JCGM 2008).

5.1.3   Environment

Accurate dimensional metrology generally requires measurements to be performed at 20 °C with variations on the order of 1 °C or less. To fulfil this requirement is often challenging in CT because the CT system is itself a heat source. To give an idea, a 50 W X-ray tube will produce around 49.8 W of energy in the form of heat. It is, therefore, not a trivial task to keep the temperature stable and constant over time and over the CT cabinet volume; even by installing a large cooling system in the X-ray tube and in the cabinet. Figure 5.8 shows the evolution of temperature at two different X-ray power levels. It can be seen that CT measurements are conducted in a dynamic environment. Temperature variations cause dimensional changes in both the workpiece and the CT system, depending on the material thermal proprieties. For example, a 1 m long bar of steel expands by approximately 0.012 mm K^{-1}, while one made of plastic expands by more than 0.05 mm K^{-1}. Temperature variations in CT modify the kinematics of all the components involved by further magnifying the geometrical errors. The temperature also influences the electronic components by generating thermal agitation inside an electrical conductor at equilibrium. The major effect of thermal agitation is to modify the detector response and to corrupt the detector calibration, usually at the expense of the accuracy of geometrical measurements. A change of 1 °C in the detector temperature can yield up to 10% deviation of the dark signal (Kuusk 2011). An enhanced temperature leads to an expansion of the CsI scintillator crystals (the thermal expansion coefficient is equal to 54×10^{-6} K^{-1}). If a CT measurement is started as soon as the workpiece is placed on the rotary table, temperature changes during measurement may produce significant measurement errors. A good solution is to

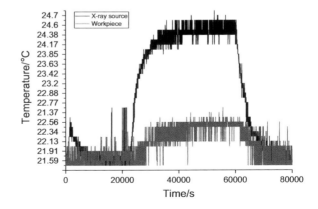

Fig. 5.8 Temperature variation in a CT equipped with a sealed micro-focus X-ray source. By using two sensors, one on the workpiece and one on the X-ray source, the temperature was sampled for 22 h (80,000 s) (courtesy of UNC)

leave the workpiece and its fixture within the CT cabinet for some time (for example one to two hours). The environment temperature is also very important to obtain better performance from CT. Manufacturers state permissible ambient temperatures ranging from 15 °C to 35 °C and the storage temperature should not exceed below 5 °C.

The usual humidity range requirement for metrology laboratories is 40% to 60% relative humidity without condensation. A stable humidity is important to prevent oxidation or rust on the measurement equipment, but also to avoid workpiece expansion. When polymer parts are exposed to moisture, water molecules will diffuse into the matrix, by passing through the open structure of the polymer. Moisture can decrease the surface free energy and simultaneously increase the volume of the polymer. Compared to other environmental considerations, humidity is a minor contribution for CT.

Shock and vibration have an impact on accuracy in the cases where they exceed the maximum levels of vibration a CT system can withstand. Vibrations are often caused by heavy-duty machine tools located in adjacent rooms. A sudden fracture of the filament can occur under the influence of shock or vibration. Vibration also modifies the alignment between the X-ray focal spot, the axis of the rotary table, and the detector, with a direct repercussion on the success of the measurements.

5.1.4 Data Processing

Data processing is an essential component of CT measurements (see Chap. 2 for more details on the various algorithms employed). The main steps to be considered are reconstruction, surface determination and data handling. Reconstruction in CT is mathematically described by the Radon transform, and practically conducted using the filtered back-projection (FBP) algorithm. FBP is a technique to correct the blurring encountered in simple back-projection in which reconstructed images appear to be blurry. The simplest filter implemented in FBP algorithms is a ramp, also known as a Lak filter (Toft 1996). This filter compensates the unwanted blurring, provided the image quality is very high. The disadvantage of the Lak filter is that it boosts high frequencies, which bring unwanted information into the final reconstruction volume. Several other low-pass filters with high frequency cut-offs have been developed (Toft 1996). The most popular filters in the reconstruction software are shown in Fig. 5.9. The Shepp-Logan and Butterworth filters yield the least smoothing at any frequency (Umbaugh and Umbaugh 2011), whereas the Hamming filter induces the strongest smoothing. Despite the smoothing effect, high frequency cut-off values have an impact on the structural resolution (see Chap. 6). Hamming and cosine filters typically reduce the structural resolution by up to 20% (Arenhart et al. 2016a, b) compared to the Ram-Lak filter.

Many reconstruction software packages replace filtering during reconstruction with a pre-reconstruction filtering, usually based on a Shepp-Logan filter. In the presence of low noise levels, all filters yield similar reconstruction outcomes and

Fig. 5.9 Five reconstruction filters that are commonly used in FBP. Apart from the ramp filter, all filters reduce the signal amplitude at high frequencies

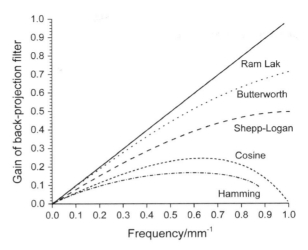

Fig. 5.10 A comparison showing differences between global (*black squares*) and local (*red circles*) threshold method. Six calibrated lengths of a polymer step gauge were considered: M1 = 2 mm, M2 = 6 mm, M3 = 10 mm, M4 = 2 mm, M5 = 14 mm, M6 = 18 mm, and M6 = 21 mm. The lengths were based on plane-to-plane distances

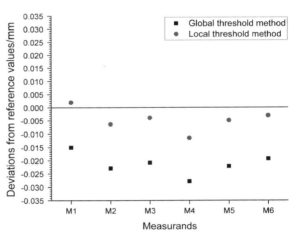

thus measurement results. Surface segmentation (see Chap. 4), defined as the separation of the image into regions, is a necessary step for image analysis and dimensional evaluation. Surface determination for dimensional metrology is usually based on thresholding methods. Global thresholding methods, based on ISO 50%, are a straightforward way to segment a CT data set and lead to accuracy of the order of the voxel used. Local thresholding methods are capable of reaching sub-voxel accuracy, even in the presence of low quality data sets (Borges de Oliveira et al. 2016). Several comparisons between global and local thresholding methods have been conducted and all indicate that local thresholding approaches lead to more repeatable measurements than global methods (Kruth et al. 2011). Figure 5.10 shows a comparison between global and local methods by using a step gauge made from aluminium. The results show that a global thresholding segmentation results in deviations up to 6 μm compared to a local approach. Tan (2015) show that local thresholding segmentation reduces down to 50% the standard deviation obtained with global thresholding approaches. Stolfi et al. (2016) reported that local thresholding approaches are highly repeatable. Borges et al. (2016) also confirmed the sub-voxel accuracy of local thresholding methods for multi-material segmentation.

All thresholding methods are influenced by the starting point, which is the grey value intensity from which the segmentation process starts. Changing the starting points, even of a few grey values, results in surface offsets because of the absence of a sharp transition between grey values of the workpiece and background, as shown in Fig. 5.11. Depending on the homogeneity of the grey value distribution, a transition can vary from a tenth of a voxel to several voxels.

Global thresholding methods are, however, more sensitive to the starting points than local methods, in which the starting point is refined locally. Other surface determination methods, based on region-based segmentation (Borges de Oliveira et al. 2016) and edge-based segmentation (Yagüe-Fabra et al. 2013), exist. Unfortunately, these region- and edge-based methods are still in a state of evolution and, therefore, the range of applications is somewhat restricted. The impact of surface determination on the measurement depends on the feature investigated. Features such as diameters and lengths are more sensitive to the determined surface than the distance between the centres of two spheres.

Voxel models (volume models) and STL models (surface models) represent the ways to handle CT information. Voxel models are usually regarded as more robust than STL models that are more prone to meshing errors, which are very difficult to identify and correct. The magnitude of the meshing errors depends on a series of factors, such as the number of triangles used, linear and angular resolution, and the quality of data sets from which the STL model is generated. Figure 5.12 shows a comparison between two STL surface models created by using a different number

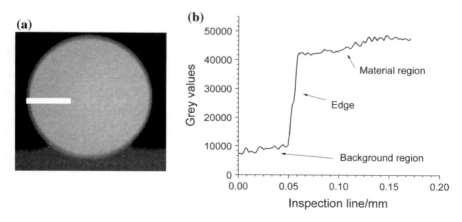

Fig. 5.11 **a** 2D slice of a voxel volume from which **b** a grey value profile was extracted over an inspection line of 0.20 mm (*white line*). The profile showing three regions, namely background plateau, edge region and material plateau. It can be seen that in this case the edge transition covers more than 0.02 mm, which is less than the voxel size (0.035 mm)

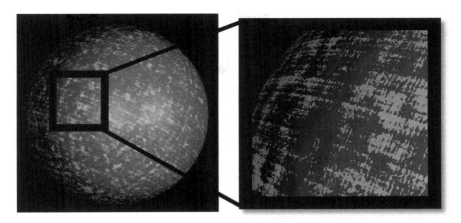

Fig. 5.12 Two STL models having different number of triangles are overlapped. It can be seen that varying the number of triangles surface quality decreases. A ruby sphere with an outer diameter of 5 mm was used as reference object

Table 5.3 Comparison between volume model and surface model, at two meshing quality, with respect to data size and generation time

	Volume model	Surface model Rough mode	Surface model Fine mode
Data size \MB	1600	2.5	116
Creation\s	\	3	120

of triangles. It can be seen that varying the number of triangles, the surfaces decrease in quality. A lower number of triangles leads to larger bias for form measurements. Linear resolution is expressed as a linear dimension and is the maximum distance that the surface of a STL model is allowed to be away from the original voxel model. The angular resolution is the angular deviation allowed between adjacent triangles. Linear and angular resolution should be kept below the target measurement uncertainty. The image quality of data sets has an impact on the information accuracy of STL models. Inspections conducted on voxel and STL models from high quality data sets generally result in similar measurement results. Inspections conducted on voxel and STL models coming from low quality data sets result in very different measurement results, especially for geometrical measurands.

The big advantage of STL models over voxel models is the relatively small size of the data to be handled due to the absence of volumetric information. Table 5.3 shows a file size comparison between a volume model, based on 1000 projections, and two surface models with increasing numbers of triangles. It can be seen that a surface model is always smaller compared to a volume model, even for fine meshing.

5.1.5   *Operator*

Workpiece fixturing and orientation, magnification, X-ray source, number of projections and image averaging, and measurement strategy are the key factors by
which an operator may influence CT performance.

5.1.5.1   Workpiece Fixturing and Orientation

Figure 5.13 shows the influence of the two fixtures, a loose fixture and a tight
fixture, on six length measurements. The results show that an increase in measurement errors between 0.3 µm and 5 µm, depending on the position of the
measurand within the measured volume, can be obtained when a loose fixture is
used. A good fixture should, therefore, be able to hold the workpiece, avoiding
displacements and rotations during imaging. A good fixture should also have a very
low absorption in order not to modify the spectrum. These two conditions are very
difficult to achieve for very low absorption materials. Fixture materials, such as
polyurethane and polystyrene, are mostly used for low absorption workpiece
materials. A disadvantage of polymeric fixtures is the material relaxation and the
thermal instability, both of which cause slippage. Epoxy resins are used for high
abortion materials as well as for heavy parts.

The orientation of the workpiece has a considerable impact on the variation of
the length of the way the X-rays pass through the object during scanning. In the
presence of significant length variation, weak and improper image reconstruction
are experienced. Villarraga-Gómez and Smith (2015) investigated the link between
orientation and measurement accuracy by using several workpieces and different

Fig. 5.13 A comparison
showing difference between a
step gauge in a loose fixture
(*black squares*) and in a tight
fixture. Six calibrated lengths
of a polymer step gauge were
considered: M1 = 2 mm,
M2 = 6 mm, M3 = 10 mm,
M4 = 2 mm, M5 = 14 mm,
M6 = 18 mm, and
M6 = 21 mm. The lengths
were based on plane-to-plane
distances

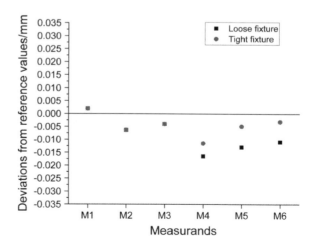

materials. Villarraga-Gómez and Smith show that angles in the range of 10° to 40° lead to smaller measurement errors, including for form measurement. Angular positions above 60° yield larger deviations. The impact of orientation becomes more and more evident for high absorption material such as, for example steel. Angel et al. (2015) reported that for steel step gauge measurement, errors at 90° are five times higher than those at 45°.

5.1.5.2   Magnification

The magnification of an X-ray CT system can be expressed as follows:

$$M = \frac{L_{SD}}{L_{SW}},$$ (5.2)

where L_{SD} is the distance between the source and the detector and L_{SW} is the distance between the source and the workpiece, as shown in Fig. 5.14.

Users obtain a small voxel size by selecting very low values for L_{SW} or high values for L_{SD} and thus a large geometrical magnification. In order to make an accurate reconstruction of the volumetric data, the entire sample must remain within the field of view during the rotation. The maximum magnification M_{max} is limited by the ratio of the effective detector width D and the sample diameter d and can be expressed as

$$M_{max} = \frac{D}{d}.$$ (5.3)

From the magnification, the voxel size V can be derived, which is also related to the detector pixel size p. The characteristic curve linking the three parameters is presented below:

$$V = \frac{M_{max}}{d}.$$ (5.4)

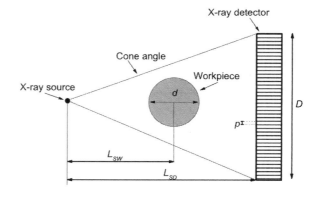

Fig. 5.14 Geometrical magnification is determined by the source-workpiece distance L_{SW}, source-detector distance L_{SD}, the detector width D, the measuring volume d, and the detector pixel size p

Fig. 5.15 A test regarding the effect of five voxel sizes on the accuracy of two different dimensional measurands D1 outer diameter (*black squares*) and D2 inner diameter (*red circles*). Two calibrated diameters of a polymer industrial part were considered: D1 = 15 mm and D2 = 4 mm. The diameters were all least square fitted using 1000 points

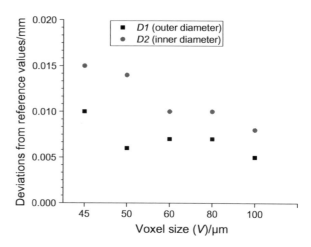

The voxel size should always be larger than the focal spot size to avoid image blurring. As a result, high absorption workpieces cannot be scanned at high magnifications. If the detector runs in binning mode (see Sect. 5.1.1.2), the characteristic curve varies, allowing the scanning of larger focal spot sizes without inducing blurring. Figure 5.15 shows a test regarding the effect of five voxel sizes, ranging from 40 μm to 100 μm, on the accuracy of two different dimensional measurands, namely an outer diameter D1 and an inner diameter D2. It can be seen from Fig. 5.15 that the deviations between CT and CMM measurements do not increase linearly with the voxel size. Voxel size has instead an impact on the detection of sharp edges.

5.1.5.3 X-ray Source Settings

The choice of the proper voltage and current of the X-ray source for a specific measurement and/or workpiece should satisfy two conditions. The first condition is that the X-rays are strong enough to penetrate the workpiece at all rotation angles. The second condition is that the X-rays must not saturate at any rotation angle in any part of the image to be used. Operators can easily satisfy these two conditions with simple workpieces, whereas they may find it difficult to achieve the same aim in workpieces having large variations in their cross-section. The voltage and current selected for a particular cross-section may not fully penetrate the workpiece in directions in which the material being penetrated is larger and simultaneously to avoid detector saturation in the directions in which there is less material. As a result, some projections can be well-exposed and others badly-exposed. In order to have a uniform projection exposure, operators usually find a compromise, based on prior experience, between the X-ray penetrability and the detector salutation. Figure 5.16 shows how twenty operators selected X-ray source settings for two items involving metal and polymer parts during an international comparison (Angel and De Chiffre

Fig. 5.16 Current against voltage for polymer part (*blue diamonds*) and for a metal part (*red square*) (Angel and De Chiffre 2014)

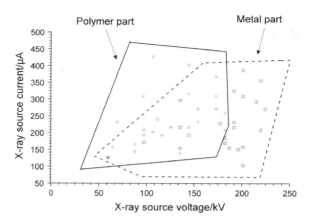

2014). It can be seen that, although the items are very different in X-ray absorption and size, very similar power levels were used for both parts.

The situation described is even more complex for multi-material assemblies that simultaneously include low absorption and high abortion materials. Low energy X-rays are sufficient for lower absorption material. High energy X-ray beams are necessary for completely penetrating the denser parts. Multi-material assemblies having parts with similar absorption coefficients can more easily be imaged.

5.1.5.4   Number of Projections and Image Averaging

The number of projections is intrinsically connected with scanning time and the amount of generated data. Performing a scan with a large number of projections not only increases the acquisition time, but also increases the reconstruction time. Moreover, a large number of projections yield a large amount of data generated which needs to be stored. It is, therefore, preferable to deal with the lowest possible number of projections providing an acceptable accuracy. Several studies have identified the minimum condition to achieve a successful reconstruction of data sets in which the dominant source of noise is photon statistics and the workpiece is a simple geometry. The minimum condition P_{min} can be expressed as follows:

$$P_{min} = \frac{\pi}{4} M^2, \tag{5.5}$$

where M is the number of pixels necessary to inscribe the workpiece. Equation (5.5) assumes that the workpiece is simple in shape and presents symmetry relative to the rotary axis. A weak correlation between the measurement accuracy and the increase of the number of projections has been reported in different studies, even for geometrical measurements. By increasing the number of projections above 800, deviations reduced by less than 5% to 10% (Weckenmann and Krämer 2013). Villarraga-Gómez et al. (2016) showed that the accuracy of measurements of lengths with CT were unchanged using between 400 and 2000 CT projections, with deviations from a reference measuring equipment below 20 μm. Villarraga-Gómez et al. also noticed that the effect of the number of projections does not change as the workpiece material changes. Image averaging represents a further parameter by which operators may influence CT performance. This parameter reduces stochastic fluctuations in the projections by averaging more projections taken at the same angular position. Image averaging increases scanning time which may represent a disadvantage if the number of projections is not reduced. To fully take advantage of image averaging, the projections to be averaged must be collected at the same temperature.

5.1.5.5 Measurement Strategy

Dimensional measurements can be performed using actual-to-nominal comparison or using substitute features. Actual-to-nominal comparison allows the highlighting of differences between a CT data set and its reference information by using a colour map. Each colour represents a different extent of variation between the scanned part and reference information, which can be obtained from a more accurate measuring instrument or from a nominal CAD model. Results of actual-to-nominal comparisons are directly influenced by the alignment process used to overlay the two data sets. Three families of alignment methods are available in inspection software packages namely: best-fit alignment, iterative closest point alignment and non-iterative alignment. Best-fit alignment is a method that globally minimises the distance of every measured point to its reference, using an iterative least-squares fitting algorithm. All points are considered to be equally important in the alignment calculation. This simple equality condition may lead to errors in the presence of data sets with low quality. Iterative closest point alignment methods define an alignment based on selected features rather than points. Closed form alignment methods resolve the alignment by specific coordinates that the user sets. Changing alignment method affects mean values, even if the repeatability is similar. In many cases, alignment differences are also dependent on the coordinates of the CT data set relative to the reference data.

Substitute features are mathematical geometries with no defect whose orientation and size are defined from the points measured on the surface of the workpiece (see Fig. 5.17). Substitute features are affected by three parameters, namely: the number of sampled points, the distribution of points and the strategy of fitting. The number of points should be sufficient to have a sound representation of the geometric feature.

By increasing the number of points used for a given measurement, the resulting measurement will generally be more accurate, especially for complex shapes. In some cases, a large number of points is counterproductive as the probability of having outliers increases. The distribution of measured points should yield a uniform coverage of the feature being measured with a distance between points smaller than the expected defects. This condition will ensure that even small defects are measured. Regular or random distributions can be used depending on the nature of the object's surface texture. The fitting strategy also plays a role in the measurement

Fig. 5.17 Two substitute features (planes) are fitted on a miniature step gauge to measure the distances between flanks. The substitute features are based on 10,000 equally distributed points. The step gauge is made from aluminium

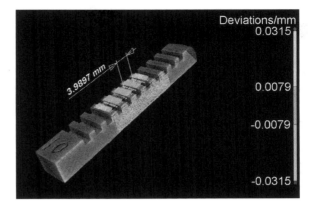

accuracy, as many inspection software packages allow users to select a percentage of the measured points. Varying the percentage of points, will significantly change the resulting measurement. For example, form measurements on a ruby sphere can change on the order of several micrometres when increasing the percentage of points from 95% to 98%. Müller et al. (2016) reported that the choice of a specific measuring strategy did not yield significant differences, provided that the data set presents a good image quality. Nevertheless, Müller et al. reported that the measurement strategy plays a role in the condition of low quality data sets. Similar results were reported by Stolfi et al. (2016) in connection to the datum system definition.

5.2 CT Artefacts in the Reconstructed Volume

In CT, the term "artefact" is applied to any systematic discrepancy causing streaking, shading, rings and bands in a reconstructed volume that are not present in the original object. No reconstructed volume can be assumed to be free of discrepancies. However, different conditions can accentuate the impact of image artefacts on reconstructed volumes. This section presents the most significant image artefacts occurring in CT.

5.2.1 Feldkamp Artefacts

The Tuy-Smith sufficiency condition (Tuy 1983) states that an exact reconstruction is possible, provided that all surfaces intersecting the object intersect the trajectory of the X-ray source at least once. CT based on a circular trajectory can only satisfy this condition within a torus region, in 3D Radon space (Toft 1996). Outside the torus, shadow zones along the z-axis of rotation exist, which make any reconstruction of the 3D object inaccurate, regardless of the detector resolution. Shadow zones can be seen as void regions in 3D Radon space for all images. Shadow zones generate rhombus-like artefacts in the reconstructed volume, commonly known as Feldkamp artefacts. Feldkamp artefacts are present at the top and bottom of reconstructed volumes. Asymmetrical distributions of Feldkamp artefacts are also possible in a CT system having local errors. The extent of Feldkamp artefacts varies with the cone angle, as shown in Fig. 5.18. It can be seen that increasing the cone angle, increases the portion of volume corrupted. However, these errors are less significant when the cone angle is small, that is, when the distance from the X-ray source to the workpiece is large in comparison with the object size. Müller et al. (2012) showed that large cone angles may result in measurement deviations from a CMM up to 30 µm. The presence of Feldkamp artefacts leads to overestimating the sphericity (how spherical a workpiece is) of a sphere up to five times, as shown in Fig. 5.19.

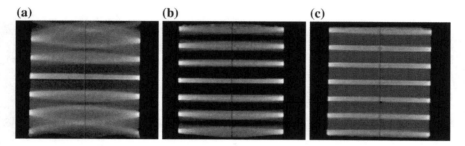

Fig. 5.18 CT reconstruction slices acquired with a cone angle of **a** 30°, **b** 11°, and **c** 5°. The artefacts increase as the cone angle increases (courtesy of QRM Quality Assurance in Radiology and Medicine GmbH)

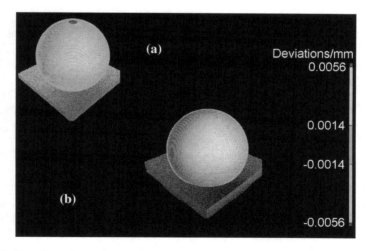

Fig. 5.19 Fit points with an 8 mm ruby sphere showing **a** strong deviations at its pole due to Feldkamp artefacts and **b** no Feldkamp artefacts. Despite the small area covered, the error introduced by Feldkamp artefacts results in worsening of the sphericity by more than five times compared to the reference value (courtesy of PTB)

Feldkamp artefacts are also caused by the longitudinal truncation owing to the limit of a cone beam in irradiating the whole measuring volume uniformly. The non-uniform coverage means that the portion of the measuring volume lying closer to the detector is entirely sampled. The portion that is closer to the X-ray source is scarcely irradiated. Because of the irradiation discrepancies, the reconstruction process cannot deliver correct results in the truncated region. The longitudinal truncation is only an issue for long workpieces. If the workpiece is spherical, cone-beam projections can penetrate the workpiece without significant data truncation.

5.2.2 Beam Hardening Artefacts

The assumption that the X-ray tube emits photons having the same energy is rarely satisfied. A real source provides an energy spectrum which has the general shape of a broad distribution between zero and a cut-off equal to the maximum energy applied to the electrons, known as the peak energy. For example, a peak energy of 120 keV produces X-rays varying between 10 kV and 120 kV with a mean energy of only one-third to one-half of the peak energy. Given that the mean energy is definitely below the peak energy, many of the photons comprising an X-ray beam will be characterised by energy levels far below the mean energy. Very low-energy X-rays do not cause any problem because they are immediately absorbed within the target and X-ray window. Low-energy X-rays are gradually removed after the first few millimetres of material to be traversed, principally because of photoelectric absorption (see Chap. 2). Consequently, the spectrum of the beam becomes richer in high-energy photons and harder to attenuate as it penetrates the workpiece (Lifton et al. 2016). This non-linear effect is referred to as beam hardening (Fig. 5.20).

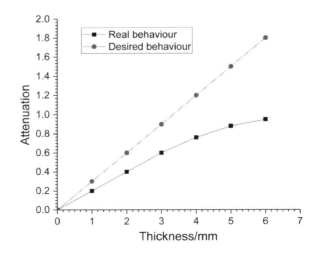

Fig. 5.20 Relation between the X-ray attenuation and the penetrated material thickness with (*line with red circles*) and without (*line with black squares*) the beam hardening effect

Fig. 5.21 A grey value profile of a multi-material sphere with a diameter of 10 mm. The half-sphere made out of steel shows the cupping effect while the grey value profile of the half-sphere made out of aluminium shows no cupping effect. Grey values are mapped ranging from 0 to 65,000 with background at 0. The data set was based on simulation by the chapter author

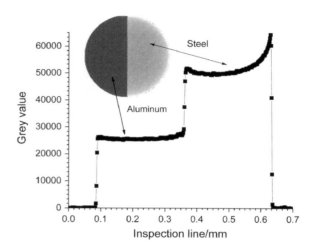

The result of beam hardening is an underestimation of the attenuation and a lowering of opacity values along the ray paths most suffering from this error. Examples of the typical artefacts in CT images arising from beam hardening are shown in Fig. 5.21.

Cupping artefacts have the effect that the apparent material absorption decreases with the depth of material penetrated. As a result, the X-ray intensity at the centre of a projection is smaller than the true values. Streak artefacts can be seen in scans including more workpieces in the measuring volume simultaneously. When an X-ray beam passes through two workpieces, it becomes harder than when it only passes through one material. The presence of beam hardening artefacts can also be seen within the grey value histogram of a CT data set as a severe distortion of the grey value peaks. Beam hardening occurs in any workpiece but its magnitude increases with the absorption and the thickness of the workpiece. Polymer workpieces are subject to an insignificant amount of beam hardening even in the presence of large wall thickness values. Arenhart et al. (2016a, b) stated that when beam hardening artefacts corrupt the datasets, a low frequency drop occurs in the MTF.

Beam hardening influences dimensional measurements by decreasing the inner dimensions and increasing outer dimensions. This diametrically opposite effect may be explained by the fact that beam hardening modifies the inner and outer grey values of background, influencing the contrast and surface determination (Müller et al. 2016). Beam hardening, however, induces larger measurement errors for outer dimensions than for inner dimensions due to the fact that the X-ray attenuation-thicknesses relationship is strongly non-linear for the first millimetres of material to be penetrated (Lifton et al. 2016). Form and bidirectional length measurements are typically influenced by beam hardening, which spreads the distribution of fitted points by increasing the likelihood of having outliers. Unidirectional length measurements are not affected by beam hardening as they are not sensitive to the workpiece material (Bartscher et al. 2014).

5.2.3 Scatter Artefacts

Most CT reconstruction algorithms assume that all X-rays travel in a straight line from the source to the detector. In practice, what happens is that, together with the primary radiation, a secondary radiation, known as scatter, reaches the detector.

Scatter comprises X-rays which are absorbed and then re-emitted, either with the same energy or with a lower energy, as shown in Fig. 5.22. Scatter is originated by the workpiece and environment (Schörner 2012). Workpiece scatter is due to the interaction between X-rays from the X-ray tube and atoms in the workpiece. Depending on the type of interaction mechanism, which is a function of X-ray energy, the direction may be completely different, or only slightly different. The most dominant interaction probability for engineering materials is Compton scattering, which takes place for energies higher than 100 keV (Lifton et al. 2016). The environment scatter is due to all the CT components which are within the cabinet. The contribution of environmental scattering depends on the size of the X-ray beam and its opening angle, because the more photons miss the workpiece, the more photons may be deviated by the environment. Investigations observed that the environment scatter contributes a large part to the total scattered radiation (Schuetz et al. 2014). Scatter results in a high background signal in all projections and in a general loss of contrast. In the reconstructed images, scatter generates artefacts that are similar to those due to beam hardening, namely cupping and streaking (Hunter and McDavid 2012), but less pronounced. Cupping artefacts appear in those cases in which scatter slowly varies during scanning. Streaking artefacts are visible in the presence of very sharp scatter fluctuations, as with a multi-material workpiece (see Fig. 5.23). The amount of total scatter present in a projection is characterised by the scatter-to-primary ratio (SPR) which is defined as

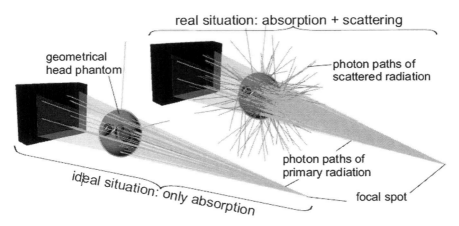

Fig. 5.22 The left part shows the paths of photons that are either absorbed or passing the sample, representing the ideal scenario. The right part gives an idea of the large amount of scattering events influencing attenuation and the detection of scattered radiation in real situations (Wiegert 2007)

(a)   **(b)**

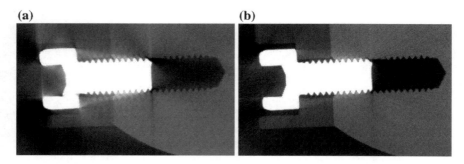

Fig. 5.23 Reconstruction image showing a metal screw in a multi-material workpiece **a** with and **b** without scatter artefacts (courtesy of GE Inspection Technologies)

$$\text{SPR} = \frac{S}{P}, \qquad (5.6)$$

where S is the total secondary radiation and P the primary radiation. For a SPR value equal to two, the background intensity is composed of one unit of primary and two units of scattered radiation. In order to quantify the SPR, the primary and secondary signal should be detected from each projection. This process can often be complex because in most cases the two signals overlap in frequencies through projections. Image contrast is also influenced by scatter as expressed in the following equation

$$C_s = \frac{C_p}{(1 + \text{SPR})}, \qquad (5.7)$$

where C_s is contrast in the presence of scatter and C_p is the primary contrast scatter, which depends on the difference in attenuation between the materials in the workpiece to be imaged. From Eq. (5.7), it can be seen that the contrast can be reduced by about 35% when the SPR is 0.5.

Scatter also influences the detector performance by modifying its DQE which is normally quantified in the absence of scatter. The DQE in the presence of scatter DQE_s can be expressed as follows

$$\text{DQE}_s = \frac{\text{DQE}_f}{(1 + \text{SPR})}, \qquad (5.8)$$

where DQE_f is quantified in a non-scatter scenario. Scanning parameters, magnification, workpiece geometry and X-ray tube voltage play a role in generating scattered radiation (Glover 1982). In particular, the effect of scatter becomes greater at high X-ray voltages. Lifton et al. (2016) presented two methods to check the presence of scatter artefacts. The first method looks at the material grey value distribution: if a Gaussian distribution with minimal skew is seen then the data is likely to contain minimal artefacts. The second method analyses the contrast of the inner and outer edge profiles from a given CT image: if the contrast of inner and outer edges is similar, the data is likely to present negligible artefacts.

5.2.4   Metal Artefacts

The presence of metal components in the workpiece being scanned can significantly deteriorate image quality by causing artefacts, known as metal artefacts. The causes of metal artefacts are the polychromatic X-ray spectrum produced by the X-ray tube, and the complete absorption of the primary radiation by metal parts. An additional cause is the discrepancies between the actual non-linear acquisition process and the ideal mathematical model used in the reconstruction process. Regardless of the causes of metal artefacts, their effects in the reconstructed images are the same. Metal artefacts are visible within the reconstructed images as dark stripes between the metal parts, and glow artefacts on the non-metal parts, in a workpiece (see Fig. 5.24). Although any metal yields artefacts, their importance strongly depends on the size and material of the metal object. High atomic number metals such as iron, steel and platinum produce more pronounced artefacts with respect to metals having low atomic numbers, such as aluminium. Moreover, low atomic number metals lead to artefacts with magnitudes that are independent of the amount of metal.

(a)   **(b)**

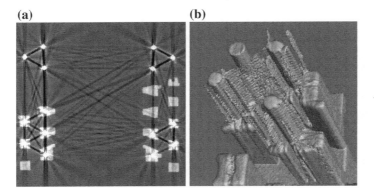

Fig. 5.24 Metal wires in a plastic connector, **a** 2D reconstructed slice showing dark stripes between all wires, **b** 3D volume depicting the metal artefacts that almost hide the actual shape of the workpiece

5.2.5 Detector Artefacts

Malfunctioning and miscalibrations of detector pixels, and impurities on the scintillator screen result in ring artefacts that are visible as patterns of circles centred on the rotation axes (see Fig. 5.25). The rings are present in all the reconstructed slices, but appear to be stronger or weaker depending upon the position of the slice relative to the rotary axis. By approaching the rotary axes, the pattern becomes more evident. Ring artefacts have also different intensities, depending on the cause originating them. Defective detector elements generate sharp rings with a width of one pixel in the reconstructions. Miscalibrated detector elements give rise to wider and less strongly modulated rings. Dust and damaged scintillator areas cause the widest ring artefacts in the reconstructed images, as they influence several detector pixels at the same time. Yousuf et al. (2010) studied the correlation between ring artefacts and scanning parameters. The results showed that the ring patterns are dependent not only on the tube voltages but also on the detector integration time. The rings may also be caused by beam hardening (Ketcham and Carlson 2001).

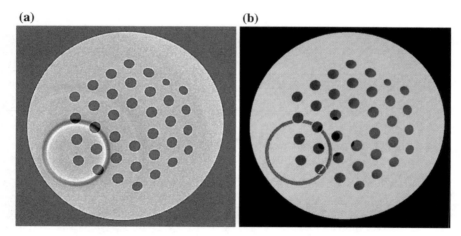

(a)   **(b)**

Fig. 5.25 Polymer workpiece affected by ring artefacts. **a** 2D reconstructed slice and **b** 3D reconstructed view that both show the pattern of circles. The workpiece has an external diameter of 3 mm and series of through holes with a diameter of 0.1 mm

5.2.6 *Noise Artefacts*

Image noise is unwanted variation of intensity across an X-ray projection. Five categories of image noise: random noise, quantum noise, electronics noise, round-off noise and reconstruction noise, can impair the CT measurement process. Random noise is due to fluctuations that have an approximately Gaussian amplitude distribution. Random noise can be neither predicted nor corrected. Thermal noise, also known as Johnson–Nyquist noise (Landauer 1989), is one example of random noise which charges carriers inside an electrical conductor, regardless of the applied voltage. Quantum noise is associated with the limited number of X-ray photons that are used for creating the voxel information. Mathematically, the number of photons N measured by a given sensor element over a time interval t is modelled by the Poisson distribution (Stigler 1982) as follows

$$P(N) = \frac{e^{-\lambda t}(\lambda t)^N}{N!}, \tag{5.9}$$

where λt is the shape parameter that indicates the average number of expected incident photons in the given time interval. Figure 5.26 shows what happens when the number of photons per voxel decreases with respect to the ideal condition. The number of photons per voxel describes the X-ray concentration for a given area. Reducing the number of X-rays per voxel causes the workpiece surface to become noisy, thereby influencing the accuracy of the surface determination process. For large numbers, the Poisson distribution approaches a normal distribution about its mean, and the elementary events (photons, electrons, etc.) are no longer individually observed, typically making quantum noise indistinguishable from random noise. Electronic noise is due to the electronic circuits that inevitably add noise to signals. Analogue circuits are much more susceptible to noise than digital circuits, as a small change in the signal can represent a significant change in the information conveyed in the signal. Round-off noise is an error due to the limited number of bits used to represent a signal digitally. For example, the product of two numbers must be rounded off to the least significant bit available in the computer representation of the number. Additional noise comes from reconstruction and from geometrical errors (see Sect. 5.1.1.3). In general, noise increases from the projection to the reconstructed image. The noise power spectrum (NPS) analysis is a useful image metric that provides a quantitative description of the amount and frequency of the noise fundamentally produced through CT imaging. In metrology, noise influences both dimensional and geometrical measurements. In particular, geometrical measurements are greatly influenced by noise, as even one outlier in the data set can significantly influence the result (Müller et al. 2016). Noise can be quantified by conducting a probing error test assessing the diameter and form deviation of a calibrated sphere having negligible form and surface error. The probing error test should be conducted in conditions reflecting the normal scanning conditions. The results of the test may be used as an indication of systematic error caused by noise.

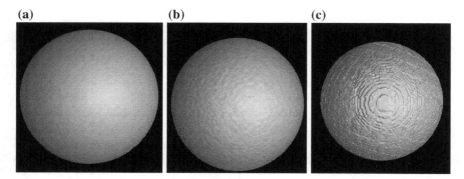

Fig. 5.26 CT volumes simulated of a sphere using **a** 3×10^9, **b** 3×10^5, and **c** 3×10^4 photons per voxel. Reducing the number of protons impinging on the surface of the sphere causes the extent of the noise to increase. Simulations conducted using a ruby sphere with a nominal diameter of 10 mm

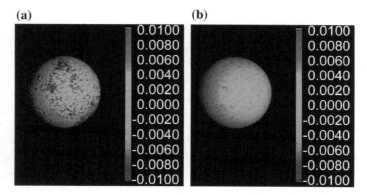

Fig. 5.27 Two nominal-actual comparison maps showing difference between two data sets scanned at two different noise levels: **a** high noise and **b** low noise and. A ruby sphere with a diameter of 5 mm was used for the comparison. Deviations refer to a CAD model and are expressed in millimetres

Figure 5.27 shows a 3D comparison between two spheres scanned at two noise levels. It can be seen that just increasing the noise of 20% causes deviations to increase by a few micrometres.

5.3 CT Artefact Correction

Many reduction techniques have been developed for the suppression of image artefacts in industrial CT. In this section, the principal approaches for reducing image artefacts presented in the previous sections are discussed.

5.3.1   Beam Hardening Correction

The relevance of the problems caused by beam hardening has resulted in a broad variety of artefact reduction methods. These methods can be organised into three classes: hardware, linearization and iterative.

5.3.1.1   Hardware Correction

Hardware correction uses a physical filter on the X-ray source's output window. The presence of the filter reduces the gap in intensity between the parts of the image formed by the X-rays penetrating large path lengths and the parts formed by X-rays not passing through the workpiece. Moreover, the filter removes low-energy photons, leaving Compton scattering as the dominant attenuation mechanism. In common CT, filter materials generally used are aluminium, copper, silver and tin as physical filters. Copper has a high photoelectric absorption which makes it an efficient filter for industrial applications. The K-shell binding energy of Cu is 9 keV, resulting in absorbing X-rays in the range of 9 keV to 30 keV. Tin has a K-shell binding energy of 29.2 keV, leading to absorbing photons of energies in the range of 30 keV to 70 keV, using photoelectric interactions. The K-edge of aluminium is 1.56 keV, resulting absorption of only low energy X-rays. Note that the K-edge is the binding energy of the K-shell electrons of an atom (Hemachandran and Chetal 1986). A photon having an energy just above the binding energy of the electron is, therefore, more likely to be absorbed than a photon having an energy just below this binding energy. The thickness of X-ray filters varies from 0.1 mm to several millimetres, depending on the material used. Since hardware correction reduces the overall number of X-rays hitting the detector, an increase in noise may occur, unless the image exposure is increased to tackle the noise. Note that the spatial resolution is also improved by the beam hardening correction (Van de Casteele et al. 2004). This improvement in resolution is due to the fact that beam hardening enhances the edges, making two objects in an image easier to be resolve.

5.3.1.2   Linearization Techniques

Linearization techniques aim to transform the measured polychromatic attenuation data into monochromatic attenuation data. The most reliable linearization method computes the linearization curve from a reference object similar to the workpiece (Van Gompel et al. 2011). The efficiency of a linearization method based on a reference object is high provided that the workpiece and reference object are made from the same material, and are both investigated under similar scanning and environment conditions. The use of linearization methods based on reference objects is limited, especially in industry, in which many different workpieces and materials are used. A second and widely used linearization method is based on polynomial curves, implemented within the reconstruction software packages. The general format describing polynomial curves is

$$Y = a(b + cX + dX^2 + eX^3 + fX^4),$$ (5.10)

where X represents the initial grey value of a pixel in an X-ray image, Y represents the linearized grey value, and a, b, c, d, e and f are the coefficients that can be automatically or manually selected in order to linearize the profile. The selection of the coefficients is often difficult because an erroneous selection of polynomial curve yields surface offset in both internal and external features. Form measurements can be influenced by the coefficients selected for the beam hardening correction. Bartscher et al. (2014) showed that the local form error of a hole plate cylinder can vary in the range of 2.8 μm to 7.0 μm for three different beam hardening settings. A second order polynomial curve is generally sufficient for low absorption materials, such as polymer and thin aluminium parts. High absorption materials typically require polynomials of higher order. In multi-material workpieces, the selection of a well-suited linearization curve is even more difficult because it requires compensation for the high absorption material without modifying the grey value distribution of the low absorption counterpart.

Reprojection is a further correction that compares the ideal projection, taken using a monochromatic source, with the real one, taken using a polychromatic source. The difference between the two projections' intensities in each pixel is used for the correction. The reprojection method shows clear advantages in reducing artefacts, especially in multi-material workpieces (Krumm et al. 2008). The accuracy of the reprojection method is influenced by the initial reconstruction and surface determination thresholding from which the method defines the correction.

5.3.1.3 Iterative Artefact Reduction

The iterative artefact reduction (IAR) method is an iterative multistage process (Kasperl et al. 2002). Several post processing steps are applied to the reconstructed volume in order to calculate a beam hardening correction independently of any reference object. The IAR is a multistage process, which combines pre- and post-processing techniques. Each iteration of the IAR algorithm delivers an improved correction function H_S^{-1}. The algorithm requires the following four steps.

Initialisation: The correction function H_0^{-1} for the first reconstruction is initialised under the (wrong) assumption that the projection data were scanned with monoenergetic X-rays

$$H_0^{-1} = -\ln(I^P/I_0).$$ (5.11)

Reconstruction: The first iteration reconstructs the initial 3D voxel data from the measured set of raw data, P_0, using a 3D inverse Radon transformation R^{-1}. This initial volume $V_0(\underline{r})$ is uncorrected and thus includes all artefacts, thus

$$V_0(\underline{r}) = R^{-1} H_0^{-1} P_0. \tag{5.12}$$

Subsequent iterations reconstruct improved volumes, $V_N(\underline{r})$, using the modified projection data, P_N, from the pre-processing, i.e. the linearization step.

$$V_N(\underline{r}) = R^{-1} P_N = R^{-1} H_N^{-1} P_0. \tag{5.13}$$

Post processing: After each reconstruction, the post processing steps, segmentation, ray summing and fitting are processed in order to create the correction function H_{N-1}^{-1} for the next iteration. These post processing steps replace the expensive calibration measurements of a reference object. In order to get the path length through the object, it is essential to know the geometry of the object. A segmentation step separates object and non-object voxels in the reconstructed volume as well as possible. The task of the ray summer is the extraction of "measuring points" $\left(I^p \times I_0^{-1}, L\right)$ for the function H, see Eq. (5.13). For all rays from the X-ray source, traced to all raster points in the detector plane, the attenuation of radiation is registered and the ray path length L through the object is calculated. As mentioned above, the data points $\left(I^p \times I_0^{-1}, L\right)$ are fitted with a sum of exponentials. The fit is carried out with a standard non-linear least-squares routine. The inverted function is used as a correction function for the linearization correction. If the difference of the two sequenced system characteristics is deemed

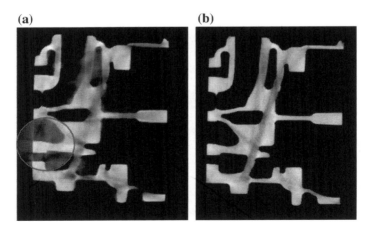

(a)   **(b)**

Fig. 5.28 Reconstruction of an aluminium workpiece, without (**a**) and with (**b**) IAR correction. Artefacts are evidently reduced (Kasperl et al. 2002)

satisfactory, the IAR process terminates, otherwise the next iteration starts with linearization. In Fig. 5.28, two reconstructions without and with an IAR correction are shown.

5.3.2 Scatter Correction

Scatter reduction techniques can be grouped in two categories: scatter reduction techniques and scatter correction techniques.

5.3.2.1 Scatter Reduction Techniques

Anti-scatter grids (ASGs) consist of a plurality of lead strips, separated by an interspace material of very low atomic number such as, carbon fibre and aluminium (see Fig. 5.29). Placing ASGs close to the detector causes them to reduce scatter, by only letting through X-rays that are parallel to lead strips. The pattern of the strips can be modified by changing the ratio between the height of the strips and the width of the interspace material. Bor et al. (2016) showed that ASGs yield an increase in image quality, and a grid ratio of twenty-five fully eliminates the scatter. Two major disadvantages of ASGs are that they only reduce scattered radiation parallel to the strips, and that they require high X-ray energies to penetrate the grid. Moreover, ASGs add some blurring on the projections because they are stationary during the acquisition. The air-gap method minimises the scattered signal by moving the workpiece away from the detector. It has been reported that the scatter reduction with air-gaps is comparable to that of anti-scatter grids (Wiegert 2007). Moreover,

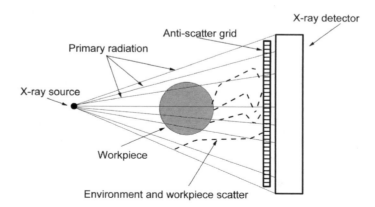

Fig. 5.29 Illustration of an anti-scatter grid placed next to the detector

the air-gap method is inexpensive because no additional component is necessary. In industrial CT, the air-gap method can be implemented by increasing magnification at the risk of stronger Feldkamp artefacts.

5.3.2.2 Scatter Correction Techniques

Scatter correction techniques remove the scatter by first estimating the scatter and then subtracting its extent from each projection. A series of methods have been developed to correct scatter by means of its direct quantification or analytical and Monte Carlo simulations (Lifton et al. 2015). Direct measurement methods require the placement of a beam stop object, for example a lead disk, between the X-ray source and the object. The beam stop object absorbs all X-rays incident upon it, thus any signal measured in the shadow of the beam stop must be due to scatter originating from the object (Lifton et al. 2016). The registered scatter is then subtracted from the workpiece projections. The scatter quantification based on a beam stop object needs to be conducted before each scan. When similar workpieces are investigated under similar scanning conditions, the scatter quantification can be performed only once. The direct measurement methods are an elegant and easy procedure to reduce scatter, although they assume that the X-ray beam does not change over time. Brunke (2016) presented a commercial scatter correction method. Brunke reported that using the developed method, an improvement by a factor of twenty-seven in penetration length can be obtained. Analytical simulations, based on Compton scattering, give the probability of a photon scattering at a given angle and the X-ray energy. Analytical simulations are typically limited to first order scattering interactions otherwise simulation times rise exponentially. Monte-Carlo simulation is the most popular simulation method for quantifying scatter (Thierry et al. 2007). Monte-Carlo methods simulate the X-ray propagation through a scattering medium by using either the CAD data or the material properties of a workpiece. A very large number of photons need to be simulated in order to obtain accurate estimates.

5.3.3 Signal-to-Noise-Ratio Improvement

Since noise is enhanced during the reconstruction, it is necessary to keep it as low as possible during scanning. Two classes of methods: signal enhancement and image filtering, can be used to reduce noise.

Signal enhancement methods rely on the fact that the quantum noise is signal-dependent (the standard deviation varies with the square root of the signal). An increase in the X-ray voltage yields stronger X-rays passing through the specimen to hit the detector, so that the signal becomes stronger. Increasing the X-ray voltage, however, reduces the contrast between the low density materials and the background noise level. The X-ray current reduces the noise, by spreading the grey value histogram over the linear range of the detector. As a result, the grey values of the background and the workpiece are spaced apart from each other, which leads to well-exposed projections at any angular position. If the voltage is high enough, then increasing the current is a better option than increasing the voltage further. It is, however, important to ensure that the selected voltage and current lead to grey values always within the linear range of the detector, at any angular position. If they are close to the linearity limit or even in the non-linear range, the information value of the projections will be reduced. A good rule of thumb is to make sure all image pixels fall within the linear range of detector.

Image averaging is a further noise reduction technique working on the assumption that noise is randomly distributed. By averaging more and more similar images, random fluctuations will be reduced by the square root of the number of images averaged, as shown in Fig. 5.30. Image averaging requires that the frames to be averaged are sampled under similar conditions of temperature and X-ray source voltage in order to successful scale down noise. Even small changes between projections can reduce the efficiency of image averaging.

Flux normalisation is a further process that compensates for variations in brightness during scanning, leading to additional noise during reconstruction. The intensity normalisation is conducted by defining a reference area on the detector on which no workpiece is projected throughout the entire measurement. The mean radiation intensity for each projection is determined with respect to the reference area.

Image filtering tends to increase the SNR by reducing the extent of the noise. The first method to reduce noise is the flat field correction (Van Nieuwenhove et al. 2015), which is a pre-processing method. The idea behind any flat field correction is to collect a number of projections, called reference projections, without the workpiece in the view, at different power levels, including a dark level in which no X-rays are produced. The reference projections captured with the sensor in the dark

Fig. 5.30 Impact of number of frames per projection on the original random noise

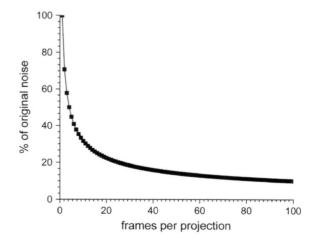

aim to minimise a component of the image noise which is the offset from the average across the imaging array at a particular setting (temperature, integration time). The reference projections taken at increasing power levels aim to correct the variation of sensitivity between pixels for a uniform level of light intensity. The reference projections acquired are subsequently used as a mask to normalise the workpiece projections. The corrected projections can also be stored and used over a long period of time, provided that environment conditions are stable. Attention should be paid to selecting an adequate number of projections in order to keep noise in the reference projections as small as possible. For example, if the reference projections and workpiece projections show similar noise levels, the corrected projections will have higher noise due to the propagation of the errors associated with the convolution. Figure 5.31 shows a comparison of two data sets acquired using two different flat field corrections. The first correction is based on eight levels and thirty-two projections per level and required ten minutes, while the second correction is based on eight levels and sixteen projections per level and required five minutes. Form measurements are significantly different between two methods. Although flat field corrections are well-suited for noise removal, they do not ensure against noise caused by variations in X-ray emission and temperature. Moreover, the flat field correction methods assume a strictly linear behaviour of the detector, which is not fully correct when increasingly large regions of the detector are used. Flat field corrections are conducted before scanning and typically require between a few minutes to a few hours depending on the scanning parameters.

A further method to reduce noise is to use digital filtering, which can be conducted in the 2D and 3D domains. 2D filtering is typically applied to acquired images, while 3D filtering is applied to the reconstructed volume. Mean filters,

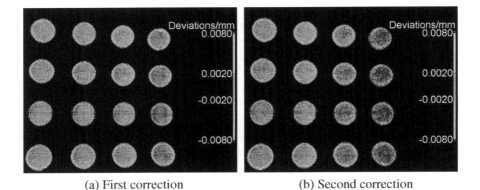

(a) First correction   (b) Second correction

Fig. 5.31 Comparison between two scans conducted using two different shading correction procedures: **a** eight levels and thirty-two projections per level and **b** eight levels and sixteen projections per level and required ten minutes. The first correction yielded lower noise and better form measurements (≈ 10 µm on average, at 95% confidence level) than the second one (≈ 18 µm on average, at 95% confidence level). 1000 points were fitted to each of sixteen ruby spheres. Same scanning parameters were used for two scans

(a)   (b)   (c)

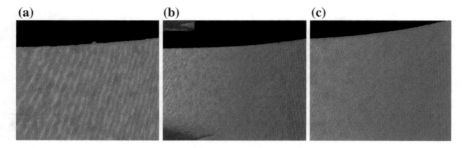

Fig. 5.32 Three 3D views showing the same surface **a** not filtered, **b** filtered using a median filter with a convolution kernel of three voxels and **c** filtered using a median filter with a convolution kernel of five voxels. It can be noticed that increasing the filtering improves the surface in both peak and valley distributions

median filters and morphological filters are the main digital filtering techniques used for this task. Median filters with small convolution matrices (less than or equal to five voxels) are a good way for minimising the noise and measurement deviation in both geometrical and dimensional measurands (Stolfi et al. 2016). Median filters with larger matrices (greater than five voxels) result in measurement errors of the same order as mean filters. Moreover, it is advisable not to simultaneously conduct 2D and 3D filters on the same data set (Bartscher et al. 2012). Figure 5.32 shows the effect of median filtering on the inner surface of a polymer workpiece. It can be noticed that increasing the convolution kernel improves the surface in both peak and valley distributions.

Improvement to the image quality can also be achieved by avoiding any focal spot drift while scanning. It is advisable to allow the X-ray tube to stabilise for an appropriate time before imaging. The stabilisation time depends upon the design of the X-ray tube and the selected power.

5.3.4 Ring Artefact Correction

The most effective way of avoiding ring artefacts is to use hardware correction methods, which eliminate artefacts from projections. The first method to reduce ring artefacts is to make the scanning and detector calibration conditions as close as possible. This method enables the complete removal of ring artefacts, although it requires calibration of the detector each time. Ring artefacts can also be removed by moving the workpiece during the acquisition (Davis and Elliott 1997). As a consequence, the characteristics of detector elements are averaged, reducing the occurrence of artefacts. The major limitation of hardware correction, based on the movement of the workpiece, is that it increases the scanning time by up to 50%.

Whenever hardware corrections cannot be used, software methods, including pre- and post-processing approaches, have been proposed and implemented. A typical pre-processing method is the flat-field correction (Sijbers and Postnov

2004). This method eliminates ring artefacts caused by non-uniform sensitivity of the detector pixels and the non-uniform response of the scintillator screen. Post-processing approaches aim to eliminate ring artefacts by means of filters. Fourier filters, wavelet filters, combined wavelet-Fourier filters, and median filters are typically used to remove ring artefacts from the sinogram or from the reconstructed projection (Anas et al. 2011). The type of filter is selected depending on the extent of the artefacts. Ring artefact corrections can be implemented using either a Cartesian coordinate system or a polar coordinate system. The polar coordinate system is more efficient for artefacts close to the centre of rotation. Matern et al. (2016) investigated the extent to which a software-based ring artefact reduction method enhances measurement accuracy. A series of features, including diameters and roundness, were taken into account within the study. Matern et al. advised to use ring artefact reduction, although the improvements were relatively limited. Compared to hardware corrections, software corrections may often result in corrupted information as they not only neutralise ring artefacts, but also workpiece information. Moreover, software corrections may produce additional secondary artefacts and loss of resolution. To minimise secondary artefacts induced by the correction, the images are usually segmented using two different thresholding values (Prell et al. 2009).

References

Abu Anas EM, Kim J, Lee S, Hasan Mdk (2011) Comparison of ring artifact removal methods using flat panel detector based CT images. Biomed Eng onLine 10:72

Aloisi V, Carmignato S, Schlecht J, Ferley E (2017) Investigation on the effects of X-ray CT system geometrical misalignments on dimensional measurement errors. In: 7th conference on industrial computed tomography (iCT), 7–9 Feb, Leuven, Belgium

Angel J, De Chiffre L, Kruth JP, Tan Y, Dewulf W (2015) Performance evaluation of CT measurements made on step gauges using statistical methodologies, CIRP J Manuf Sci Technol 11:68–72

Angel J, De Chiffre L (2014) Comparison on computed tomography using industrial items. CIRP Ann Manuf Technol 63(1):473–476

Arenhart F, Baldo CR, Fernandes TL, Donatelli GD (2016a) Experimental investigation of the influencing factors on the structural resolution for dimensional measurements with CT systems. In: 6th conference on industrial computed tomography, Wels, p 12.

Arenhart FA, Nardelli VC, Donatelli GD (2016b) Comparison of surface-based and image-based quality metrics for the analysis of dimensional computed tomography data. Case Stud Nondestruct Test Eval

Bartscher M, Sato O, Härtig F, Neuschaefer-Rube U (2014) Current state of standardization in the field of dimensional computed tomography. Meas Sci Technol 25(6)

Bartscher M, Staude A, Ehrig K, Ramsey A (2012) The influence of data filtering on dimensional measurements with CT. 18th WCNDT—World conference on nondestructive testing, pp 16–20

Bequé D, Nuyts J, Bormans G, Suetens P, Dupont P (2003) Characterization of pinhole SPECT acquisition geometry. IEEE Trans Med Imaging 22(5):599–612

Bor D, Birgul O, Onal U, Olgar T (2016) Investigation of grid performance using simple image quality tests. J Med Phys 41(1):21–8

Borges de Oliveira F, Stolfi A, Bartscher M, De Chiffre L (2016) Experimental investigation of surface determination process on multi-material components for dimensional computed tomography. Case Stud Nondestruct Test Eval 6(Part B):93–103. doi:10.1016/j.csndt.2016.04. 003

Brunke O (2016) Recent developments of hard-and software for industrial CT systems. In: 6th conference on industrial computed tomography (iCT), 9–12 Feb, Wels, Austria

Carmignato S, Aloisi V, Medeossi F, Zanini F, Savio E (2017) Influence of surface roughness on computed tomography dimensional measurements. CIRP Ann Manuf Technol 66(1):499–502. doi:10.1016/j.cirp.2017.04.067

Davis GR, Elliott JC (1997) X-ray microtomography scanner using time-delay integration for elimination of ring artefacts in the reconstructed image. Nucl Instrum Methods Phys Res Sect A 394(1-2):157–162

De Chiffre L, Carmignato S, Kruth JP, Schmitt R, Weckenmann A (2014) Industrial applications of computed tomography. CIRP Annals 63(2):655–677. doi:10.1016/j.cirp.2014.05.011

Ferrucci M, Leach R, Giusca C, Carmignato S, Dewulf W (2015) Towards geometrical calibration of X-ray computed tomography systems—a review. Meas Sci Technol 26(August):92003. doi:10.1088/0957-0233/26/9/092003

Glover GH (1982) Compton scatter effects in CT reconstructions. Med Phys 9(6):860–867. Available at: http://www.ncbi.nlm.nih.gov/pubmed/7162472. Accessed 23 May 2016

Hemachandran K, Chetal AR (1986) X-ray K-absorption study of copper in malachite mineral. Physica Status Solidi (B) 136(1):181–185. Available at: http://doi.wiley.com/10.1002/pssb. 2221360120. Accessed 24 Oct 2016

Hiller J, Maisl M, Reindl LM (2012) Physical characterization and performance evaluation of an X-ray micro-computed tomography system for dimensional metrology applications. Meas Sci Technol 23(8):85404. Available at: http://stacks.iop.org/0957-0233/23/i=8/a=085404?key= crossref.f16b74da17fdf2dcb54f5caf3bc9722e. Accessed 26 Apr 2016

Hunter A, McDavid W (2012) Characterization and correction of cupping effect artefacts in cone beam CT. Dentomaxillofac Radiol 41(3):217–223. Available at: http://www.birpublications. org/doi/abs/10.1259/dmfr/19015946. Accessed 26 Apr 2016

IEC (2003) IEC 62220-1 medical electrical equipment—characteristics of digital X-ray imaging devices—Part 1: determination of the detective quantum efficiency, Geneva, Switzerland. Available at: http://www.umich.edu/~ners580/ners-bioe_481/lectures/pdfs/2003-10-IEC_ 62220-DQE.pdf. Accessed 24 May 2016

Ihsan A, Heo SH, Cho SO (2007) Optimization of X-ray target parameters for a high-brightness microfocus X-ray tube. Nucl Instrum Methods Phys Res Sect B 264(2):371–377

JCGM (2008) JCGM 200 : 2008 international vocabulary of metrology—basic and general concepts and associated terms (VIM) vocabulaire international de métrologie—concepts fondamentaux et généraux et termes associés (VIM). In: International organization for standardization Geneva ISBN, 3(Vim), p 104. Available at: http://www.bipm.org/utils/ common/documents/jcgm/JCGM_200_2008.pdf

Kasperl S, Bauscher I, Hassler U, Markert H, Schröpfer S (2002) Reducing artifacts in industrial 3D computed tomography (CT). In: Conference: proceedings of the vision, modeling, and visualization conference 2002, Erlangen, Germany, 20–22 Nov, pp 51–57

Ketcham RA, Carlson WD (2001) Acquisition, optimiziation and interpretation of X-ray computed tomography imagery: applications to the geosciences. Comput Geosci 27:381–400

Konstantinidis A (2011) Evaluation of digital X-ray detectors for medical imaging applications. Ph.D. thesis.University College London

Krumm M, Kasperl S, Franz M (2008) Reducing non-linear artifacts of multi-material objects in industrial 3D computed tomography. NDT E Int 41(4):242–251. Available at: http:// linkinghub.elsevier.com/retrieve/pii/S0963869507001478. Accessed 26 Apr 2016

Kruth JP, Bartscher M, Carmignato S, Schmitt R, De Chiffre L, Weckenmann A (2011) Computed tomography for dimensional metrology. CIRP Annals 60(2):821–842. doi:10.1016/j.cirp. 2011.05.006

Kumar J, Attridge A, Wood PKC and Williams MA (2011) Analysis of the effect of cone-beam geometry and test object configuration on the measurement accuracy of a computed tomography scanner used for dimensional measurement. Meas Sci Technol 22(3)

Kuusk J (2011) Dark signal temperature dependence correction method for miniature spectrometer modules. J Sens 2011:1–9. doi:10.1155/2011/608157

Landauer R (1989) Johnson-nyquist noise derived from quantum mechanical transmission. Physica D: Nonlinear Phenom 38(1–3):226–229. Available at: http://linkinghub.elsevier.com/retrieve/pii/0167278989901978. Accessed 26 Apr 2016

Lazurik V, Moskvin V, Tabata T (1998) Average depths of electron penetration: use as characteristic depths of exposure. IEEE Trans Nucl Sci 45(3):626–631. Available at: http://ieeexplore.ieee.org/lpdocs/epic03/wrapper.htm?arnumber=682461. Accessed 17 July 2016

Lifton JJ, Malcolm AA, McBride JW (2015) A simulation-based study on the influence of beam hardening in X-ray computed tomography for dimensional metrology. J X-ray Sci Technol 23 (1):65–82

Lifton JJ, Malcolm AA, McBride JW (2016) An experimental study on the influence of scatter and beam hardening in X-ray CT for dimensional metrology. Meas Sci Technol 27(1):15007

Matern D, Herold F, Wenzel T (2016) On the influence of ring artifacts on dimensional measurement in industrial computed tomography. In: 6th conference on industrial computed tomography (iCT) 2016, 9–12 Feb 2016

Mehranian A, Ay MR, Alam NR, Zaidi H (2010) Quantifying the effect of anode surface roughness on diagnostic X-ray spectra using Monte Carlo simulation. Med Phys 37(2):742

Müller P, Hiller J, Cantatore A, Tosello G, De Chiffre L (2012) New reference object for metrological performance testing of industrial CT systems. In: Proceedings of the 12th euspen international conference

Müller P, Hiller J, Dai Y, Andreasen JL, Hansen HN, De Chiffre L (2016) Estimation of measurement uncertainties in X-ray computed tomography metrology using the substitution method. CIRP J Manuf Sci Technol 7(3):222–232. Available at: http://linkinghub.elsevier.com/retrieve/pii/S1755581714000157

Prell D, Kyriakou Y, Kalender WA (2009) Comparison of ring artifact correction methods for flat-detector CT. Phys Med Biol 54(12):3881–3895

Schuetz P, Jerjen I, Hofmann J, Plamondon M, Flisch A, Sennhauser U (2014) Correction algorithm for environmental scattering in industrial computed tomography. NDT E Int Volume 64:59–64, ISSN 0963-8695, http://dx.doi.org/10.1016/j.ndteint.2014.03.002

Schörner K (2012) Development of methods for scatter artifact correction in industrial X-ray cone-beam computed tomography. Technische Universität München. Available at: http://mediatum.ub.tum.de/doc/1097730/document.pdf. Accessed 24 May 2016

Sijbers J and Postnov A (2004) Reduction of ring artefacts in high resolution micro-CT reconstructions. Phys Med Biol 49(14):N247–N253

Stachowiak GW, Batchelor AW (1993) Engineering tribology. Elsevier

Stigler SM (1982) Poisson on the poisson distribution. Stat Probab Lett 1(1):33–35. Available at: http://linkinghub.elsevier.com/retrieve/pii/0167715282900104. Accessed 24 May 2016

Stolfi A, Thompson MK, Carli L, De Chiffre L (2016). Quantifying the Contribution of Post-Processing in Computed Tomography Measurement Uncertainty. Procedia CIRP 43:297–302. doi:10.1016/j.procir.2016.02.123

Stolfi A, Kallasse M-H, Carli L, De Chiffre L. (2016). Accuracy Enhancement of CT Measurements using Data Filtering. In: Proceedings of the 6th Conference on Industrial Computed Tomography (iCT 2016)

Tan Y (2015) Scanning and post-processing parameter optimization for CT dimensional metrology. Ph.D. thesis, KU Leuven, Science, Heverlee, Belgium

Thierry R, Miceli A, Hofmann J (2007) Hybrid simulation of scattering distribution in cone beam CT. In: DIR 2007—International symposium on digital industrial radiology and computed tomography, 25–27 June, Lyon, France

Toft P (1996) The radon transform theory and implementation. Ph.D. thesis. Technical University of Denmark

Tuy HK (1983) An inversion formula for cone-beam reconstruction. SIAM J Appl Math 43(3):546–552

Umbaugh SE, Umbaugh SE (2011) Digital image processing and analysis: human and computer vision applications with CVIP tools. CRC Press

Van de Casteele E, Van Dyck D, Sijbers J, Raman E (2004) The effect of beam hardening on resolution in X-ray microtomography. In: Fitzpatrick JM, Sonka M (eds) roc. SPIE 5370, Medical Imaging 2004: Image Processing, 2089. International society for optics and photonics, pp 2089–2096

Van Gompel G, Van Slambrouck K, Defrise M, Batenburg KJ, Mey J, Sijbers J, Nuyts J (2011) Iterative correction of beam hardening artifacts in CT. Med Phys 38(S1):S36

Van Nieuwenhove V, De Beenhouwer J, De Carlo F, Mancini L, Marone F, Sijbers J (2015) Dynamic intensity normalization using eigen flat fields in X-ray imaging. Opt Express 23 (21):27975. Available at: https://www.osapublishing.org/abstract.cfm?URI=oe-23-21-27975. Accessed 24 May 2016

Verein Deutscher Ingenieure (2008) VDI/VDE 2630 Blatt 1.2: Computertomografie in der dimensionellen Messtechnik. Einflussgrößen auf das Messergebnis und Empfehlungen für dimensionelle Computertomografie-Messungen, pp 1–15

Villarraga-Gómez H, Smith ST (2015) CT measurements and their estimated uncertainty: the significance of temperature and bias determination. In: Proceedings of 15th international conference on metrology and properties of engineering surfaces, University of North Carolina—Charlotte, USA, pp 509–515

Villarraga-Gómez H, Clark D, Smith S (2016) Effect of the number of radiographs taken in CT for dimensional metrology. In: Proceedings of euspen's 16th International Conference & Exhibition

Weckenmann A, Krämer P (2013) Predetermination of measurement uncertainty in the application of computed tomography. Prod Lifecycle Manag: Geom Var 317–330

Weiß D et al (2012) Geometric image distortion in flat-panel X-ray detectors and its influence on the accuracy of CT-based dimensional measurements. In: Conference on industrial computed tomography (iCT), Wels, pp 173–181

Welkenhuyzen F (2016) Investigation of the accuracy of an X-ray CT scanner for dimensional metrology with the aid of simulations and calibrated artifacts. Ph.D. thesis, KU Leuven, Science, Heverlee, Belgium

Wenig P, Kasperl S (2006) Examination of the Measurement Uncertainty on Dimensional Measurements by X-ray Computed Tomography, Proceedings of 9th European Congress on Non-Destructive Testing (ECNDT 2006)

Wiegert J (2007) Scattered radiation in cone beam computed tomography: analysis, quantification and compensation. Publikations server der RWTH Aachen University

Xi D et al (2010) The study of reconstruction image quality resulting from geometric error in micro-CT system. In: 2010 4th international conference on bioinformatics and biomedical engineering, pp 8–11

Yagüe-Fabra JA, Ontiveros S, Jiménez R, Chitchian S, Tosello G, Carmignato S (2013) A 3D edge detection technique for surface extraction in computed tomography for dimensional metrology applications. CIRP Ann Manuf Technol 62(1):531–534. doi:10.1016/j.cirp.2013.03.016

Yousuf MA, Asaduzzaman M (2010) An efficient ring artifact reduction method based on projection data for micro-CT images. J Sci Res 2(1):37–45

Chapter 6
Qualification and Testing of CT Systems

**Markus Bartscher, Ulrich Neuschaefer-Rube, Jens Illemann,
Fabrício Borges de Oliveira, Alessandro Stolfi
and Simone Carmignato**

Abstract This chapter focuses on system verification and conformance to specifications. System qualification is carried out to ensure that the system and its
components achieve the best performance—usually corresponding to the specifications made by the manufacturer. Acceptance and reverification testing are
undertaken on the overall integrated system to check whether the system performs
as specified.

Several steps are required to ensure that the best performance of a complex measurement system is achieved and maintained. During the purchasing process, the
user has an important role in the selection and performance verification of the new
system. The selection is mainly based on the specification statements provided by
the manufacturers. For this reason, fair comparability between systems through
using appropriate technical data (i.e. specifications) is highly required.

In order to ensure that the system performs according to the specifications (i.e.
the best performance is achieved), the manufacturer and sometimes the user conduct a series of experiments in which specific error conditions are tested and
parameters are obtained for the fine adjustment/correction of the system. The tests
and fine adjustments implemented to ensure that the system and its components
achieve the best performance are known as *system qualification* (see Sect. 6.1).
Additionally, before accepting the system, the customer carries out an overall test,

M. Bartscher (✉) · U. Neuschaefer-Rube · J. Illemann · F. Borges de Oliveira
Physikalisch-Technische Bundesanstalt (PTB),
Bundesallee 100, 38116 Braunschweig, Germany
e-mail: Markus.Bartscher@ptb.de

A. Stolfi
Department of Mechanical Engineering, Technical University of Denmark,
Produktionstorvet Building 425, 2800 Kongens Lyngby, Denmark

S. Carmignato
Department of Management and Engineering, University of Padova,
Stradella San Nicola 3, 36100 Vicenza, Italy
e-mail: simone.carmignato@unipd.it

© Springer International Publishing AG 2018
S. Carmignato et al. (eds.), *Industrial X-Ray Computed Tomography*,
https://doi.org/10.1007/978-3-319-59573-3_6

executed to check that the integrated system as a whole performs as stated in the specifications. The overall test is called the *acceptance test* (see Sect. 6.2). Subsequently, *reverification tests* are conducted to a reduced extent to check the system performance periodically. For the implementation of dimensional measurement tests, *reference objects* are needed; examples of such objects for testing computed tomography (CT) measuring systems are described in Sect. 6.3. An upcoming requirement for CT is the dimensional measurement of small structures, e.g. edges and slits. In Sect. 6.4, an overview of *structural resolution* is therefore given. In addition, the current state of *standardization* for CT is summarized in Sect. 6.5. Finally, it is worth mentioning that, for the quality assurance (QA) of a measurement laboratory, it is necessary that QA audits are undertaken periodically. A substantial component of proving measurement capability is performing *inter-laboratory comparisons*. Selected international comparisons are presented in Sect. 6.6.

6.1 System Qualification and System Parameter Set-Up

When a complex measurement device is manufactured/set up, a series of tests and adjustments is performed on the system by the manufacturer before delivering it to the customer. The system is then shipped to the customer. After the preliminary set-up of the device, a series of test measurements is carried out to perform fine adjustments of the system ensuring that its best performance can be achieved. These fine adjustments are called system qualification. There are qualification steps that are only performed by the manufacturer on a very rare basis (for example, only once when the system is being installed or after a system accident, e.g. collision) and the user has limited access to such procedures. In contrast, there are qualification steps that are performed by the user on a regular basis. It is important to remark that a qualification test requires the system and its components to be in operation, and it should be able to provide parameters for the fine adjustment/correction of the system.

For industrial CT, system qualification consists of a series of measurements in which each main component of a CT system (i.e. manipulator system, detector and X-ray source) is tested and specific error conditions are investigated. Usually, the specific error conditions are specified by tolerances. When the specific errors are within the tolerance range specified, this means that the system is well adjusted and its best performance can be achieved.

Some qualification tests require the assistance of dedicated reference standards or additional measurement devices, while other tests only need the CT components. Several approaches are used to perform system qualifications in different CT systems (i.e. each manufacturer has its own approach). An example of a workflow is presented in this chapter for an arbitrary industrial cone-beam CT system featuring

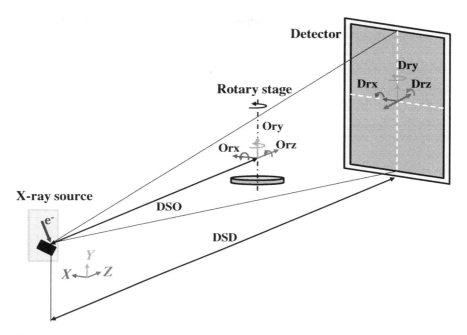

Fig. 6.1 Geometry of sample cone-beam CT system to be qualified. Geometric parameters of distance X-ray source detector (DSD), distance X-ray source rotary axis (DSO, where "O" stands for "object"), rotary axis tilt around X-axis (Orx), rotary axis tilt around Y-axis (Ory), rotary axis tilt around Z-axis (Orz), detector tilt around X-axis (Drx), detector tilt around Y-axis (Dry) and detector tilt around Z-axis (Drz) are represented

an X-ray source, a detector, three translation stages and one rotary stage. See Fig. 6.1 for the orientation of the coordinate system. The X-, Y- and Z-axes are defined along the X-direction linear stage (X-LS), the Y-direction linear stage (Y-LS) and the Z-direction linear stage (Z-LS), respectively.

The geometry of the manipulator (i.e. the relative orientations and positions of the linear stages' axes and the rotary stage's axis), as well as the focal spot of the X-ray source and the intensity of the detector, need to be known, and therefore qualified as detailed below:

- Z-direction linear stage alignment;
- Y-direction linear stage alignment and Z-direction linear stage position qualification;
- Detector alignment;
- Rotary stage alignment;
- Focal spot qualification (size and shape);
- Flat panel detector intensity qualification.

6.1.1 Z-Direction Linear Stage Alignment

The orientation of the Z-LS around the X-axis (Zrx) and around the Z-axis (Zrz) is discussed in this section. The Z-LS alignment will be used as a reference for some of the following qualification steps.

Most of the cone-beam CT systems are installed in such a way that the Z-LS is horizontal, which means that the travel direction of the manipulator along the Z-axis is orthogonal to the direction of gravity. Therefore, a potential approach for checking the orientation of the Z-LS is relating it to the direction of gravity, by placing a digital tiltmeter on the rotary stage and driving the manipulator along the Z-axis. For Zrx and Zrz qualification, the tiltmeter should be placed along the Z-axis and X-axis, respectively. If the alignment of the Z-LS is not within the specifications, usually an adjustment of the manipulator is performed. The Z-LS alignment is normally performed when the CT system is installed for the first time or if a severe collision occurs.

Furthermore, a similar set-up may be used to qualify the orientation of the X-LS around the Z-axis, driving the manipulator along the X-LS with the tiltmeter along the X-axis. For common cone-beam CT using circular trajectory Feldkamp-Davis-Kress (FDK)-based reconstruction algorithms, the qualification of the X-LS may not be required. However, for CT scans using special trajectories (e.g. mosaic reconstruction), the qualification of the X-LS may be of importance.

6.1.2 Y-Direction Linear Stage Alignment and Z-Direction Linear Stage Position Qualification

The following geometrical characteristics should be qualified in order to fulfil the reconstruction algorithms' requirements (such as the parallelism between the rotary axis and the detector centre column): the alignment of the Y-direction linear stage around the Z-axis (see Yrz in Fig. 6.2), the alignment of the Y-LS around the X-axis (see Yrx in Fig. 6.2), and the relative position between the X-ray source and the rotary axis (see DSO in Fig. 6.1).

When the Y-LS is not orthogonal to the Z-axis, a change in the magnification at different height positions of the rotary stage occurs. When the Y-LS is not orthogonal to the X-axis, the rotary axis projected on the detector might be out of the detector centre column at different height positions. Whereas, when there is an offset of the Z-LS position with respect to the input value to the reconstruction algorithm, a scaling error is present.

A potential approach for qualifying the Yrz is the use of an engineer's square and a dial indicator. The engineer's square is placed on the rotary stage, while the dial indicator is attached to a holder and kept in contact with the square. The read-outs of the dial indicator are taken at several positions along the Y-axis. The solution

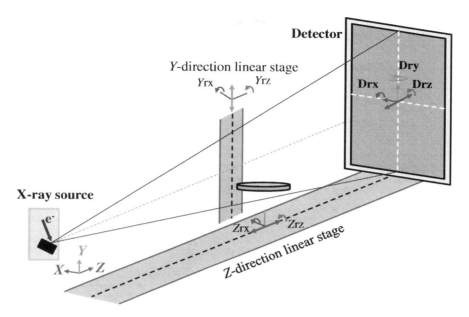

Fig. 6.2 Z-LS misalignment around the Z-axis (Zrz) and around the X-axis (Zrx), and Y-LS misalignment around the Z-axis (Yrz) and around the X-axis (Yrx)

using an engineer's square and a dial indicator is also valid for the alignment of the *Yrx*.

For qualifying the Z-LS position and the alignment of the *Y*-LS, a laser interferometer as a length standard may be used. Several mounting configurations are possible in this approach; e.g. in one of these, the laser interferometer is placed on the rotary stage and a reflector is attached to the X-ray source (see Fig. 6.3). An array of measurement points at several magnifications and different height positions should be assessed using a laser interferometer and a high quality optical block reflector (i.e. with low flatness error). The laser interferometer read-outs are compared with the embedded CT system read-outs and provide a set of parameters that allows the performance of a fine adjustment in the system.

In most of the closed X-ray source CT systems, however, the laser interferometer method is not enough for determining the distance between the X-ray source and the rotary axis. The position and shape of the focal spot are power dependent, i.e. higher power demands larger spot sizes due to the increase of the temperature in the X-ray tube. The increase of the temperature may also change the focal spot position at different X-ray settings (i.e. current and voltage). This means that, for every X-ray setting, a new focal spot position might be obtained, consequently changing the position between CT components. Thus, performing additional scaling factor corrections during the use of the CT system is required.

Performing scaling corrections and, thus, determining the voxel size of a CT scan can be approached in several ways. The most common methods use length

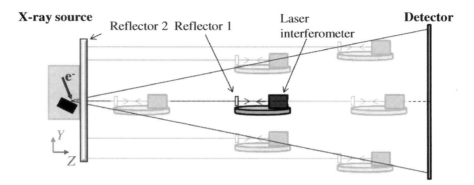

X-ray source Reflector 2 Reflector 1 Laser interferometer **Detector**

Fig. 6.3 Laser interferometer set-up at different positions along the CT volume for geometric qualification

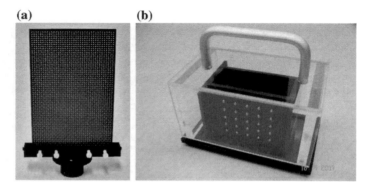

(a) **(b)**

Fig. 6.4 a Structured layer for the 2D-like approach published by Illemann et al. (2015) and **b** double-layer ball plate presented by Weiss (2005) (Courtesy of Carl Zeiss, Oberkochen, Germany)

reference standards (e.g. hole plates, ball plates, multi-sphere standards) measured as unidirectional centre-to-centre lengths. Centre-to-centre lengths are used as the measurement results are not affected by the surface determination process (for more details on length reference standards, see Sect. 6.3.1). Alternatively, there are two-dimensional (2D) image-based approaches for qualifying the relative DSO and DSD positions of the CT system, where a limited number of X-ray images are necessary. Prospective benefits of 2D methods are the significant reduction of time and costs. However, the 2D-based methods cover a limited number of possible alignment qualifications (e.g. rotary axis wobbling may not be covered by the 2D-based methods) when compared to the reference-standard-based methods.

Illemann et al. (2015) proposed a 2D grid-like approach, where a structured layer (see Fig. 6.4a) is used to calculate the voxel size from two 2D X-ray images. The method also allows quantifying the rotary axis misalignment around the X-axis (see Sect. 6.1.4). Weiss (2005) presented a proprietary approach where a calibrated two-layer ball plate is used (Fig. 6.4b) to qualify the voxel size. The method is based on radiographic images from the two-layer ball plate and the geometric parameters are calculated using the sphere centres.

6.1.3 Detector Alignment

Reconstruction algorithms assume that the detector plane is perpendicular to the Z-LS and the detector's centre column is parallel to the rotary axis. In order to meet the detector-related orientation assumptions, detector misalignments around the X- Y- and Z-axes are qualified in this section (see the Drx, Dry and Drz in Fig. 6.1). When Drx, Dry and Drz qualifications do not meet the reconstruction algorithms' requirements, a gradient of the magnification along the workpiece in the Y-direction, a gradient of the magnification in the X-direction and a distortion of the three-dimensional (3D) volume are observed, respectively.

Drx and Drz qualifications require, as a prerequisite, the Y-LS to be qualified already. Drx and Drz qualifications may be performed using X-ray images of a sphere-like tip. The manipulator is moved along the Y-axis, in a low magnification position, and the resulting X-ray images of the sphere-like tip should feature unchanged size. In addition, the tip image projected on the detector should not be out of the centre column of the detector along the Y-direction positions (see Fig. 6.5).

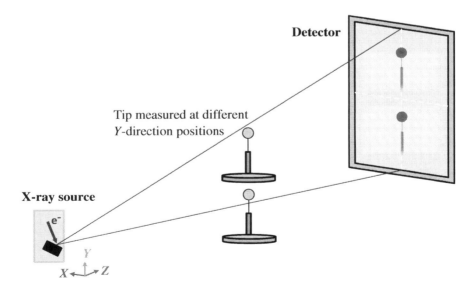

Fig. 6.5 Sphere-like tip set-up as an example of detector alignment qualification

For the *Dry*, a similar set-up as the one used for the *Drz* and *Drx* can be employed. However, for the *Dry*, the manipulator is driven along the *X*-axis and the sphere-like tip size should feature unchanged size at different *X*-LS positions. Detector-misalignment-related qualification does not require periodic repetition; usually it is performed when the system is set up or when an unexpected event occurs (e.g. collision).

6.1.4 Rotary Stage Alignment

According to the requirements of reconstruction algorithms, the rotary axis should be parallel to the detector's centre column, and orthogonal to the *X*- and the *Z*-axis. Furthermore, the misalignment of the rotary axis may also be caused by wobble, when a tilt of the rotary axis to the *X*- or *Z*-direction is present.

The tilt of the rotary axis in the *Z*-direction causes a gradient of the magnification along the workpiece (i.e. the part closer to the detector has a smaller magnification than the part closer to the X-ray source); this effect is also called trapezoidal distortion. In contrast, the tilt and offset of the rotary axis in the *X*-direction generate double edges in the reconstructed volume. Wobble misalignments cause a different position of the workpiece in every angular position within a complete revolution.

A potential approach to qualifying the rotary axis around the *X*-axis is the use of a thin linear structure (e.g. copper wire) placed in the centre of the rotary stage. Two-dimensional projections assessed this thin linear structure and will be measured at different angular positions and at different magnifications. Wobble misalignments of the rotary stage could be included in the wire experiment, too. However, the wobbling should be small enough, compared to other potential errors in the CT set-up, to be a prerequisite for the qualification steps, as the misalignment caused by wobbling of the rotary stage is usually qualified by the rotary stage manufacturer and/or by the CT manufacturer. On the other hand, some CT manufacturers provide implemented software tools which are able to perform rotary axis qualification automatically. Furthermore, there are tools available on the market [e.g. Siemens CERA reconstruction software (Siemens 2017)] which are able to correct rotary-stage-related distortions.

Rotary axis qualification is required more often than other qualification procedures, e.g. Z-LS qualification. Some manufacturers recommend qualifying the rotary axis weekly or when the contour of the object presents double edges, while others recommend qualifying the rotary axis within every CT scan.

6.1.5 Focal Spot Qualification

Ideally, the focal spot is a point. However, due to the limitations of radiation physics, the focal spot is a laterally narrow limited volume in the target material.

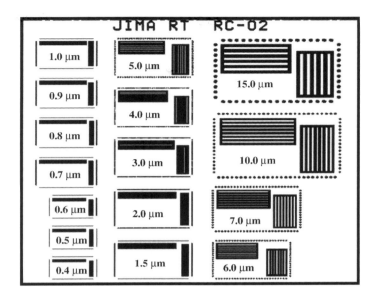

Fig. 6.6 Example of a JIMA mask layout: JIMA RT RC-02 (JIMA Mask 2006) with grid distances from 0.4 to 15 μm

See Chap. 2 for more information on the focal spot. A non-ideal focal spot (i.e. with a size larger than zero and/or with a non-symmetric shape) may cause problems such as blurring in the X-ray images, consequently affecting the CT results. Thus, some methods to qualify the focal spot size and shape are available.

A potential method of qualifying the focal spot is to place a small detailed structure (e.g. an electronic chip) in front of the X-ray source with the highest magnification possible, acquiring X-ray projection images. The JIMA masks are well-known reference standards used to qualify the focal spot size and shape, ranging grid distance from 0.1 to 50 μm with different layouts (see example Fig. 6.6). Another possibility is the Siemens star (which is a pattern structure of a planar carrier plate). For focal spots smaller than 5 μm, the QRM Micro CT Bar Pattern NANO Phantom (Möhrendorf, Germany) is a possibility available on the market.

6.1.6   Flat Panel Detector Intensity Qualification

There is no practical way to ensure that every single detection element (pixel) in the detector is working as designed. Although the number of irregularly working or defective pixels should be much smaller in comparison to the number of properly working pixels, the defective pixels may influence the CT results significantly, if not corrected. Some manufacturers provide built-in solutions for flat panel detector qualification.

In general, two procedures exist to qualify the X-ray intensity of flat panel detectors: defective pixel correction (also called "bad" pixel correction) and intensity variation correction (i.e. flat-field and gain correction, also named shading correction). For defective pixel correction, a series of X-ray images of the free detector (i.e. no pre-filter and no object in the field of view) is taken. A map of the defective pixels is created and corrected based on the average of the neighbouring good pixels. This procedure is typically not performed very often; it is recommended that defective pixel correction is performed e.g. every half year, or when defective pixel-related artefacts (e.g. ring artefacts) are observed in the reconstructed volume (see Chap. 5 for more details).

Intensity variation correction, to the contrary, compensates for the variation response of the detector pixels when illuminated with constant X-ray conditions. Most of the CT manufacturers provide built-in intensity variation correction routines in their software packages and they are usually performed for almost every CT scan. The procedure is based on acquiring several X-ray images from the detector using at least two levels of image intensity (i.e. no intensity and maximum intensity used for that CT scan). This procedure has to be repeated when changing X-ray and exposure conditions (i.e. tube voltage, tube current, tube focusing, pre-filter, detector exposure and gain settings).

6.2 Procedures for Performance Verification: Acceptance and Reverification Testing

This section provides an overview of the current status of the methodology and the principal approaches to performance verification and acceptance testing. At present, only a national guideline (VDI/VDE 2630-1.3) exists on this topic and there is no international standard for CT. The working group that is responsible for the development of international ISO 10360 standards—the ISO Technical Committee 213 Working Group 10 (ISO/TC 213/WG 10)—is working on filling this gap. Details about the implementation of the procedures in published standards and guidelines can be found in Sect. 6.5.

6.2.1 Methodology

The aim of acceptance tests is to check whether the metrological performance of a delivered coordinate measuring system (CMS) fulfils the specifications stated by the manufacturer. The test results are normally relevant to payments and warranty claims. Reverification tests are required to ensure a reliable performance during the operation of the CMS. They should be carried out by the user of the CMS at regular intervals and after special events (e.g. collisions, changes of the environmental conditions, etc.). Acceptance and reverification tests should reflect the typical use of the CMS in such a way that they comprise the complete measurement process (from data acquisition to the final result). Acceptance and reverification tests are complemented with interim checks—being simplified tests to ensure the reliable operation of the CMS during the longer intervals between the acceptance and reverification or multiple reverification tests.

Industrial CT systems used for dimensional measurements should be considered as CMSs. Therefore, the approach and—if technically possible—the specific procedures for system verification and acceptance testing should be similar to the corresponding procedures for other (optical and tactile) CMSs. These tests had been first developed for tactile coordinate measuring machines (CMMs), but they were later adapted to optical CMSs and special kinds of CMSs (e.g. multisensor CMMs, laser trackers for measuring point-to-point distances).

All dominant error sources of the tested measurement system should be taken into account. Therefore, tests for CMSs with CT sensors should show, for example, the influence of the geometry and the material of the workpiece on the measurement result. The tests must be performed under the rated operating conditions given by the manufacturer, which may include limitations, e.g. regarding the allowed system parameters, characteristics of the measured workpieces (e.g. material, maximum roughness), the environmental conditions and the measurable quantities. It is assumed that the measurement system to be tested was previously qualified (see Sect. 6.1) for the specific measurements under test, and that all corrections are applied after the qualification of the system. The result of the tests provides characteristics describing the three-dimensional error behaviour of the overall system. Therefore, specific error sources (e.g. errors of single components of the CMS under study) cannot necessarily be identified.

In the relevant standards and guidelines (ISO 10360 series and VDI 2630-1.3:2011, see also Sect. 6.5), there are two complementary types of tests for CMSs:

1. The local P-test (probing error test), which tests system performance to measure a surface locally, i.e. in a very small measurement volume. This test was originally defined to test the probing system of tactile CMMs (Salesbury 2012). The respective characteristics "probing error for form" (P_{Form}) and "probing error for size" (P_{Size}) are determined from measurements of ideally perfect spheres (or half spheres).

2. The global E-test (length measurement error test) which tests system performance in the entire measurement volume. This test was originally defined to test the CMM kinematics and, thus, the translation axes of tactile CMMs (Salesbury 2012). The characteristic "length measurement error" (E) is determined from point-to-point length measurements.

The characteristics can be used to specify the performance of CMSs in data sheets (which can be used for the marketing of CMSs). In the data sheets, "maximum permissible error" (MPE) values are specified. During acceptance and verification tests, the specified MPE values are used to check the conformance of the determined performance. The proof of the conformance with specifications is not simply a comparison of the determined characteristics with the MPE values, but must take into account also the "test value uncertainty" (U_{Test}), which is the uncertainty of the determined characteristics caused by the tester and the testers' equipment (e.g. reference object). ISO 14253-5:2015 describes the general methodology and criteria for the test value uncertainty.

Important contributions to U_{Test} are geometrical imperfections (e.g. form errors) of the reference objects, the uncertainty of the calibration of the reference objects, errors caused by missing knowledge of the thermal expansion of the reference standards in use when testing thermally compensated CMSs, and "tester errors" due to the misalignment and fixing of the reference objects. Further specific guidance to determine U_{Test} for tactile CMSs is given in ISO/TS 23165:2006. To determine the conformance and non-conformance zones (e.g. using ISO 14253-1 as a decision rule), the particular party that performs the test must be taken into account: when the test is performed by the manufacturer, U_{Test} must be subtracted from the specified (MPE) value, while it must be added to the specified value when the test is performed by the customer.

If there are no restrictions on operating parameters (e.g. voltage, current, position, test specimens [material, size, position in the measurement volume, etc.]) and no details about functions for calculating measuring points given by the manufacturer in the data sheet, they can be chosen freely by the tester. Data filtering methods used during practical measurements of workpieces must also be used during the tests. CMSs with CT sensors can be used in different operation modes which may lead to different specifications. Testing procedures may be described for dedicated measurement modes or principles (e.g. cone-beam CT featuring a circular trajectory). Other modes of operation (e.g. helical CT systems) can often be tested based on mutual agreements between the manufacturer and the user.

6.2.2 Local Error Testing—Probing Error Tests

In these kinds of tests, spheres or spherical caps or arrangements of spheres and spherical caps are measured. The test specimens must be made of a suitable material (e.g. stable dimensions, appropriate attenuation coefficient) which is not excluded by the manufacturer. The diameter must be calibrated and the roughness and the form error must be negligibly small.

The Gaussian fitted sphere of the measured points is determined without any constraints on its position and size. The diameter and form error are then evaluated. The probing errors are calculated as follows:

1. The full span of the radial deviations of the measured points from the calculated sphere is the "probing error for form" (P_{Form}):

$$P_{\text{Form}} = R_{\text{max}} - R_{\text{min}}, \qquad (6.1)$$

which is always positive. In Eq. 6.1, R_{max} and R_{min} are the maximum and minimum distances of the measured points from the centre of the fitted sphere.

2. The difference between the measured and the calibrated sphere diameter is the "probing error for size" (P_{Size}):

$$P_{\text{Size}} = D_{\text{a}} - D_{\text{c}}, \qquad (6.2)$$

which can be positive or negative. In Eq. 6.2, D_{a} is the diameter of the fitted sphere and D_{c} is the result of the calibration of the sphere diameter.

There are different implementations of the probing errors. To achieve comparability to the probing error of a tactile CMM, determined according to ISO 10360-5:2010, in some characteristics only 25 measurement points at one hemisphere of the sphere are taken into account. These points can be determined as "representative points" from several single localized measurement points, e.g. by the evaluation of patches (see Fig. 6.7). Additionally, there are also characteristics considering the large amount of measurement points determined, e.g. by CMSs equipped with optical line sensors, area sensors or CT. Examples are "probing dispersion", taking 95% of the measured points into account, for analysing form and "probing size error All" for analysing size.

Especially when optical CMSs and CMSs with CT sensors are tested, a structural resolution test should be implemented to identify data filtering which can artificially retouch the P-test result especially for form. As an example, low pass filtering of the data mostly reduces the form errors, hence improving the probing error for form, while worsening the structural resolution. In Sect. 6.4, the state of the art and

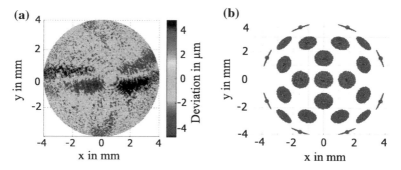

Fig. 6.7 Approaches of local behaviour assessment determining radial errors of Gaussian fit to measurement data—P-test using large number of points **a** probing dispersion analysing 95% of the points and probing size error analysing all points, as well as probing error size and form **b** using 25 representative points (given in *red*) based on 25 patches (given in *blue*) of all probed points (Reproduced from Borges de Oliveira et al. 2015)

current research activities regarding structural resolution are described in detail. At present, stating the structural resolution for dimensional measurements is optional, but may become mandatory in the future.

6.2.3 Global Error Testing—Length Measurement Error Tests

In length measurement error tests, calibrated length standards are measured in different orientations and at different positions in the measurement volume.

Suitable test specimens are, e.g. ball rods with two or more spheres, ball rails, gauge blocks, step gauges, or ball plates (see Sect. 6.3).

To achieve comparability with tactile CMMs, bidirectional lengths should be measured. As illustrated in Fig. 6.8, bidirectional lengths are defined as distances between two points on different faces, assessed in opposite probing directions. Again, measurement points can be computed internally by several single points, e.g. by the evaluation of patches.

Bidirectional lengths should be measured because unidirectional lengths do not include all error sources. For instance, errors caused by the surface determination process are only included in bidirectional lengths. In contrast, unidirectional lengths also have practical relevance, e.g. as the distance of bore holes, and can be used to separate surface-determination-related errors from errors coming from guiding errors of mechanical axes in use, for instance.

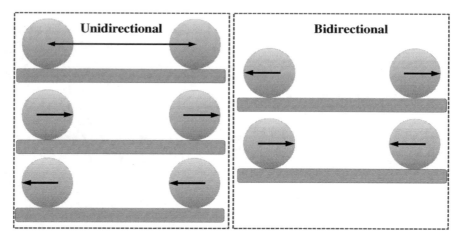

Fig. 6.8 Global behaviour assessment—*E*-test tested as uni- and bidirectional length measurements (Reproduced from Borges de Oliveira et al. 2015)

If it is not possible to determine bidirectional lengths and therefore unidirectional lengths are used, this must be stated. For example, this is the case when ball rods are used as length standards.

In these cases, there are two principal possibilities to determine equivalent bidirectional length measurement errors based on unidirectional length measurement errors:

Method A:

In this method, the probing errors (of size and form) are added to the unidirectional length measurement error to obtain a value assessing the equivalent bidirectional length measurement error. The probing errors of size and form must be added with the proper sign: i.e. the sign that maximizes the absolute value of the resulting bidirectional length measurement error E. In this way, a worst-case approximation of the real bidirectional length measurement error is performed.

Method B:

In this more precise method, an additional bidirectional measurement of a short calibrated length (e.g. a gauge block or a two-point distance at a sphere) is performed and the length error of this measurement is added (with sign) to the unidirectional length measurement error. Thus, an increase or decrease of the absolute value of the error may occur.

6.3   Reference Objects

This section describes reference objects needed for system qualification, for acceptance and reverification testing and for testing the system characteristics when specific measurement tasks are performed. The reference objects must be dimensionally stable in the long term and calibrated with an appropriate calibration uncertainty. Many reference objects used in tactile and optical coordinate metrology are not suitable for CT system qualification because they are made of high-absorbing materials (e.g. hard metal or steel) or assembled from different highly absorbing materials. The material and the X-rays penetration length must be adapted to the measurement parameters. Materials usually adopted for reference objects in CT include metals with low X-ray absorption like aluminium and titanium, as well as low absorbing ceramics and crystals like ruby. The connections between single objects are often realized with carbon-fibre reinforced polymer (CFRP) as it is fairly X-ray transparent, it has a low expansion coefficient and it often has an acceptable geometrical stability.

6.3.1 Length Standards

Length standards used for system qualification to determine the scale factor (see Sect. 6.1) and for the determination of the length measurement error E (see Sect. 6.2) mostly contain spheres, calottes or cylinders.

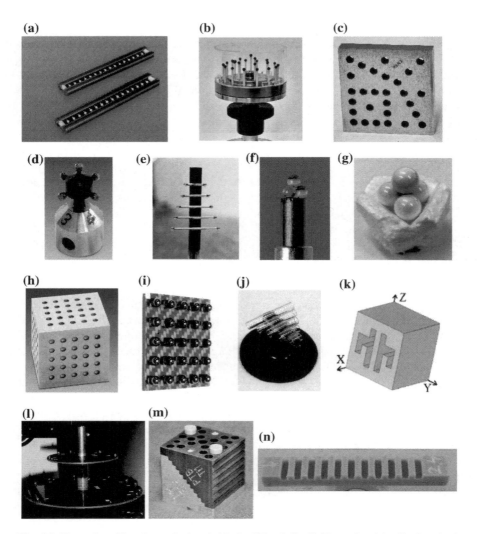

Fig. 6.9 Examples of length standards suitable for CT: **a** ball rail (Trapet Precision Engineering), **b** multi-sphere standard (Bartscher et al. 2016), **c** hole plate (Bartscher et al. 2016), **d** star probe (Bartscher et al. 2008), **e** CT tree (Müller 2012), **f** sphere tetrahedron (Bartscher et al. 2008), **g** sphere tetrahedron (Léonard et al. 2014) **h** calotte cube (Bartscher et al. 2008), **i** ball plate (Müller 2012), **j** pan flute standard (Carmignato 2012), **k** cactus standard (Kiekens et al. 2011), **l** sphere disk (Courtesy of Nikon Metrology, Tring, UK), **m** multi-material hole cube developed by PTB and **n** miniature step gauge (Cantatore et al. 2011)

Sphere distances are well suited to the correction of CT scale factors, because the measurements of sphere centre positions are only minimally affected by the surface determination process and beam hardening (depending on the evaluated area of the sphere surface). Spheres are, in principle, also suited to test the surface determination procedure, through their diameter and form measurements. For the mounting of the spheres, low absorbing material should be used to reduce its influence on the sphere measurements. For this reason, CFRP is frequently used to produce reference standards, as shown, for example, in Fig. 6.9a, d, e. Assemblies of spheres can be glued together (Fig. 6.9f, g), or calottes can be used (Fig. 6.9h).

Ball rails (Fig. 6.9a) are commercially available in different sizes and versions. They can be calibrated with a tactile CMM. The ball rail shown in Fig. 6.9a (length: 120 mm) allows the measurement and calibration of bidirectional lengths, which is not the case for ball bars containing more than two spheres which are mounted on a single rod.

Multi-sphere standards containing ruby spheres fixed on CFRP or ceramic shafts (Fig. 6.9b) are commercially available in different sizes and containing e.g. up to 27 spheres. The outer diameter of the standard shown in Fig. 6.9b is 72 mm. Often the limited long-term stability of the embodied lengths requires a tactile recalibration in short intervals (e.g. in the order of half a year). Furthermore, any mechanical contact with the spheres may ruin the calibration and must be avoided.

The hole plate design shown in Fig. 6.9c is the result of a cooperation project between the National Metrology Institute of Japan (NMIJ) and Germany (PTB). The holes are arranged in differently oriented lines, providing a large number of lengths in different orientations. Two variants (6 mm × 6 mm × 1 mm made of steel, 48 mm × 48 mm × 8 mm made of aluminium) and multiple specimens were manufactured by electrical discharge machining (EDM).

Star probes (Fig. 6.9d) with CFRP shafts can be easily manufactured in various designs. When the sphere diameters and lengths between the spheres are determined from CT surface data, the evaluated surface areas should be symmetrical to the sphere centre to achieve precise results. The horizontal distance of the spheres of the star probe shown in Fig. 6.9d is 10 mm.

The CT tree (Fig. 6.9e) developed by the Technical University of Denmark (DTU) consists of five ball bars of different lengths (normally 16 mm to 40 mm). The ruby balls (⌀ 3 mm) are glued onto CFRP rods. The results of the tactile calibration are sphere coordinates, distances of spheres belonging to one rod and form errors of the spheres.

Sphere tetrahedrons (Fig. 6.9f, g) were realized by gluing four spheres with diameters from 0.127 mm to 10 mm on a shaft made of amorphous carbon or by supporting four spheres in a pyramidal polystyrene holder. Tetrahedrons with spheres of sub-mm diameters are the smallest realized 3D-length standards usable to qualify and test CT-CMSs. Glued tetrahedrons with sphere diameters down to 0.5 mm have been successfully calibrated with tactile CMMs (Bartscher et al. 2008). A tetrahedron with a sphere diameter of 0.5 mm is shown in Fig. 6.9f. The diameter of the spheres of the tetrahedron shown in Fig. 6.9g is 14.29 mm.

The sphere calotte cube (Fig. 6.9h) is a length standard developed by PTB. At three side faces of the cube, a grid of spherical calottes was manufactured by EDM. One size (edge length equal to 10 mm and calotte diameter equal to 0.8 mm) is commercially available in a solid and a hollowed version. It is made of titanium to be measurable by typical industrial CT systems. Due to the large number of calottes, the cube embodies 2775 lengths. This allows a precise determination of scale factor errors and the assessment of the reproducibility of the measured lengths (Bartscher et al. 2008).

The ball plate (Fig. 6.9i) developed by DTU features a regular 5 × 5 array of ruby spheres with a nominal diameter of 5 mm glued on a 2 mm carbon fibre plate. The nominal pitch between the sphere centres is 10 mm. The calibration was carried out with a tactile CMM and the achieved expanded uncertainties of the sphere coordinates were 1.4 μm or 1.6 μm (in plane) and 3.1 μm (off plane).

The "pan flute" standard developed by the University of Padova, Italy (Fig. 6.9j) has five bidirectional calibrated glass tubes of different lengths (ranging from 2.5 to 12.5 mm). Each tube has three calibrated dimensions: length, inner diameter and outer diameter (Carmignato 2012).

The cactus standard developed by KU Leuven, Belgium (Fig. 6.9k) has eight internal parallel surfaces in a prismatic aluminium part (edge length: 45 mm). The lengths to be measured are the horizontal distances between these surfaces.

Nikon Metrology developed a length standard which consists of two CFRP disks and a CFRP cylinder with glued-on ruby spheres (Fig. 6.9l). The stable fixing of the spheres to enable a tactile calibration with sufficiently small uncertainty is a challenge. The diameter of the biggest disk shown in Fig. 6.9l is 160 mm.

Additionally, PTB developed the multi-material hole cube standard (MM-HC), see Fig. 6.9m. It has a size of 30 mm × 30 mm × 30 mm featuring 17 holes inside and 12 "V"-shaped grooves outside. The design consists of two symmetric parts made of different materials The MM-HC also features a step-like "cut" shape enabling different multi-material ratios along the standards' height (Borges de Oliveira et al. 2017a).

The miniature step gauge (length 40 mm, see Fig. 6.9n) made of bis-acryl material was produced by DTU using a replica process and has 11 grooves. The embodied lengths are the distance between the parallel planes.

Further standards have been used in intercomparisons. These standards are shown in Sect. 6.6.

6.3.2 Reference Spheres

Reference spheres are used to determine the probing error and to test, e.g. the surface determination procedure. Depending on the application, the diameter should be calibrated and the form error should be either calibrated or negligibly small. ISO 3290-1:2014 defines criteria for diameter and form deviations of spheres in the form of grades. Grade 10 spheres with a maximum form deviation of 250 nm are

(a)   (b)

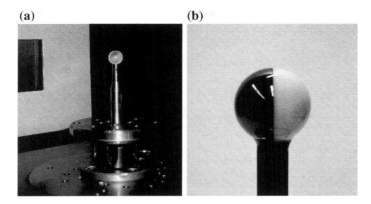

Fig. 6.10 a Werth® universal calibration sphere as an example of a sphere suited to CT testing (Christoph and Neumann 2011); **b** example of a multi-material sphere developed by PTB for testing CT systems

recommended. An example of a special reference sphere—made of glass—is shown in Fig. 6.10a. This sphere is suitable for testing a CMS equipped with tactile probes, optical sensors and CT sensors due to the material and the diffuse reflecting surface.

Another option is to calibrate at least one of the spheres at the length standards shown in Fig. 6.9b, d, g, i.

Additionally, PTB developed a multi-material sphere for testing CT systems consisting of half spheres of different materials, see Fig. 6.10b (Borges de Oliveira et al. 2017b).

6.3.3 Step Cylinders

Step cylinders (see Fig. 6.11, possibly with a central or stepped bore hole inside) are used to assess the errors of external (and internal) measurements simultaneously. Additionally, step cylinders with a central bore hole are mentioned in VDI/VDE 2630-1.3 to investigate material- and geometry-dependent effects (see Sect. 6.2). Effects, dependent on material-specific absorption and penetration thickness, can be investigated by analysing the diameter and the form deviations of the measured steps. The maximum penetration length can be determined.

The step cylinder shown in Fig. 6.11a was one of the objects measured in a test survey organized by ISO TC 213 WG 10 (Bartscher et al. 2016). From the results, it can be concluded that a penetration thickness dependence can be observed. However, the behaviour appears to be inconsistent. A beam-hardening correction enlarged the errors at the outer diameters while the error of the inner diameters remained unchanged.

(a)   (b)

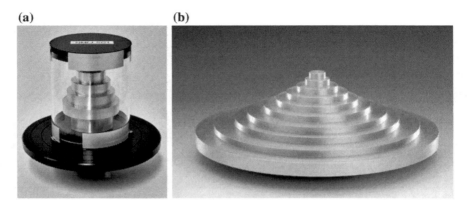

Fig. 6.11 Examples of step cylinders (max diameters: **a** 50 mm, standard of NMIJ, Japan, with a central bore hole **b** 250 mm, standard of PTB without a central bore hole) (Bartscher et al. 2016, 2008)

Fig. 6.12 Chamfer standard made of steel showing examples of an application-related reference object (Neuschaefer-Rube et al. 2012)

6.3.4 Application-Related Reference Objects

Application-related reference objects should help to determine the measurement deviations of CT occurring at a specific measurement task. The best option would be to use calibrated workpieces or to apply multisensor measurements (VDI/VDE 2630-1.3:2011) but that is often not possible.

Alternative solutions are reference objects embodying geometries similar to the workpieces. Figure 6.12 shows an example of a chamfer standard (Neuschaefer-Rube et al. 2012). It features a "roof edge" which is increasingly flattened along the edge. Laser marks define positions where profiles rectangular to the edge are investigated. The standard tests the ability to measure the size of chamfers occurring in the automotive industry. Since the object is made of steel, it was divided into small segments, to reduce the material penetrated by the radiation and, hence, to facilitate the measurability by CT. Moreover, the segments can be CT scanned at a higher magnification.

6.4 Structural Resolution

Resolution concepts originally come from the field of optics. Different criteria were proposed to define an optical resolution (see e.g. Barrett and Myers 2004): the classical approach uses the minimal distance—given in μm—of line or point structures that can just be distinguished using a given criterion. More modern approaches use the *optical transfer function* (OTF) (its absolute value is the *modulation transfer function* (MTF), the *detection quantum efficiency* (DQE) and the *noise power spectrum* (NPS) (see e.g. Illers et al. 2005). From MTF and NPS, a prognosis can be given as to which pattern is visible or not. Relevant and comprehensive information about the spatial resolution and the use of the MTF for CT can be found in the relevant standards on *non-destructive testing* for CT (ISO 15708-1/-2).

For *dimensional measurements* with CT systems, the information obtained is a (discrete sampled) two-dimensional surface in the three-dimensional space—or multiple surfaces, if the object is segmented in different regions or topologies. In this context, the following definition of the term *structural resolution for dimensional measurements* is given in the guideline VDI/VDE 2630-1.3:2011: it "describes the size of the smallest structure that can be measured dimensionally". This guideline proposes taking the limiting diameter of the smallest sphere for which the measurement system is able to determine a diameter with a manufacturer-stated error relative to the calibrated value. This definition cannot describe spatial anisotropy of resolution. Moreover, it is not transferrable to other coordinate measurement systems, and has no unambiguous criterions. To overcome the aforementioned restrictions, two further approaches are given in this section.

For the first approach, we focus the discussion on cases where—in a region of interest—only one single curved surface element exists. Topologically it can be described by its local curvature or by its spatial frequency components (Illemann et al. 2014; Fleßner et al. 2014, 2015; Arenhart et al. 2015). The transition from the real profile to the measured profile is described by a *curvature transfer function* or by the MTF. In the case of the MTF, the height value over a model function (e.g. line, circle) is treated as the functional value. A single length measure is utilized to describe these transfer functions. This measure should be denoted as the *metrological structural resolution for dimensional measurements* (MSR) (see Sect. 6.4.2). The MSR statement is an appropriate measure to compare CMSs— even with different sensor principles—with respect to the measurement capability for small structures consisting of a single topology.

The second approach targets the special ability of CT systems to measure inner surfaces/interfaces. That means the case where different surfaces converge very closely. Similar to optical concepts, the minimal distance of parallel planes is determined in which they can just be distinguished using a given criterion. It is denoted in the following text as *interface structural resolution* (ISR) (see Sect. 6.4.3).

At the time this book was published, the international standardization of acceptance and reverification tests for coordinate measuring systems using CT sensors—in respect of a new definition of structural resolution—was still under discussion (see Sect. 6.5). Therefore, the following subsections are confined to introducing known concepts and tests. The following subsection lists influence factors and demands on the definitions and procedures, followed by a section each for MSR and ISR.

6.4.1   Influence Quantities on Structural Resolution and Demands on Definitions

This section is focused on industrial cone-beam CT using flat panel detectors and performing measurements based on X-ray absorption contrast. For synchrotron instrumentation, phase-contrast methods and fan-beam CT, the contents of this section can be applied, too—if appropriate. Table 5.1 contains a list of general influence quantities, which affect the results of CT. Table 6.1 shows a more detailed list of influence quantities which impair structural resolution, along with their typical effect on structural resolution in respect to position and direction dependence. Basically, the causes of resolution loss can be sorted into technological, physical and mechanical reasons. The prominent technological causes are the focal spot size and the detector pixel size. When reviewing the resolution, it is important to note that the pixel size does not necessarily describe a realistic resolution of the detector: even for a needle-fine X-ray beam hitting one pixel on the detector, there can be crosstalk to the neighbouring pixels. This is described by the point-spread function (PSF). The PSF strongly depends on the type of the detector's active material. In thick unstructured scintillators (e.g. Gd_2O_2S-based scintillators), it is dominated by visible light diffusion. The limitation in structured scintillators (CsI-based) is only given for physical reasons (backscatter and X-ray fluorescence). Only for the case of structured scintillator-based detectors can it be estimated that the pixel size is the same as the detector resolution.

An important physical cause of structural resolution loss is the photon shot noise. For a CT using absorption contrast, the number of absorbed photons in a voxel has to exceed the square root of the passing photons through the empty voxel to overcome the shot-noise limit from Poisson statistics. An improvement of the signal-to-noise ratio can be achieved by the prolongation of the integration time or by powering up the X-ray source. Countering that, the structural resolution depends explicitly on the integration time and the number of images, but also on the attenuation coefficient of the absorbing material.

Table 6.1 Typical causes of structural resolution loss

Cause	Direction/position dependence	Comments
Technological causes		
Focal spot size	Possibly strongly anisotropic because of elliptic spot	Dominates at high power or high voltage using reflection targets
Focal spot drift	Possibly anisotropic, at reflection target also magnification dependent	Depends on warm-up state, sensitive to long-term use and high power
Detector pixel size	Isotropic	Dominates at low power, low magnification and with binning use
Detector PSF	Isotropic	Light diffusion in scintillator, X-ray fluorescence, backscattering
Faulty detector pixels	Radial and polar	Ring artefacts, decrease radial, no tangential component
Detector read-out noise	Isotropic	At short integration time, radiation sensitivity of preamplifier
Physical causes		
Photon shot noise	Isotropic	At low power, short integration time or high magnification
Scattering in the object	Isotropic	Minor effect as similar in neighbouring voxel
X-ray diffraction	Polar direction	At hardened X-ray in low magnification, effect is averaged radially
Cone-beam artefacts	Polar direction	Feldkamp artefact produces excessive noise in gracing incidence
Beam hardening	Radial direction	Corners around hard shadowed areas are forged
Mechanical causes		
Rotary axis wobble and eccentricity	Radial distance dependent, height dependent	Only non-sine part contributes, increasing with mounting height
Azimuth angle jitter	Radial distance dependent, only in tangential direction	Non-constant velocity of gear, too low no. of projections
Dimensional stability of metrological frame	Possibly anisotropic	Analogue as focal spot run-out
Fixation of object	Possibly anisotropic	Typically in radial direction for high magnification

A higher total measurement time increases effects from the drift of the mechanical components and of the focal spot, if they are not compensated for by downstream software procedures. Drift manifests itself in blurring of the reconstructed voxels. From the anisotropic character of some of influence quantities listed in Table 6.1, it is also clear that structural resolution might also be dependent on the direction and on the position in the CT reconstructed volume. Figure 6.13 shows the effect of the electron beam current and the filament heating current on single projection images as an example. With higher beam currents, the focal spot size increases equally in the horizontal and vertical directions. An insufficient thermionic emission causes peripheral electrons to contribute to and to reduce drastically the image resolution only in one direction. If horizontal, as seen, it is obvious that for a circular CT, the structural resolution is mainly affected in the radial and tangential, but not in the axial direction. This is just one example where the structural resolution shows a spatial anisotropy. Further effects which might lead to a position dependency of the resolution in the reconstructed volume are deviations of the rotary stage during revolution. Here a sine-like wobble or an eccentricity does not contribute as it would be equivalent to an object's tilt or shift —only the irregular parts contribute. Different distances between object points and the wobble pivot create different blurring in the projected image. What is easily overseen is a jitter of the azimuth angle of the rotary stage positions. This produces a jitter of the object positions similar in effect to a jitter in the X-ray source position, but proportional to the radial distance and acting in the tangential direction. The effect of sampled angular positions which are too low in number is similar.

There are two different approaches to evaluate structural resolution: deductive and empirical. Deductive means that these single influence quantities are experimentally determined and are combined to a single value using a theory—e.g. with numerical simulation software. Thus for different standardized tasks, a value of structural resolution can be given. This is a good tool for optimization, development and system verification. Typically, the focal spot size in combination with the PSF of the detector can be obtained by measurements of electron-lithographically produced masks (e.g. Siemens star or JIMA masks). More information about geometrical stability, that also degrades the structural resolution, can be obtained by test objects (Sire et al. 1993; Smekal et al. 2004; Weiss 2005; Hiller et al. 2012; Illemann et al. 2015).

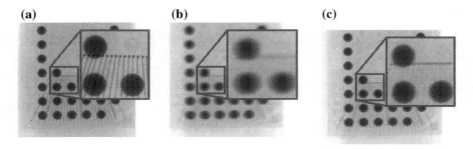

(a)   **(b)**   **(c)**

Fig. 6.13 An integrated circuit with a ball grid array (pitch 500 μm) as a test chart for the focal spot size: **a** optimal setting with a low 35 μA current and properly set filament heating, **b** at 350 μA with filament heating that is too low and **c** at 350 μA with correct filament heating

For the acceptance test of an industrial CT system, this approach only has a restricted value. The reconstruction and surface determination software is, in general, not disclosed and the complete measurement chain has to be benchmarked. These software components can affect the structural resolution, but may also enhance the data quality by using other (e.g. redundant) information to correct certain influence quantities. Modern algebraic (iterative) reconstruction algorithms are highly non-linear so that the results are not predictable. Thus, the empirical approach is recommended. Possible definitions of structural resolution—both metrological as well as interface structural resolution—should be based on a well-defined mathematical model, but the procedures to realize them should refer to the specific circumstances of the measurement. That means they should be sensitive to typical influence quantities. Particularly the reference standards should allow the measurement of the structural resolution at different positions, in different orientations and with different (typical) absorbing materials.

6.4.2 Metrological Structural Resolution (MSR)

The precision of the determination of a single point of a surface in its normal direction is the positional resolution (PR). MSR—in contrast—is a property of local form measurement, i.e. the correlation between neighbouring points on the surface. The PR certainly limits the MSR, but systematic lateral effects contribute, too.

Figure 6.14 illustrates the direction dependence of the PR and therefore of the MSR. It shows a cross section of a cylindrical workpiece (cf. Fig. 6.15). The workpiece diameter is 40 mm; it was scanned with a voxel size of 35 μm, an X-ray tube current of 370 μA, an X-ray tube voltage of 200 kV, and a copper filter of 1 mm thickness. The measurement was done with the sub-optimal set focal spot producing distinct image-blurring as given in Fig. 6.13b. At four different locations and directions, a grey-value profile orthogonal to the surface was extracted. The nearly vertically (collinear to the rotation axis) oriented profile (4) has the steepest profile by a factor of about two in comparison to the more horizontal profile (3), but about the same noise. The position noise is the slope times the grey-value noise visible in Fig. 6.14b. Hence, the surface position can be determined (statistically) with a precision factor of two better at profile (4) compared to profiles 1–3. The noise-related shift of the surface is symmetric, so position noise can be eliminated by lateral averaging without a systematic shift. That is different to tactile measurements, for example, where—due to the morphological filtering of tactile probes —noise always causes a non-symmetric shift.

Figure 6.15 shows one possible reference standard, made by CERTI, Brazil, for the determination of the MSR (Jusko and Lüdicke 1999; Arenhart et al. 2015). In a diamond-turning process on an aluminium cylinder, the superposition of five different sinusoidal height profiles is worked on, thus, creating a multi-wave standard (MWS). The idea is to Fourier-transform the measured (averaged in the axial direction) circumferential profile. From the ratio between the so-determined discrete

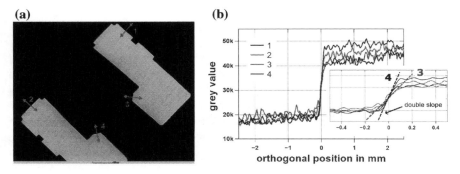

Fig. 6.14 Centrally cut view of a cylindrical workpiece with four density profiles orthogonal to the marked surface positions (**a**) and respective detailed view (**b**). The mean slopes of profiles three and four are shown in a magnified profile view: profile four has a double slope

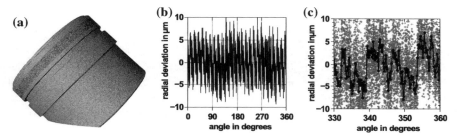

Fig. 6.15 CT of a multi-wave standard **a** with extracted angular topography—the *colour scale* represents radial deviation from −10 μm to +15 μm. Diagram **b** shows the circular profile averaged in the axial direction. The detailed view **c** additionally shows all extracted points of the non-averaged tracks as *red dots*

amplitude values of a reference measurement in comparison to the CT measurement, an amplitude transfer function is obtained.

Figure 6.16a shows the spectral analysis of the profile normalized to a tactile reference measurement with a small probe (radius $R = 50$ μm). The circumference of the MWS is about 125 mm, so at an interpolated wavenumber of 290 (corresponding to a wavelength of 430 μm), the amplitude is reduced to about 50%. This is proposed as a value of the MSR in Arenhart et al. (2015). A tactile reference measurement with a 1 mm radius sphere would be insufficient (Fig. 6.16b), as it reduces higher frequencies. The reason for this is that the MWS is partially too steep, meaning that this large probe cannot touch all surface points.

Another concept for the definition of the MSR is proposed in Illemann et al. (2014). It is based on the transfer of local curvature rather than of extended sine waves. Figure 6.17 shows the principle of this curvature-based structural resolution. It is assumed that a CMS, i.e. a CT in this case, will ideally lead to a Gaussian broadening (full width S) of the surface points' coordinates. Thus, a curved edge is

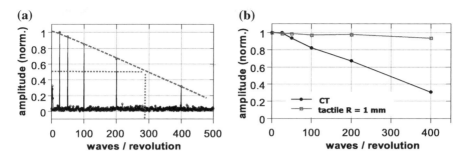

Fig. 6.16 Spectral components of the CT measurement **a** of the multi-wave standard normalized to a tactile calibration (probe radius $R = 50$ μm). Comparison between CT measurement and tactile measurement using an $R = 1$ mm probe (**b**): a tactile measurement with a bigger probe significantly decreases amplitude with increasing spatial frequency

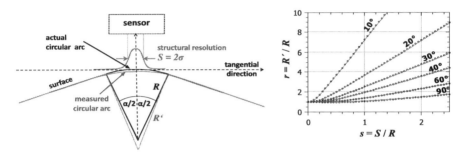

Fig. 6.17 Principle of curvature-based structural resolution. The nomogram shows Eq. 6.3 in dependence of the angle α

transferred with a change in the radius of curvature. The multiplicative factor is given in Eq. 6.3:

$$r \approx erf\left(\frac{\tan \alpha/2}{s/\sqrt{2}}\right)^{-1} \approx 1 + b \cdot e^{-1.06/b}, \quad b = \frac{0.628 \cdot s}{\tan \alpha/2} \tag{6.3}$$

The structural resolution in length units is $S = s \cdot R$, with s a dimensionless resolution measure, and can be read from the nomogram in Fig. 6.17. The input value is the ratio between the measured radius R' and the radius R known from calibration.

Figure 6.18 shows a realization of a cylindrical reference standard and its CT measurement result. Its diameter is 2 mm and its length 4 mm. The radii $R_1 - R_6$ of the reference standard are in the range between 1 μm and 10 μm, and are respectively concave and convex. The rounded triangle-shaped profile has an obtuse angle of $140°$. The structural resolution is calculated from the radii of curvature measured.

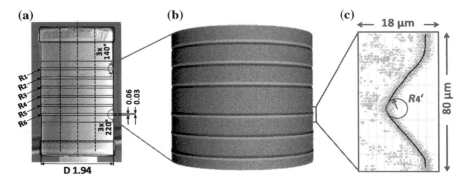

Fig. 6.18 Reference standard for local curvature determination. **a** Dimensional drawing overlaid over an electron-microscopic image, **b** rendered CT, **c** exported data detail with fitted radius

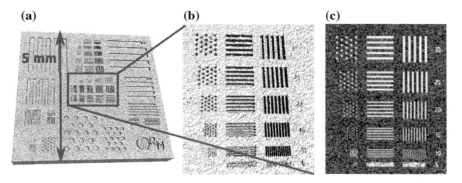

Fig. 6.19 Commercially available 3D test pattern. **a** Surface rendered CT, **b** magnified central area, **c** density cut in 50 μm depth. The *red* contours are the surface profiles found

Finally, another approach shall be mentioned—the two-spheres method (Fig. 6.20). This method, originally introduced with a simplified analysis (Carmignato et al. 2012b, cf. Sect. 6.4.3), assesses shape variations due to loss of resolution at the very small contact point of two spheres. In the area around the contact point, the geometry to be measured is changing from convex to concave (cf. Fig. 6.21 right). In the concave region of the contact zone where the two separate surfaces of the sphere still have a significant distance—and where there is no risk of the two surfaces mixing—a loss of resolution can be measured as a geometrical height error (Zanini and Carmignato 2017). This test is easy to perform and inexpensive because calibrated reference spheres are readily available.

Close to the very contact point of the two spheres, two surfaces next to each other exist. The resolution value obtained from separating two nearly parallel surfaces is linked to—per the definition of this book—a different aspect of resolution—the Interface Structural Resolution (ISR), which will be discussed in the following section.

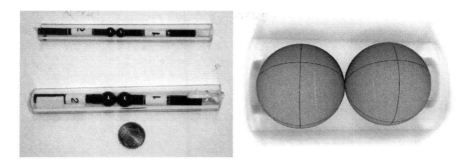

Fig. 6.20 Two-spheres test assembly. A pair of identical ceramic spheres (glued onto carbon fibre stems) is touching each other at a point. They are simply put in plastic tubes

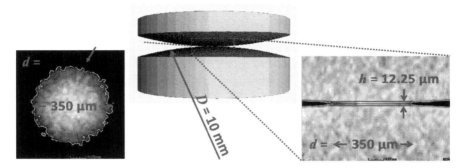

Fig. 6.21 Evaluation of two-spheres test. The contour line in the symmetry plane of the two spheres (diameter D) is fitted with a circle with a diameter d. The neck's height h is calculated

6.4.3 Interface Structural Resolution (ISR)

For practical work, an important question is what the minimum distance of two (nearly) parallel surfaces comes to, where these surfaces can still be separately recognized with a sufficient positional resolution. This minimum distance is the ISR.

Figure 6.19 shows a commercially available test pattern (QRM GmbH, Germany). Deep grooves with steep flanks are etched in silicon. It contains groups of equidistant line patterns in vertical and horizontal directions and circular dots. The structure size in the magnified area ranges from 5 μm to 30 μm. The minimum pitch measure can be seen, where contours are still observable, but in discrete steps. The example shown in Fig. 6.19c allows the determination of horizontal and vertical line structures down to 10 μm, but that of circular hole-structures, only down to 15 μm. In materials testing, this kind of test pattern is proposed to test resolution on the level of grey values inside a measured volume [see also EN 16016-3 (2012)]. There, only the modulation depth of the grooves is relevant.

Carmignato et al. (2012) proposed the two-spheres test (Fig. 6.20) in 2012 to test structural resolution. In the vicinity of the contact point, this two sphere assembly could—per the definition of this section—be used as a continuous scale for measuring the ISR. The magnified CT reconstruction of the contact point is shown in Fig. 6.21. The two spheres with a diameter D form a neck with a diameter d to be determined in the symmetry plane. The calculated height h of the neck is correlated with the structural resolution:

$$h = D - \sqrt{D^2 - d^2} \approx \frac{d^2}{2D} \tag{6.4}$$

Further investigations are needed to study the influence of different sphere diameters, different orientations and beam hardening. Recently, further work has been performed to extend the interpretation of the two-spheres method and to study some important aspects of the influence factors (Zanini and Carmignato 2017).

6.5 The Current State of Standardization

Computed tomography has existed since the early 1970s as a medical imaging technique and for nearly as long as a non-destructive testing technique. Standardization has taken place in both of these fields of application of CT. While medical imaging is outside the scope of this book, it is worth noting that the current standardization in the field of non-destructive testing, EN 16016, as a relevant example of standardization in the field of non-destructive testing, contains a series of four standards. These cover the following: (1) terminology; (2) principle, equipment and samples; (3) operation and interpretation; and (4) qualification. The reading of these standards may provide an introduction to cases where CT is used for non-destructive testing as well as to dimensional measurements.

Historically, standardization in the field of dimensional CT started after standardization in non-destructive testing. Current activities in the standardization of dimensional metrology using CT are distributed over several levels: (i) national guidelines, (ii) national standards and (iii) international standards. In contrast to the well-established standardization in medical CT, dose aspects and aspects of exposure security have not been discussed in technical applications of dimensional CT up to now. This is mainly due to the much higher dose resistance of technical workpieces compared to living human/animal patients or biological samples. The dose aspect may be technically relevant to sensitive materials and cases where technical parts (e.g. reference standards) are exposed repeatedly to high doses. The key points of work on standardization at the moment are: terminology, influence factors, specifications and testing procedures, as well as measurement uncertainty. This section mainly focuses on the general methodology of standards in the field of dimensional computed tomography which is relevant to both industrial standard CT systems applied to dimensional measurements and dedicated metrological CT

CMSs—being mainly used e.g. by research institutes, metrology service providers and national metrology institutes.

6.5.1 National Standardization in Germany

Up to now, only one national guideline describing acceptance and reverification tests for dimensional CT has existed: the German guideline VDI/VDE 2630-1.3:2011. This guideline is part of a framework of guidelines that introduces concepts and terminology. It describes influence factors, specifications and the respective acceptance tests as well as a comparison of CT with other measurement technologies. The series ends with an assessment of the measurement uncertainty and the process suitability of series of parts inspected by CT-based CMSs. The different parts of the guideline are presented below.

VDI/VDE 2630-1.1:2016-05: Computed tomography in dimensional measurement—Fundamentals and definitions.

This first part of the series details the fundamental terminology to be used in the entire series.

VDI/VDE 2630-1.2:2016-07: Computed tomography in dimensional measurement—Influencing variables on measurement results and recommendations for computed tomography dimensional measurements.

This second part of the series details the influence quantities that are relevant to dimensional CT measurements. The use of this part is important when considering error sources and the uncertainty of CT measurements, as well as the correction of errors. Furthermore, it contains a user guide to performing dimensional CT measurements. This guide tries to create sensitivity to everyday issues and problems when using CT. A special focus is placed on actual/nominal comparisons and the measurement of sculptured surfaces (free-form surfaces).

VDI/VDE 2630-1.3:2011-12: Computed tomography in dimensional measurement—Guideline for the application of ISO 10360 for coordinate measuring machines with CT sensors.

This third part of the series aims to create specifications and testing guidance for performing acceptance and reverification tests. In particular, it details the tests for: probing errors for size and form, length measurement errors, material- and geometry-dependent errors, as well as structural resolution. This German national guideline is the only one that is attributed to two series of guidelines. VDI/VDE 2630-1.3 is identical to VDI/VDE 2617-13. VDI/VDE 2617 is a series of guidelines dedicated to the acceptance and reverification testing of CMMs. This double identification is to stress that CT-based CMSs, which are dedicated to performing dimensional CT measurements, act as CMMs.

VDI/VDE 2630-1.3 describes the reference standards to be used in acceptance testing, the testing procedures and conditions, the data analysis of tests and the conformance or non-conformance with specifications. As described in Sect. 6.2, the test is subdivided into a probing error test and a length measurement error test.

What is specific to VDI/VDE 2630-1.3 is a description of how to assess the further influence of CT, the so-called, material- and geometry-dependent effects. These effects can be measured using calibrated workpieces or, as detailed in the guideline, by using measurements of a calibrated step cylinder—as a simplified workpiece. For the case of the step cylinder, new characteristics are defined, measuring the size and the form of the outer cylinder and the inner bore hole of the step cylinder as well as the straightness of the inner bore hole.

Several concepts new to dimensional CT are introduced in VDI/VDE 2630-1.3 and are worth being described here. First, there are the concepts of performing measurements in two principal different modes of operation: stationary measurements and translational measurements. The first type indicates measurements where reconstructed volumes are created and evaluated with a procedure that uses no data stitching; i.e. a set of projections is assessed and then directly reconstructed. For the second mode, a data stitching process occurs; either based on projections or reconstructed volumes. These stitching processes may impair the precision of the measurement as errors may occur at the interfaces of stitched projections or reconstructed volumes. Consequently, VDI/VDE 2630-1.3 distinguishes between stationary (mode TS) and translatory (mode TT) measurements and provides testing guidance for both regimes. The two modes—if applicable—are applied to both probing error testing and length measurement error testing.

A second concept is based on the observation that today CT-based systems are able to work with different measuring field sizes, i.e. these systems apply different magnifications to perform the measurements. Consequently, the metrological performance of the CMS is tested using two distinctively different magnifications.

A third concept is based on the observation that bidirectional calibrated reference standards for length measurement testing may not be available as desired. In this case, unidirectional calibrated reference standards must be used. Here, VDI/VDE 2630-1.3 applies a concept which was originally introduced for optical CMMs in VDI/VDE 2617-6.1 (for CMMs with optical sensors for lateral structures) and in VDI/VDE 2617-6.2 (for CMMs with optical distance sensors). Two approximate methods are introduced to create a measure which is comparable to a bidirectional error statement: firstly, the use of a short end gauge (as a bidirectional standard) plus measurements of unidirectional standards of length, and secondly, adding the probing error for form and the probing error for size to measurements of unidirectional standards of length (see Sect. 6.2 for a description of the concept). For the second approach, special attention has to be given to the correct numerical sign when adding the two probing error values. VDI/VDE 2630-1.3 provides a mathematical distinction for this in the given formula.

As introduced in Sect. 6.2, VDI/VDE 2630-1.3 requests a conformant acceptance test to measure the probing error for form as well as the probing error for size. VDI/VDE 2630-1.3 describes testing conditions and requirements as to which spheres can be used to perform the test.

Comparable to ISO 10360-2, VDI/VDE 2630-1.3 describes how length measurement error testing has to be performed using standards with a (linear) thermal expansion coefficient (CTE) of at least 2×10^{-6} K^{-1}. In cases where this condition is not fulfilled, a substitute calculation to a "normal" CTE between 2×10^{-6} K^{-1} and 13×10^{-6} K^{-1} has to be performed. As an alternative, a measurement of one single length using a standard made of a normal CTE material has to be performed.

VDI/VDE 2630-1.4:2010-06: Computed tomography in dimensional metrology— Measurement procedure and comparability.

This fourth part of the series tries to illustrate measurement technology CT in contrast to CMSs using other sensor principles—ranging from tactile CMMs to different optical CMSs. This version of VDI/VDE 2630-1.4 does not solve, in general, the problem of comparing measurements performed with different sensor technologies. The reader is advised that strong systematic differences may occur as the sensors assess surface data in different ways.

VDI/VDE 2630-2.1:2015-06: Computed tomography in dimensional measurement—Determination of the uncertainty of measurement and the test process suitability of coordinate measurement systems with CT sensors.

This fifth and currently last part of the series describes several methods of assessing the measurement uncertainty of dimensional CT measurements. The measurement uncertainty will then be used to calculate the process suitability of a measurement process of a series of parts from an industrial production process. One method of assessing the measurement uncertainty that is explained in more detail is an experimental approach based on ISO 15530-3. Here, a calibrated workpiece is required which is measured at least 20 times with the CT-based CMS under test. The results are analysed on a statistical basis. The resulting uncertainty statement is then attributed to a series of parts that is going to be measured with the CT-based CMS. The guideline also describes verification procedures to check that the uncertainty statement achieved is valid over a certain time (repeated testing of the hypothesis that the real behaviour is in conformance with the predetermined uncertainty statement).

The German national series of guidelines VDI/VDE 2630 is constantly being revised by the Standardization Committee VDI/VDE-GMA 3.33 "Computed tomography in dimensional metrology". It will be adapted to new developments in ISO/TC 213/WG 10—the international standardization committee that is at present developing an ISO 10360 standard for the acceptance and reverification testing of CT-based CMSs. One of the future activities will be the development of criteria for the simulation of dimensional CT measurements. Due to the policy of VDI/VDE, in the past guidelines were either withdrawn when an ISO standard emerged which described the same topic, or a so-called application guideline was created—which provided further assistance in the use of the (new) ISO standard.

6.5.2 The Current State of National Standardization in Japan

In Japan, one national guideline on the terminology of dimensional CT has been published (JIS B7442). The scope of this standard is comparable to VDI/VDE 2630-1.1. Further national Japanese standards dedicated to dimensional CT are currently not publicly available.

6.5.3 The Current State of National Standardization in the US

In the US, no specific national guidelines exist for dimensional CT. There is a series of national guidelines under the framework of the ASTM International (previously the American Society of Testing and Materials) that discusses aspects of non-destructive testing and that may partly be of interest also to users of dimensional CT. These guidelines are introduced below. ASTM E 1441 describes the theory and general aspects of CT-based imaging. ASTM E 1672 describes the choice of a CT system. ASTM E 2767 describes the data format to be used for data exchange. ASTM E 1695 describes a test method for assessing the performance of a technical CT system. ASTM has started working on a guideline related to metrology applications. The results are not yet available to the public.

The American Society of Mechanical Engineers (ASME) started developing a national guideline for "CT Measuring Machines" under the identification B89.4.23. The results are not yet publicly available.

6.5.4 The Current State of Standardization at ISO Level

ISO/TC 213/WG 10 started work on the part of ISO 10360 dedicated to the acceptance and reverification testing of CMSs using the principle of computed tomography in the year 2010. The aim is to create an ISO 10360 standard which is comparable to the scope of VDI/VDE 2630-1.3 and which includes the current ISO/TC 213/WG 10 principles for an ISO 10360 standard (cf. Sect. 6.2). The first few years of work on this new topic were dedicated to answering technical questions related to influence factors. ISO/TC 213/WG 10 has initiated two technical test surveys to answer specific questions. The first test survey performed in 2013/2014 focused on the applicability of hole plates as standards to assess length measurement errors. As a result, hole plates appear, in general, to be feasible to perform the test, while testing conditions still have to be studied further. A second test survey was performed in 2015 to provide information on the material thickness influence to be present in length measurement testing. The question to be answered was whether a test using a hole plate is sufficient or whether a test consisting of two parts, the test of a multi-sphere standard and the test of a step cylinder, is required. The test survey showed that the use of a hole plate seems to be sufficient. Testing results created input for the future work of ISO/TC 213/WG 10. See e.g. Bartscher et al. (2016) for details of the test survey and for specific results of one participant.

In mid-2017 a first official draft was submitted to ISO/TC 213 as a New Work Item Proposal (NWIP, or in short NP) and members were asked for their comments. ISO member states supported the draft and voted for a four years development time frame. The next step is publishing a Committee Draft (CD) for a future ISO 10360 standard on CT acceptance testing. The preliminary assignment is ISO 10360-11— but this notation may change depending on ISO/TC 213 requirements. The regulations of ISO require a standard to be released or rejected within the selected development time frame. The next steps prior to a published standard are a Draft International Standard (DIS) and a Final Draft International Standard (FDIS). It is worth noting that this international standard will conform to the notation rules of ISO that are implemented for any new ISO standard. The ISO 80000 series requires symbols of characteristics to consist of one leading character followed by subscripts. This terminology has been explained in (Bartscher et al. 2014), but it will be constantly revised. Concerning the technical solutions described in any publicly available draft, it is important to note that there is no guarantee that the approach to testing and the current technical implementation will stay unchanged.

6.5.5 Special Topic: Current Standardization Activities for Structural Resolution

In the previous Sect. 6.4, the "metrological structural resolution" concept was introduced and detailed for the case of dimensional CT measurements. The current standardization for dimensional CT describes this concept in VDI/VDE 2630-1.3. The problem when using this national guideline is that the test described here for structural resolution testing provides only a partial solution for the reader. VDI/VDE 2630-1.3 requires the diameter of the smallest sphere, which can be measured by the CMS under study, to be the measure for the metrological structural resolution. A manufacturer conforming to this definition has to make a statement about the measurement error, but precise criteria for a measurement limit of this smallest sphere are missing. A link to any large-scale specifications of the CMS under study—e.g. MPE values for probing errors or length measurement errors—is also missing. Thus, it is difficult for a CT user to transfer this resolution statement to any measurement of practical interest.

Currently, several other concepts to assess metrological structural resolution are under discussion and are being tested and detailed (see Sect. 6.4). A further concept that has already been proposed for standardization is "small-scale fidelity". This concept was introduced first for the field of surface topography. "The measurement

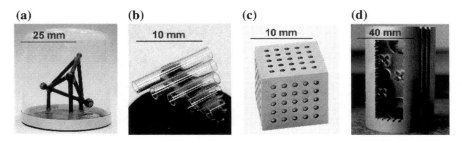

Fig. 6.22 The four CT Audit samples: **a** CT tetrahedron provided by the University of Padova, **b** pan flute gauge provided by the University of Padova, **c** calotte cube provided by the Physikalisch-Technische Bundesanstalt, **d** QFM cylinder provided by the Institute of Quality Management and Manufacturing Metrology of the University of Erlangen-Nuremberg (Carmignato 2012)

is considered to fulfil the 'small-scale fidelity' criteria when the amplitude of a structure of the fitted reference shape deviates no more than 10% from the reference amplitude. The value of 10% is a suggestion for a possible acceptance criterion" (Seewig et al. 2014).

6.6 Interlaboratory Comparisons

Due to the fact that dimensional CT is a relatively new topic in metrology, only a limited number of comparison measurements have been performed under a formal framework (intercomparisons, round robins). These *interlaboratory comparisons* cover different topics and not all of them are published in detail. Furthermore, it has to be stressed that these comparisons have not reached the level of international intercomparisons as presently attained by BIPM key comparisons (cf. http://kcdb. bipm.org and ISO/IEC 17043:2010). Nevertheless, certain important observations have been made. Thus, the comparisons are useful for describing the state of the art and for focusing on problems in the use of dimensional CT in coordinate metrology.

6.6.1 "CT Audit" Intercomparison (2009–2012)

The "CT Audit" intercomparison was the first international round robin in the field of dimensional CT, which was open to the public and which was fully published (Carmignato 2012). This intercomparison was organized and coordinated by the University of Padova, Italy. The circulation was prepared in the year 2009 and was carried out in the period from March 2010 to March 2011, involving 15 CT systems used for dimensional metrology by research institutions and companies from different countries in Europe, America and Asia. Four calibrated samples were circulated in a sequential participation scheme (i.e. from one participant to the next). The samples are shown in Fig. 6.22 and include a variety of dimensions, geometries and materials. During the circulation, all the samples were protected in thin plastic sealed boxes to reduce the risk of damage, limit contamination and avoid measurements with other sensors.

More than 5000 individual dimensional measurement results were collected from the participants and finally analysed by the coordinator. The results demonstrated that sub-voxel accuracy is clearly reachable in CT dimensional measurements: for specific measurements of size (mainly unidirectional lengths), errors in the order of down to 1/10 of the voxel size were obtained by most of the participants. However, it was shown that only few participants were able to perform length measurements with errors below their CT systems' specifications. Larger errors were found for form measurements, due to the presence of significant noise in CT reconstructions. Furthermore, it was demonstrated that only less than half of the participants' measurement results were provided with a valid measurement uncertainty statement. In fact, most of the participants failed to satisfy the proficiency assessment criterion $|E_n| < 1$, in which the E_n number is defined according to ISO/IEC 17043:2010 and is expressed by the following equation

$$E_n = \frac{x - X}{\sqrt{U_{lab}^2 + U_{ref}^2}}, \tag{6.5}$$

where x is the participant's measurement result, X is the reference value from calibration, U_{lab} is the expanded measurement uncertainty of the participant's measurement result and U_{ref} is the expanded calibration uncertainty of the reference value. In conclusion, the CT Audit intercomparison pointed out that the traceability of CT dimensional measurements is still a major challenge, and that international standards are urgently needed to establish proper procedures for measurement uncertainty evaluation and for the metrological performance verification of CT systems (Carmignato 2012).

6.6.2 Intercomparison on Structural Resolution in ISO TC 213 WG 10 (2011)

ISO/TC 213/WG 10—the international technical committee which is working on a future ISO standard on dimensional CT (see Sect. 6.5)—is interested in including a structural resolution test in a future part of ISO 10360. The first intercomparison on this topic in 2011 formed the starting point of this work. Two standards were used: (i) a JIMA mask (being a 2D standard, see Fig. 6.6a) provided by NMIJ, Japan and (ii) a micro-tetrahedron (being a 3D standard, see Sects. 6.3 and 6.4.3) provided by PTB, Germany. Both standards were measured by different committee members (two members for the JIMA, five members for the tetrahedron). The JIMA standard

(a)   (b)   (c)   (d)   (e)

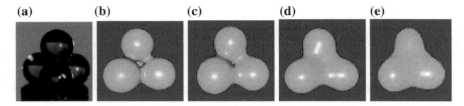

Fig. 6.23 Series of measurements of a micro-tetrahedron featuring four ruby spheres of a diameter of 0.5 mm mounted on a glass carbon shaft; measurement data of PTB. **a** Photography; **b–e** series of CT measurements with decreasing magnification (magnification 6, 3, 2 and 1.34, resp.)

(a)   (b)

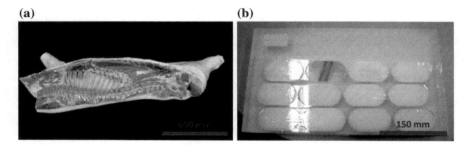

Fig. 6.24 **a** Real pig carcass and **b** synthetic phantoms (Angel et al. 2014)

was measured using projections only. Thus, no 3D structural resolution statement was achieved describing the performance of the whole measurement system. The 3D CT measurements of the micro-tetrahedron performed with different resolutions of the CMSs showed a clear visual indication of the loss of resolution with decreasing magnification (see Fig. 6.23 as an example of a result of one of the participants). While the results showed the general dependency, no specific decision was made to use the underlying concept (multiple surfaces are blurred into one surface) as a criterion for the metrological structural resolution. It is worth noting that the tetrahedron features at the intersections of the sphere concave structures only and that the topology type changes when losing resolution. Thus, resolution statements may be restricted. A further observation is that the tetrahedron, being a composite of four spheres, always has glue at its interfaces. Thus, precision statements may be impaired by the presence of the glue.

6.6.3   CIA-CT Intercomparison on CT for Industrial Applications in the Slaughterhouses (2011–2012)

A first intercomparison on medical CT for industrial applications in slaughterhouses was coordinated by the Centre for Geometrical Metrology (CGM), Department of

(a)   (b)

Fig. 6.25 The two items used in the CIA-CT interlaboratory comparison: **a** a plastic Lego brick and **b** a metal component from a medical device (Angel and De Chiffre 2014)

Mechanical Engineering, Technical University of Denmark, DTU, and carried out within the project "Centre for Industrial Application of CT Scanning—CIA-CT". The main aim of the comparison was to prove the applicability of these types of CT systems for volume measurement on pig carcasses. Two synthetic phantoms including several polymer materials were used as substitutes for biological tissues (see Fig. 6.24). In this comparison, seven laboratories from four countries were involved. The circulation took place between May 2011 and May 2012. Out of a total of 42 single results obtained by the participants using CT scanners, 31% of the measurements yield satisfying $|E_n|$ values smaller than 1 (see Eq. 6.5), with 69% being larger than 1. The reason for the poor agreement between the participants and calibration values, provided by the coordinator, could be due to the following two factors. (1) The segmentation areas between multi-materials were composed of mixed pixel values, making it difficult to evaluate which material they should belong to; (2) the lack of understanding of how to outline uncertainty budgets and the idea of implementing uncertainty budgets. As a general conclusion, medical CT systems will be suitable to become a powerful tool for volume measurement on pig carcasses after further development.

6.6.4 CIA-CT Intercomparison on Industrial CT for Measurement Applications (2012–2013)

The second CIA-CT comparison was organized by the DTU Mechanical Engineering Department, Denmark and carried out within the project "Centre for Industrial Application of CT Scanning—CIA-CT". The main goals of this comparison were to test the applicability of CT for the dimensional measurement of small objects, commonly measured in industry, which are more representative than artificial reference standards, and to evaluate the impact of instrument settings and operator decisions on the measurement of items of two different materials and geometries. Figure 6.25 shows the two commercial items used for the comparison, a plastic Lego brick and a metallic component of a medical insulin injection device.

(a)   (b)

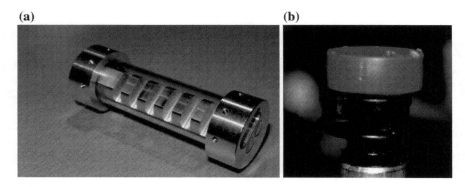

Fig. 6.26 The two items selected in the InteraqCT comparison: **a** assembly 1 consisting of an aluminium part inside a glass tube having a total length of 60 mm and **b** assembly 2 consisting of a volume data set produced from a physical assembly of plastic parts having a total length of 15 mm (Stolfi and De Chiffre 2016)

Different measurands are considered, encompassing diameters, roundness, and lengths. In the comparison, 27 laboratories from eight countries were involved. The parallel circulation of sets of items took place in spring 2013 and the final report was available in September 2013. All the items were calibrated using two tactile CMMs before and after the circulation, showing a good dimensional stability over time. Out of a total of 167 single results obtained by the participants using CT scanners, 55% of the measurements yield satisfying $|E_n|$ values less than 1 (see Eq. 6.5), and 45% larger than 1. Systematic errors were detected for some participants for all measurands investigated. The roundness measured by the participants for both plastic and metal items again showed the largest deviations (cf. 6.6.1). A clear influence from the surrounding wall thickness on the measurement of roundness was documented for the metal item reflecting the change of X-ray penetration thickness.

6.6.5 InteraqCT Comparison on Assemblies (2015/2016)

The InteraqCT interlaboratory comparison on assemblies is the most recent inter-comparison activity organized by the DTU Mechanical Engineering Department, Denmark, and carried out within the EU Marie Curie ESR Project InteraqCT which stands for "International Network for the Training of Early stage Researchers on Advanced Quality control by Computed Tomography". The InteraqCT comparison was organized to test the applicability of CT for measurements of assemblies along with their dimensions as well as of materials commonly used in industry. The comparison involved 20 laboratories from seven countries. In contrast to earlier comparisons, the InteraqCT comparison also introduced a test data set representing a scan produced by the coordinator and distributed electronically to all participants.

Figure 6.26 shows the two items used for the comparison: a physical step-gauge-in-tube assembly (assembly 1), which was produced for the comparison and distributed to the participants in parallel, and a volume data set (assembly 2) produced from a physical assembly of plastic parts. Various mono-material measurands were considered on both parts, encompassing lengths, diameters, roundness and coaxiality. A multi-material length, defined as the gap distance between the top of the first tooth of the aluminium step gauge and the inner diameter of the glass tube, was also selected on assembly 1. Two different scanning approaches were considered for physical assembly 1. The first approach, labelled as "own choice", did not apply any scanning restrictions on any of the scanning parameters. The second one, labelled as "fast scan", introduced a series of scanning limitations, including the scanning time. Twenty samples of assembly 1 and a set of virtual data sets of assembly 2 were circulated in winter 2015/2016 and a final report was available in September 2016. All the items were calibrated several times during the course of the comparison using a tactile CMM, showing a good dimensional stability over a total interval of ten months.

Measurements on assembly 1 showed that out of a total of 200 measurement values obtained by the participants, 70% of those measurements yielded satisfying $|E_n|$ values less than 1. No significant $|E_n|$ value differences were observed between mono-material and multi-material measurements of assembly 1, giving evidence that under certain circumstances, CT can deliver multi-material length measurements at the same level of accuracy as mono-material length measurements. By comparing the two different scanning approaches, it was noticed that most of the participants were able to reduce their scanning time by more than 70% without affecting the measurement accuracy. The sizes of uncertainties were found to be consistent for all measurands, although some confusion on the different methods to estimate the measurement uncertainty was observed among the participants. For example, none of the participants who stated a measurement uncertainty according to ISO 15530-3 respected the conditions of material and measurand similarity.

Measurements on assembly 2 showed that increasing the complexity of the measurand dramatically increases the range of variation among the participants and the systematic errors from the reference values provided by the coordinator. For example, the standard deviation of coaxiality measurements, representing the most challenging measurand in the present comparison, was found to be approximately three times larger than the one of diameter measurements. As a general conclusion, the post-processing activities may have a tangible influence on the accuracy of a single measurement and thus CT users need to establish procedures to control such activities. No differences were observed among different inspection software packages for the diameter measurements, while some differences were observed for roundness and concentricity measurements, which are measurands whose evaluations depend on intrinsic software filtering and how the datum system needs to be defined in a particular software package.

References

Angel J, De Chiffre L (2014) Comparison on computed tomography using industrial items. CIRP Ann 63:473–476. doi:10.1016/j.cirp.2014.03.034

Angel J, Christensen LB, Cantatore A, De Chiffre L (2014) Inter laboratory comparison on computed tomography for industrial applications in the slaughterhouses: CIA-CT comparison. CIA-CT technical report, 76 p

Arenhart FA, Nardelli VC, Donatelli GD (2015) Characterization of the metrological structural resolution of ct systems using a multi-wave standard. In: Proceedings of the XXI IMEKO world congress "measurement in research and industry", Prague, Czech Republic, 2015, online: http://www.imeko.org/publications/wc-2015/IMEKO-WC-2015-TC14–282.pdf

ASTM E 1441 (2011) Standard guide for computed tomography (CT) imaging

ASTM E 1672 (2012) Standard Guide for Computed Tomography (CT) System Selection

ASTM E 1695 (1995) Standard test method for measurement of computed tomography (CT) system performance

ASTM E 2767 (2013) Standard practice for digital imaging and communication in nondestructive evaluation (DICONDE) for X-ray computed tomography (CT) test methods

Barrett HB, Myers KJ (2004) Foundations of image science. Wiley, Hoboken. ISBN 0-471-15300-1

Bartscher M, Hilpert U, Härtig F, Neuschaefer-Rube U, Goebbels J, Staude A (2008) Industrial computed tomography, an emerging coordinate measurement technology with high potentials. In: Proceedings of NCSL 2008 international workshop & symposium. ISBN 1–584-64058-8

Bartscher M, Illemann J, Neuschaefer-Rube U (2016) ISO test survey on material influence in dimensional computed tomography. Case Stud Nondestruct Test Eval. doi:10.1016/j.csndt.2016.04.001

Borges de Oliveira F, Bartscher M, Neuschaefer-Rube U (2015) Analysis of combined probing measurement error and length measurement error test for acceptance testing in dimensional computed tomography. In: Proceedings of DIR 2015 in NDT.net, online: www.ndt.net/events/DIR2015/app/content/Paper/31_BorgesdeOliveira.pdf

Borges de Oliveira F, Bartscher M, Neuschaefer-Rube U, Tutsch R, Hiller J (2017a) Creating a multi-material length measurement error test for the acceptance testing of dimensional computed tomography systems. In: Proceedings of iCT 2017 conference, Leuven, Belgium, in NDT.net, online: http://www.ndt.net/events/iCT2017/app/content/Extended_Abstract/57_BorgesdeOliveira_Rev2.pdf

Borges de Oliveira F, Stolfi A, Bartscher M, Neugebauer M (2017b) Creating a multi-material probing error test for the acceptance testing of dimensional computed tomography systems. In: Proceedings of iCT 2017 conference, Leuven, Belgium, in NDT.net, online: http://www.ndt.net/events/DIR2015/app/content/Paper/31_BorgesdeOliveira.pdf

Cantatore A, Andreasen JL, Carmignato S, Müller P, De Chiffre L (2011) Verification of a CT scanner using a miniature step gauge. In: Proceedings of 11th EUSPEN international conference, Como, Italy

Carmignato S (2012) Accuracy of industrial computed tomography measurements: experimental results from an international comparison. CIRP Ann Manuf Technol 61–1:491–494. doi:10.1016/j.cirp.2012.03.021

Carmignato S, Pierobon A, Rampazzo P, Parisatto M, Savio E (2012) CT for Industrial Metrology - Accuracy and structural resolution of CT dimensional measurements, Proc. of iCT 2012 in NDT.net, online: http://www.ndt.net/article/ctc2012/papers/173.pdf

Christoph R, Neumann H-J (2011) X-ray tomography in industrial metrology. Süddeutscher Verlag onpact GmbH. ISBN 978-3-86236-020-8

DIN EN 16016-3:2012-12 (2012) Non destructive testing—radiation methods—computed tomography—Part 3: Operation and interpretation. German version EN 16016-3:2012

EN 16016-1:2011-12 (2011) Non destructive testing—radiation methods—computed tomography—Part 1: Terminology. Trilingual version

EN 16016-2:2012-01 (2012) Non destructive testing—radiation methods—computed tomography—Part 2: Principle, equipment and samples
EN 16016-3:2012-12 (2012) Non destructive testing—radiation methods—computed tomography—Part 3: Operation and interpretation. German version EN 16016-3:2011
EN 16016-4:2012-01 (2012) Non destructive testing—Radiation methods—Computed tomography—Part 4: Qualification
Fleßner M, Vujaklija N, Helmecke E, Hausotte T (2014) Determination of metrological structural resolution of a CT system using the frequency response on surface structures. In: Proceedings MacroScale, Vienna, Austria
Fleßner M, Helmecke E, Staude A, Hausotte T (2015) CT measurements of microparts: numerical uncertainty determination and structural resolution. In: Proceedings of SENSOR 2015. doi:10.5162/sensor2015/C8.2
Hermanek P, Carmignato S (2016) Reference object for evaluating the accuracy of porosity measurements by X-ray computed tomography. Case Stud Nondestr Test Eval. doi:10.1016/j.csndt.2016.05.003
Hiller J, Maisl M, Reindl LM (2012) Physical characterization and performance evaluation of an X-ray micro-computed tomography system for dimensional metrology applications. Meas Sci Technol 23:1–18. doi:10.1088/0957-0233/23/8/085404
Illemann J, Bartscher M, Jusko O, Härtig F, Neuschaefer-Rube U, Wendt K (2014) Procedure and reference standard to determine the structural resolution in coordinate metrology. Meas Sci Technol 25:6. doi:10.1088/0957-0233/25/6/064015
Illemann J, Bartscher M, Neuschaefer-Rube U (2015) An efficient procedure for traceable dimensional measurements and the characterization of industrial CT systems. In: Proceedings of DIR 2015 in NDT.net, online: www.ndt.net/events/DIR2015/app/content/Paper/46_Illemann.pdf
Illers H, Buhr E, Hoeschen C (2005) Measurement of the detective quantum efficiency (DQE) of digital X-ray detectors according to the novel standard IEC 62220-1. Radiat Prot Dosimetry 114(1–3):39–44. doi:10.1093/rpd/nch507
INTERAQCT (2016) International network for the training of early stage researchers on advanced quality control by computed tomography. http://www.interaqct.eu
ISO 10360-2 (2009) Geometrical product specifications (GPS)—acceptance and reverification tests for coordinate measuring machines (CMM)—Part 2: CMMs used for measuring linear dimensions. International Organization for Standardization
ISO 15530-3:2011-10 (2011) Geometrical product specifications (GPS)—coordinate measuring machines (CMM): technique for determining the uncertainty of measurement—Part 3: Use of calibrated workpieces or measurement standards
ISO/TS 23165 (2006) Geometrical product specifications (GPS)—guidelines for the evaluation of coordinate measuring machine (CMM) test uncertainty
JIMA Mask (2006) Japan Inspection Instruments Manufacturers' Association. Micro resolution chart for X-ray JIMA RT RC02B, online exhibition catalogue. Accessed 29th Sept 2016: http://www.jima.jp/content/pdf/catalog_rt_rc02b_eng.pdf
Jusko O, Lüdicke F (1999) Novel multi-wave standards for the calibration of form measuring instruments. In: Proceedings of 1st EUSPEN international conference, Aachen, Germany, vol 2, pp 299–302. ISBN 3-8265-6085-X
Kiekens K, Welkenhuyzen F, Tan Y, Bleys P, Voet A, Kruth J-P, Dewulf W (2011) A test object with parallel grooves for calibration and accuracy assessment of industrial computed tomography (CT) metrology. Meas Sci Technol 22:115502
Kingston A, Sakellariou A, Varslot T, Myers G, Sheppard A (2011) Reliable automatic alignment of tomographic projection data by passive auto-focus. Med Phys 38:4934. doi:10.1118/1.3609096
Léonard F, Brown S, Withers P, Mummery P, McCarthy M (2014) A new method of performance verification for x-ray computed tomography measurements. Meas Sci Technol 25(6):065401
Müller P (2012) Doctoral dissertation. Technical University of Denmark

Neuschaefer-Rube U, Bartscher M, Bremer H, Birth T, Härtig F (2012) Lösungsansätze zur Messung von Kanten und Radien mit Computertomographie, presentation at the "XIII Internationales Oberflächenkolloquium. Chemnitz, Germany

QRM, Quality Assurance in Radiology and Medicine GmbH, Möhrendorf, Germany, online exhibition catalogue. Accessed 29th Sept 2016: http://www.qrm.de/content/pdf/QRM-MicroCT-Barpattern-Phantom.pdf , http://www.qrm.de/content/pdf/QRM-MicroCT-Barpattern-NANO.pdf

Salesbury JG (2012) Developments in the international standardization of testing methods for CMMs with imaging probing systems. NCSL International Workshop & Symposium. http://www.ncsli.org/i/c/TransactionLib/REG_2012.MAN.874.1569.pdf

Sasov A, Liu X, Salmon PL (2008) Compensation of mechanical inaccuracies in micro-CT and nano-CT. In: Proceedings of SPIE, vol 7078. Developments in X-ray tomography VI. doi:10.1117/12.793212

Seewig J, Eifler M, Wiora G (2014) Unambiguous evaluation of a chirp measurement standard. Surf Topogr Metrol Prop 2:045003. doi:10.1088/2051-672X/2/4/045003

Siemens OEM (2017) CERA—Software for high-quality CT imaging. Available at: http://www.oem-products.siemens.com/software-components. Accessed 17th Aug 2017

Sire P, Rizo P, Martin M (1993) X-ray cone beam CT system calibration. In: Proceedings of SPIE, vol 2009, pp 229–239

Smekal L, Kachelrieß M, Stepina E, Kalender WA (2004) Geometric misalignment and calibration in cone-beam tomography. Med Phys 31(12):3242–3266. doi:10.1118/1.1803792

Stolfi A, De Chiffre L (2016) Selection of items for "InteraqCT Comparison on Assemblies". In: Proceedings of iCT 2016 in NDT.net, online: http://www.ndt.net/article/ctc2016/papers/ICT2016_paper_id70.pdf

VDI/VDE 2617-6.2:2005-10 (2005) Accuracy of coordinate measuring machines - Characteristics and their testing—guideline for the application of DIN EN ISO 10360 to coordinate measuring machines with optical distance sensors

VDI/VDE 2617-6.1:2007-05 (2007) Accuracy of coordinate measuring machines—characteristics and their testing—coordinate measuring machines with optical probing—code of practice for the application of DIN EN ISO 10360 to coordinate measuring machines with optical sensors for lateral structures

VDI/VDE 2630-1.3: 2011–12 (2011) Computed tomography in dimensional measurement—guideline for the application of DIN EN ISO 10360 for coordinate measuring machines with CT-sensors

Weiss D (2005) Verfahren und eine Anordnung zum Kalibrieren einer Messanordnung, European Patent, EP1760457 (A2)

Zanini F, Carmignato S (2017) Two-spheres method for evaluating the metrological structural resolution in dimensional computed tomography. Meas Sci Technol, in press

Chapter 7
Towards Traceability of CT Dimensional Measurements

Massimiliano Ferrucci

Abstract There can be no discussion on the accuracy of a measurement result without its proven traceability. Traceability is a fundamental property of a measurement that ensures the measured quantity is comparable to the international definition of the unit with which it is expressed. The task of achieving traceability requires a thorough understanding of the measurement procedure and, in particular, the operating principle of the measuring instrument. In this chapter, the concept of traceability is introduced as it applies to dimensional measurements. The role of instrument calibration in establishing traceability of its measurements is presented for earlier coordinate measuring systems, to serve as a precursor to the discussion on how CT instruments can be calibrated. Finally, methods for assessing task-specific measurement uncertainty are discussed.

7.1 Introduction

CT is a powerful tool not only for qualitative testing (e.g. inspection of defects and structural failures) but for quantitative analysis; that is, the extraction of numerical quantities from the acquired data. CT is particularly interesting for performing dimensional measurements of the scanned object. Segmentation of the voxel-based attenuation map gives a 3D surface model, which can be subsequently converted to a cloud of three-dimensional point coordinates by applying a surface sampling algorithm. The ability to extract both internal and external surface coordinates from scanned data has led many to dub CT as the third generation of coordinate measuring systems (CMSs), table-top coordinate measuring machines (CMMs) being the first generation and portable CMSs—the second generation. Despite the recognized potential of CT for coordinate measurement, its relatively recent emergence means that the accuracy of CT coordinate measurements is largely unknown.

M. Ferrucci (✉)
Department of Mechanical Engineering, KU Leuven, Leuven 3000, Belgium
e-mail: massimiliano.ferrucci@kuleuven.be

© Springer International Publishing AG 2018
S. Carmignato et al. (eds.), *Industrial X-Ray Computed Tomography*,
https://doi.org/10.1007/978-3-319-59573-3_7

In the field of metrology—measurement science—the accuracy of a measurement is proven when the result can be traced back to the definition of its corresponding unit and any uncertainty collected along the way is assessed and stated alongside the measurement result. The concept of measurement traceability might seem trivial at a first glance. In practice, though, achieving traceability of measurements can be an arduous task. Traceability is an important property of a measurement and has several economic and practical implications, particularly in manufacturing and in commerce. For this purpose, the significant efforts needed to establish traceability of measurements can be justified.

In this chapter, traceability is introduced in the context of dimensional measurements. Qualification and testing of CT systems (see Chap. 6) are important requirements, but are not sufficient for achieving measurement traceability. Instrument calibration, *i.e.* measurement of task-independent error sources, and assessment of task-specific measurement uncertainty are two concepts that must be clearly understood for the practical achievement of measurement traceability. Given that CT is a coordinate measuring system, it is appropriate to first discuss how the concept of instrument calibration applies to measurements by other coordinate measuring technologies. Then, the concept of instrument calibration is presented as it pertains to CT systems. Finally, methods for assessing task-specific measurement uncertainty are discussed. In the interest of conciseness, the focus of this chapter will be mainly on cone-beam CT systems.

7.2 Measurement Traceability

The International Vocabulary of Metrology (VIM, abbreviated from the French equivalent) defines *traceability* as the

> property of a measurement result whereby the result can be related to a reference through a documented unbroken chain of calibrations, each contributing to the measurement uncertainty. (BIPM 2012)

In other terms, measurement traceability provides a certain level of confidence about the validity of a result. This confidence is achieved by establishing a relationship between the result of a measurement and a common reference. A peek into the history of traceability provides context for its importance today.

7.2.1 History

The concept of traceability was born out of the need to standardize measurements. Influential breakthroughs in measurement technology coincided with the industrial revolution of the 18th and 19th centuries. A distinct change in manufacturing was the adoption of interchangeable components. Products that were once individually custom-made by a skilled craftsman and in a single manufacturing process could now be made in separate but parallel manufacturing steps; the final product was completed by assembling the various components. Parallelization of the manufacturing process significantly increased production rates. Also, there was no longer the need to have a skilled craftsman who knew how to make all components of a final product. Each manufacturing step could be manned by a technician, whose sole responsibility was to manufacture one component repeatedly. The elimination of skilled labour meant cost-saving for the manufacturer.

Interchangeability means that any component chosen at random from the same manufacturing step will fit the final assembly. However, the delegation of production to separate manufacturing steps introduced a new challenge: ensuring that all the components indeed fit in the final assembly of the product. Each technician had to ensure that their components conformed to their specifications. In order to achieve this conformance, measurements performed during quality control needed to be uniform among all manufacturing processes. In other words, all measurements needed to be consistent with a common reference. Traceability was introduced as a property of a measurement to demonstrate this consistency.

The increasingly global nature of manufacturing and trade during the 19th century meant that the issue of traceability was no longer limited to a single factory or manufacturing chain. Representatives from several industrialized states agreed on the need for international uniformity of measurements. The *Metre Convention* of 1875 established the International Bureau of Weights and Measures (BIPM, abbreviated from the French equivalent), which was tasked with defining and disseminating the International System of Units (SI) (BIPM 2006). The definition of each SI unit of measurement serves as the ultimate reference for traceable measurements of the corresponding physical quantity throughout the world.

7.2.2 The Standard Unit of Length

The international reference for dimensional measurements is the SI unit of length— the metre. The definition of the metre and the method with which it is practically realized have changed several times in history. The first international metre was defined in 1889 as the length of a physical object: the International Prototype Platinum-Iridium metre bar (Fig. 7.1).

In 1960, the metre was re-defined in terms of the wavelength of light emitted by the excitation of Krypton-86 atoms. The definition of the metre in terms of a physical phenomenon—in this case, the wavelength of light—meant that the standard unit of length could be realized in a laboratory by way of interferometric methods (NBS 1975). The most recent definition of the metre was accepted in 1983 and is given by

the length of the path travelled by light in vacuum during a time interval of 1/299 792 458 of a second. (BIPM 2006)

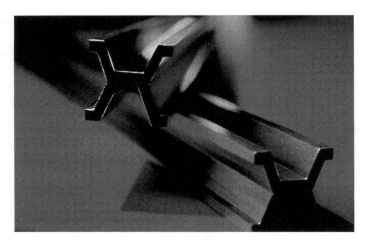

Fig. 7.1 The international prototype metre bar (Image courtesy of NIST)

This definition sets the speed of light in vacuum as a constant and ties the definition of the metre to the definition of the SI unit of time—the second. In practice, this definition meant that the realization of the standard unit of length by interferometric methods was no longer limited to light emitted by Krypton-86. The BIPM provides a list of recommended sources of light that provide a stable wavelength with narrow bandwidth. Helium-Neon (He–Ne, wavelength $\lambda_{\text{He–Ne}} = 632.8$ nm) lasers are one of the more commonly used light sources for interferometry.

7.2.3 Achieving Traceability of Dimensional Measurements

Traceability of dimensional measurements means that the unit expressed in the result is consistent with the definition of the metre. The measurement result is linked to the physical realization of the metre by one or more calibrations. While performance verification of measuring instruments (discussed in Chap. 6) contributes to the achievement of measurement traceability, it is not sufficient on its own. *Calibration* is the measurement of a physical quantity, such as the length of a feature on a test object or the distance between indications on a test instrument, by comparison to a traceable reference. The path of calibrations linking a measurement result to the definition of the metre is known as the *traceability chain*. An example of a traceability chain for CT measurements is illustrated in Fig. 7.2, including how the various calibration steps are realised. Each calibration step contributes to uncertainty of the final measurement, which must be assessed for the measurement to be considered traceable.

The result of a measurement is an estimate of the quantity being measured, i.e. the measurand, and *measurement uncertainty* is the degree of confidence with

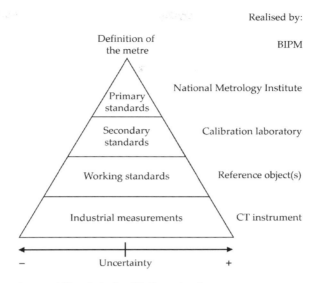

Fig. 7.2 A sample traceability chain for CT dimensional measurements

which the estimate can be attributed to the measurand. Consistent with its formal definition in the VIM, uncertainty U in the measurement of Y is expressed as a dispersion of values symmetrical to the measurement result y, i.e. $Y = y \pm U$. A common misconception is that a measurement can be "more traceable" if there are fewer calibration steps separating it from the definition of the unit or if its uncertainty is comparatively lower. Traceability is a nominal property (BIPM 2012) of measurement and, therefore, does not have a magnitude. A measurement is either traceable or it is not traceable, irrespective of the number of calibration steps linking it to the definition of the SI unit or its uncertainty.

A measuring instrument serves as the reference for establishing traceability of its measurements. The reference is typically provided by the instrument scale, for example the length indications on Vernier callipers and micrometers. In these simple cases, instrument calibration refers to comparing the positions of instrument's indications to a traceable reference and assessing the uncertainty in the comparison step. Gauge blocks are commonly used reference objects for calibrating the scales of simple one-dimensional end-to-end measuring instruments. Calibration of more complicated measuring systems requires a thorough understanding of how the scale is realized in the instrument. Section 7.3 is dedicated to the discussion of instrument calibration for other three-dimensional coordinate measuring systems. Instrument calibration of CT systems is covered in Sect. 7.4.

A measurement is the comparison of a measurand to the instrument scale. A quantity is assigned to the measurand as a result of this comparison. Calibrating the scale of an instrument alone does not ensure that its measurement results are traceable. Similarly to all other calibration steps in the traceability chain, the uncertainty in the comparison between instrument scale and measurand must be evaluated and stated alongside the assigned quantity to ensure its traceability. There are various standard procedures for calculating task-specific uncertainty in measurement, which will be discussed in Sect. 7.5.

7.3   Calibration of Coordinate Measuring Systems

The field of coordinate measurement falls within the general domain of dimensional metrology.When discussing coordinate measurements, the performance of CT is often compared to the performance of other CMSs (see Chap. 6). A CMS measures the dimensions of an object by sampling its surface as a collection of points, each defined by its three-dimensional coordinate position in the measurement volume. Subsequent dimensional analysis on the surface points includes, for example, fitting of geometrical primitives or comparison to reference CAD models. Achieving traceability of coordinate measurements entails establishing a link between the measured three-dimensional point coordinates—whether Cartesian, spherical, or cylindrical—and the instrument's calibrated scale, and evaluating the uncertainty in the coordinate positions. The method with which the instrument scale is realized depends on the measurement principle and therefore varies among measuring technologies.

Tactile CMMs

It is logical to begin this section by discussing first-generation tactile CMMs. The operating principle of CMMs is among the oldest among all CMSs. CMMs extract surface coordinates by physically touching the test object with a contact probe. Three mutually orthogonal mechanical guideways control the position of the probe and form the axes of a Cartesian coordinate frame. Several configurations of the guideways are available, each providing benefits for particular measurement tasks (Pereira 2012). The most common configuration is the moving bridge CMM, a diagram of which is shown in Fig. 7.3. CMMs can also include a rotary stage for

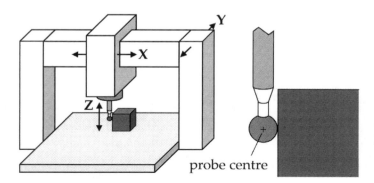

Fig. 7.3 Bridge-type CMM and contact probe

rotating the workpiece; rotation can be useful for measuring workpieces with cylindrical symmetry.

Each guideway is equipped with positioning sensors, such as linear encoders or displacement transducers, to track the position of the probe with respect to a pre-defined reference (Pereira 2012). The probe typically consists of a sphere of known radius and low form errors mounted at the end of a stylus. Various probe types can be mounted interchangeably on more modern CMMs. Surface points are collected by physically contacting the workpiece with the probe. The position corresponding to physical contact is given when the contact force detected by the probing system exceeds a specified value (Weckenmann and Hoffmann 2012). The contact force is chosen to ensure repeatable and non-destructive probing of the workpiece. The coordinate position of the probe centre is subsequently acquired and stored. Some measurement tasks demand the coordinate position of the contact point on the probe surface as opposed to the coordinate position of the probe centre. In these cases, the direction of approach of the probe can be used to compensate for the probe radius and provide an estimate for the coordinate position of the contact point. More advanced probing systems estimate the contact point by measuring the direction of stylus deflection upon contact.

The coordinate position of the probe centre is given by the readouts from each of the three axis positioning sensors. Ideally, a change in axis readouts should correspond to the actual displacement of the probe centre in the measurement volume. However, imperfections in the CMM construction and in the operation of mechanical guideways can result in a discrepancy between actual probe displacement and sensor readouts. As a carriage moves along a linear guideway, it can exhibit six error motions as a function of readout position: positioning error along the axis of motion, two straightness errors along orthogonal axis directions, and a rotation about each of the three axes (Schwenke et al. 2008). The six degrees of freedom are shown for the X guideway in Fig. 7.4. A further three squareness parameters describe angular misalignment between axes. For a typical Cartesian CMM, a total 21 kinematic error parameters—6 degrees of freedom per guideway plus 3 squareness errors—can be used to describe the possible error motions of the linear guideways. A list of these 21 kinematic error parameters for typical Cartesian CMMs is provided in Table 7.1.

Fig. 7.4 Six degrees of freedom of the X-guideway. Error parameters are described in Table 7.1

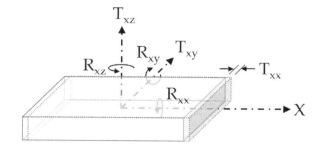

Calibration of CMMs consists of measuring the 21 kinematic error parameters as functions of the axis readouts. This procedure is known as CMM error mapping, which can be performed using reference instruments, for example interferometers, or by measuring specially designed reference objects. Typically, the reference instrument or object is chosen such that its uncertainty is at least one order of magnitude smaller than the expected kinematic errors. The measured kinematic error parameters can then be used to determine the actual coordinates of the probe centre given the readout of each axis. A common method for reducing the effects of kinematic errors is to incorporate the kinematic errors into a CMM software correction algorithm that modifies the sensor outputs to provide corrected probe centre coordinate positions (Hermann 2007).

Table 7.1 Kinematic error parameters of a three-axis coordinate measuring machine

Errors	Guideway		
	X	Y	Z
Translation along X	T_{XX}	T_{YX}	T_{ZX}
Translation along Y	T_{XY}	T_{YY}	T_{ZY}
Translation along Z	T_{XZ}	T_{YZ}	T_{ZZ}
Rotation about X	R_{XX}	R_{YX}	R_{ZX}
Rotation about Y	R_{XY}	R_{YY}	R_{ZY}
Rotation about Z	R_{XZ}	R_{YZ}	R_{ZZ}
Squareness errors			
Squareness error between X and Y axes, S_{XY}			
Squareness error between X and Z axes, S_{XZ}			
Squareness error between Y and Z axes, S_{YZ}			

Table 7.2 Other significant sources of error in CMM measurements

Elastic compression of probe and surface due to probing force (Fig. 7.5)	Error in calibrated probe radius
Errors in the position of the probe centre relative to instrument coordinate frame	Error in determining point of contact on probe surface
Limitations in surface sampling due to finite size of the probe (Fig. 7.6)	Thermal expansion of the instrument and workpiece

Fig. 7.5 Elastic compression of contact probe and measured surface under probing force. Image reproduced from NIST Engineering Metrology Toolbox

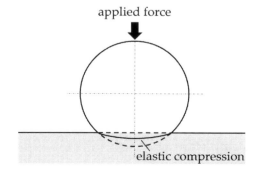

Fig. 7.6 Limitations in surface sampling due to the finite size of the probe

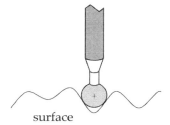

surface

While CMM calibration establishes traceability of the instrument scale, other sources of error must also be considered for traceability of coordinate measurements (Schwenke et al. 2008). A list of significant error sources for CMM measurements by contact probe is provided in Table 7.2. Note that uncertainty in sampling strategy and processing of individual coordinates, e.g. fitting of geometrical primitives, are not included in this list. Note that other probe types, e.g. non-contact probes, can be used on CMMs, further broadening the application of the instrument to different measurement tasks.

Articulated arm CMMs

Articulated arm CMMs (AACMMs, also informally known as articulated arms or robot-arms) consist of a column and at least two fixed-length shafts connected by articulating joints. Articulated arms are smaller and lighter than CMMs, making them a portable alternative for contact coordinate measurements. The diagram in Fig. 7.7 illustrates an AACMM with two fixed-length shafts and six articulating joints. A contact probe is attached by a final articulating joint at one end of the articulated arm, while the column at the other end is fixed on a platform. Each joint is equipped with an angular encoder that measures the rotation angle of the joint θ_i. Given the known lengths of the shafts, the coordinate position of the probe centre can be determined by the set of encoded angles. Note that other probe types, e.g. non-contact probes, can be used on AACMMs, further broadening the application of the instrument to different measurement tasks.

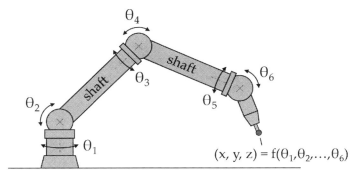

Fig. 7.7 Articulated-arm CMM

Each rotating joint exhibits error motions and angular encoders suffer from indexing errors due to, for example, encoder eccentricity. Additionally, misalignments can exist between coupled rotation axes; these misalignments can be considered analogous to squareness errors in linear axes of CMMs, for example. The parameterization of these geometrical error sources typically consists of assigning local Cartesian coordinate frames to each joint. Each coordinate frame has six degrees of freedom—three translational errors and three rotational errors (Santolaria and Aguilar 2010; Sladek et al. 2013). A common method for measuring geometrical errors is to track the position of the probing system with a reference instrument, for example a laser tracker (Santolaria et al. 2014). Alternatively, geometrical errors can be determined by least-squares fitting the parameters of a geometrical error model to observed errors in the measurement of reference objects. The significant sources of error in coordinate measurements in Table 7.2 apply for AACMMs with a contact probe.

Laser trackers

Laser trackers measure the coordinate position of a cooperative target by employing a laser-based range measuring device and two rotational axes (Muralikrishnan et al. 2016). A mirror directs the laser light along a horizontal angle (azimuth) θ, while the head can be rotated to control the direction of the laser along a vertical angle (zenith) φ (Fig. 7.8). Angular encoders track the angular position of each rotary axis. The cooperative target is a spherically-mounted retroreflector (SMR), which reflects light in a direction parallel to, but offset from, the incident beam (Fig. 7.9). The reflected light returns to the tracker assembly, where it is analyzed to determine the range ρ of the target. Range-measurement in laser trackers is achieved by interferometry or by absolute distance measurement (ADM) techniques, such as time-of-flight, amplitude modulation, or by employing multiple optical frequencies of the emitted light.

The name laser tracker is given because the instrument is designed to track, i.e. follow, the position of the SMR. A portion of the returned light is directed to a

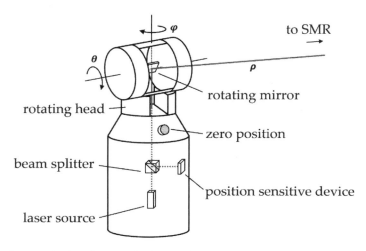

Fig. 7.8 Opto-mechanical diagram of a laser tracker

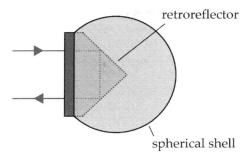

retroreflector

spherical shell

Fig. 7.9 Spherically-mounted retroreflector (SMR)

Fig. 7.10 Magnetic nest for repeatable SMR positioning

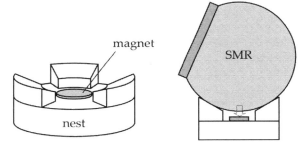

magnet

SMR

nest

position sensing device (PSD). When the laser is centred on the SMR, the returning light is centred on the PSD. As the SMR is moved, the measured light will deviate from the centre of the PSD. The laser tracker applies appropriate rotations to the mirror and head to bring the beam back to the centre of the PSD. A magnetic 'nest' (Fig. 7.10) on the body of the laser tracker is used to establish the zero position of the instrument's coordinate system. When mounted on an ideal nest, the position of the target centre does not change with the orientation of the SMR. Nests can therefore be used elsewhere in the measurement volume to provide repeatable positioning of targets, e.g. for reference measurements.

The coordinate position of the SMR is determined by the angle readouts θ and φ from the angular encoders and the rangefinder readout ρ. Misalignments between

Table 7.3 Opto-mechanical error sources of laser trackers and scanners

Tilts and offsets between beam path and transit axis (beam path from source to mirror is ideally coincident with transit axis)	Tilts and offsets between mirror and transit axis
Angular encoder eccentricity (applies to both vertical and horizontal encoders)	Angular encoder second-order scale error (applies to both vertical and horizontal encoders)
Zero range offset	Zero vertical angle offset
Tilt and offset between standing and transit axes	Time-varying tilt error motions (wobble) of rotating components

the various components contribute to errors in the measured coordinates. A list of opto-mechanical errors for laser trackers is shown in Table 7.3. A dedicated set of test procedures can be applied to map geometrical errors of laser trackers (Muralikrishnan et al. 2009). These procedures consist of measuring both a calibrated length and a non-calibrated length from a series of defined positions and orientations of the laser tracker. A geometrical error model parameterizes each error source and establishes an analytical relationship to errors in measured coordinates. Observed coordinate errors from the test procedures are used to solve for the various error parameters in the geometrical model by least-squares fitting.

Laser scanners

Laser scanners are similar in construction to laser trackers. A spinning mirror scans the laser along a vertical angle θ, while a rotating head directs the laser along a horizontal angle φ (Fig. 7.11). Laser scanners do not require cooperative targets as they are designed to measure surfaces directly. Backscattered light from the measured surface is detected by a photodetector within the instrument. Various techniques are employed to determine the distance travelled by the laser beam. In the phase-shift method, the optical amplitude of the laser is modulated at various temporal frequencies. The superposition of the modulated signals upon returning to the instrument creates a temporal interference signal that is used to determine the path travelled by the light (Petrov et al. 2011). Time-of-flight instead determines the distance travelled by the light by multiplying the speed of light by the time difference between the emergence of the signal from the instrument and its detection upon return.

While laser trackers are designed to 'hold' the beam at any angular position of the rotating axes, the mirrors in laser scanners typically spin continuously about a horizontal transit axis. Vertical angle measurement can be achieved either by employing an angular encoder or by calculating the angle from the rotational speed of the spinning mirror and the time stamp for each measurement point. The system head rotates about a vertical standing axis, the position of which is tracked by an angular encoder.

Instrument calibration therefore consists of measuring misalignments and offsets between the laser steering components (Muralikrishnan et al. 2015). Calibration

Fig. 7.11 Laser scanner

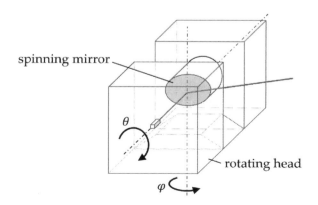

procedures for laser scanners are similar to calibration procedures for laser trackers. The opto-mechanical error sources provided in Table 7.3 also apply to laser scanners. Other sources of error that should be considered for ensuring measurement traceability include interactions of the laser beam with the measured surface, for example penetration of the laser prior to reflection.

7.4 CT Instrument Calibration

CT earned the reputation of establishing a new paradigm of coordinate measurement techniques and is recognized for its ability to non-destructively measure inner and outer surface coordinates by employing penetrating radiation (see Chap. 1). To understand how traceability of CT measurements can be achieved, it is necessary to understand how the instrument scale is realized and transferred to the measured object as a set of surface coordinates. CT measurements differ from measurements by other CMSs in that the coordinates of a surface point are not directly given by the indexed position of the instrument's kinematic axes. The imaging nature of the radiographic data acquisition step is further complicated by the tomographic reconstruction of the measurement volume and subsequent thresholding to convert the volumetric data to surface coordinates. In this section, calibration will be discussed in the context of CT data acquisition and tomographic reconstruction steps. The influence of post-processing of CT volumetric data is not covered here but is discussed, for example in Lifton et al. (2015), Stolfi et al. (2016), and Moroni and Petrò (2016).

The measurement volume is defined as a three-dimensional distribution of voxels centred at the intersection of the magnification axis and the axis of rotation. The shape of the voxels is typically cuboid but not necessarily cubic. For simplicity, measurement volumes with cubic voxels are discussed here. The length of the voxel's sides, namely the voxel size, is generally a function of detector pixel size and the magnification factor, as shown in Eq. 7.1.

$$voxel\ size = \frac{pixel\ size}{M} \tag{7.1}$$

where M is given by the ratio of the source-to-detector distance SDD and the source-to-rotation axis distance SRD, as shown in Eq. 7.2.

$$M = \frac{SDD}{SRD} \tag{7.2}$$

Substituting Eq. 7.2 into Eq. 7.1,

$$voxel\ size = pixel\ size \left(\frac{SRD}{SDD}\right) \tag{7.3}$$

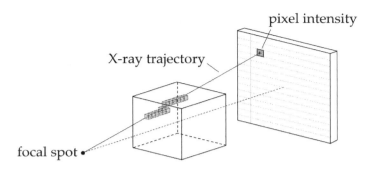

Fig. 7.12 The backprojection step in the tomographic reconstruction algorithm traces the X-ray trajectories from each pixel, through the measurement volume, and to the focal spot. Discrepancies between the backprojection geometry and the actual geometry in the acquisition of radiographs will result in errors of the reconstructed volume

It should be noted that the relationship provided in Eq. 7.3 is only a general rule; the physical size and shape of the voxels can be specified differently in the tomographic reconstruction step.

The goal of tomographic reconstruction is to populate the voxel space with values proportional to the X-ray attenuation incurred within the volumetric extent of each voxel. The measured X-ray intensity at each pixel is back projected as a linear trajectory from the corresponding pixel position to the X-ray focal spot. Each trajectory traces a path through the measurement volume and intersects a series of voxels along that path (Fig. 7.12). Attenuation values for each voxel are calculated from the collection of radiographic intensities whose back projected trajectories intersect that voxel for all angular positions of the volume. Accurate reconstruction of the measurement volume therefore relies on accurate knowledge of the instrument geometry and of the detector response to incident X-rays, namely the imaging system. In this section, the concept of instrument calibration is introduced as it pertains to industrial CT systems. Calibration of the CT geometry is presented separately from the calibration of the imaging system. Furthermore, the algorithm applied to perform tomographic reconstruction can also introduce errors. A short discussion on characterizing the effects of the reconstruction algorithm is provided. The examples presented here pertain to cone-beam CT systems. However, the topics can be applied to other CT instrument architectures provided appropriate adaptations are made.

7.4.1 CT Geometrical Calibration

The purpose of geometrical calibration is to measure the actual CT radiographic acquisition geometry by comparing it to a traceable reference and to assess uncertainty in the measured geometrical parameters. While instrument manufacturers have methods for measuring the CT geometry, standardized methods do not yet exist. It should be noted that compensation of measured geometrical errors, for example by adjustment of the physical components or by applying software correction, does not fall within the scope of calibration. Compensation is an optional step that can be performed after calibration to reduce the measured geometrical errors. A subsequent calibration would confirm whether or not the geometrical errors were indeed reduced. In the absence of error compensation, the measured geometrical errors must be included in the calculation of CT measurement uncertainty.

CT acquisition geometry

The CT acquisition geometry of a typical cone-beam system can be described for a given angular position α of the voxel space by a set of 10 parameters (Fig. 7.13). This parameterization applies to the 'static' CT geometry, not considering drifts and error motions of the components. It should be noted that various parameterizations of CT systems are possible and that the following is not unique.

The X-ray focal spot S is parameterized as an infinitesimally small point source and is the origin of a global right-handed Cartesian coordinate system, i.e. $S = (0, 0, 0)$. The global Y axis is parallel to the axis of rotation and the global Z axis is defined as the line from the source that intersects the rotation axis orthogonally. The global X axis follows the right-hand screw rule. A coordinate position $R = (0, 0, z_R)$ is assigned to the intersection of the global Z axis and the axis of rotation; R coincides with the centre of the voxel space.

The detector is assumed to be perfectly flat and its position is defined by the position of its geometrical centre, $D = (x_D, y_D, z_D)$. Detector orientation is given by its pixel row unit vector $\hat{u} = (u_X, u_Y, u_Z)$ and pixel column unit vector $\hat{v} = (v_X, v_Y, v_Z)$. The detector normal $\hat{w}$ is given by the cross product of the row and column vectors, i.e. $\hat{w} = \hat{u} \times \hat{v}$. Many studies on geometrical calibration parameterize the orientation of the detector as three Euler rotation angles (Ferrucci et al. 2015). This parameterization introduces the need to specify rotation conventions, *i.e.* extrinsic and

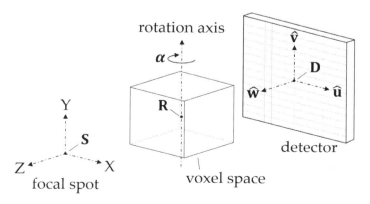

Fig. 7.13 Parameterization of the CT geometry

intrinsic, and rotation sequence. By defining the orientation of the detector by its row and column vectors, there is no need for specifying such conventions.

Nominal alignment

Most commercially-available tomographic reconstruction algorithms assume a nominal alignment of the CT components, which is illustrated in Fig. 7.14 for a typical cone-beam CT system and is defined as follows. The detector pixel row vector $\hat{u}$ is parallel to the global X-axis, while the detector pixel column vector $\hat{v}$ is parallel to the global Y-axis. The global Z-axis intersects the detector at its geometrical centre D and the detector normal vector $\hat{w}$ coincides with the global Z-axis. The imaging magnification axis is defined as the X-ray path normal to the detector. The intersection of the magnification axis with the detector is known as the principal point. Therefore, the principal point in an aligned system coincides with the detector geometrical centre D. Detector pixels are assumed to be equally-sized and equally-spaced across the detector. The parameters of the static CT geometry, *i.e.* at a fixed position of the rotation stage, and the corresponding aligned values are presented in Table 7.4.

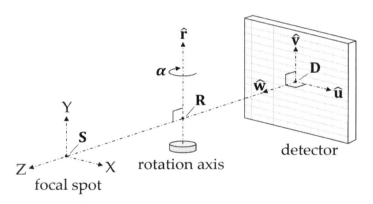

Fig. 7.14 Typical alignment of a cone-beam CT system

Table 7.4 Cone-beam CT static geometrical parameters and corresponding aligned values

Feature		Parameters	Aligned values
Detector	Position	$D = (x_D, y_D, z_D)$	$D = (0, 0, SDD)$
	Orientation	$\hat{u} = (u_X, u_Y, u_Z)$	$\hat{u} = (1, 0, 0)$
		$\hat{v} = (v_X, v_Y, v_Z)$	$\hat{v} = (0, 1, 0)$
		$\hat{w} = \hat{u} \times \hat{v}$	$\hat{w} = (0, 0, 1)$
Rotation axis	Position	$R = (0, 0, z_R)$	$R = (0, 0, SRD)$

Measurement of CT geometrical parameters

The geometrical parameters of a test system can be determined by imaging a reference object placed on the stage at one or more rotation positions α and analysing the projections. Commonly used reference objects consist of several small radiographically opaque spheres. Each sphere centre has a known coordinate position in a local coordinate frame $\mathbf{C}_{\text{local}} = (x_C, y_C, z_C)$. The pixel coordinate position (u_C, v_C) of each projected sphere centre can be estimated by applying appropriate image processing techniques. A common method for finding the projected centre involves fitting an ellipse to the projected sphere edge. In some studies, the pixel coordinate corresponding to the centre of the fitted ellipse is taken to be the coordinate of the projected sphere centre (Clackdoyle and Mennessier 2011). A more recent study by Deng et al. (2015) provides an equation for a more accurate estimation of projected sphere centre coordinates using the combination of the fitted ellipse centre and the fitted elliptical supports, i.e. major and minor ellipse axis lengths, and ellipse orientation.

The CT geometry can be determined by solving for the forward projection operator that relates the sphere center coordinates in the global frame $\mathbf{C}_{\text{global}}$ (determined by applying a translation and rotation to $\mathbf{C}_{\text{local}}$) to (u_C, v_C) for all rotation positions α.

Forward projection of the sphere centres onto the detector space consists of first parameterizing straight lines from the X-ray focal spot $\mathbf{S}$ to the global coordinate of each sphere centre $\mathbf{C}_{\text{global}}$. The set of points $\mathbf{L}$ along a straight line containing two points x_1 and x_2 can be parameterized using Eq. 7.4.

$$\mathbf{L} = x_1 + t(x_2 - x_1) \tag{7.4}$$

where t is the parametric variable denoting the location along the line with respect to point x_1; when $t = 1$, $\mathbf{L} = x_2$. Substituting x_1 with $\mathbf{S}$ and x_2 with $\mathbf{C}_{\text{global}}$, the straight lines from the source focal spot to the sphere centre are parameterized by Eq. 7.5.

$$\mathbf{L} = \mathbf{S} + t(\mathbf{C}_{\text{global}} - \mathbf{S}) \tag{7.5}$$

The projection of the sphere centres on the detector correspond to the intersection point between the parameterized lines and the detector plane, which is parameterized as follows. Given a point on the plane $\mathbf{P}_0$ and a vector normal to the detector plane $\hat{\mathbf{n}}$, all points $\mathbf{P}$ on the plane are given by Eq. 7.6.

$$\hat{\mathbf{n}} \cdot (\mathbf{P} - \mathbf{P}_0) = 0 \tag{7.6}$$

Substituting $\mathbf{P}_0$ with the detector centre point $\mathbf{D}$ and substituting $\hat{\mathbf{n}}$ with the detector normal $\hat{\mathbf{w}}$ gives Eq. 7.7.

$$\hat{\mathbf{w}} \cdot (\mathbf{P} - \mathbf{D}) = 0 \tag{7.7}$$

The intersection of line $\mathbf{L}$ and plane $\mathbf{P}$ occurs when $\mathbf{P} = \mathbf{L}$. The point of intersection I_C in the global coordinate frame is determined by first substituting Eq. 7.5 into Eq. 7.7 as follows. Note that the variable C in Eqs. 7.8–7.10 corresponds to the sphere center coordinate positions in the global frame $\mathbf{C}_{\text{global}}$, which was simplified for better visualization of the formulae.

$$\hat{\mathbf{w}} \cdot (\mathbf{S} + t(\mathbf{C} - \mathbf{S}) - \mathbf{D}) = 0 \qquad (7.8)$$

Then, the parametric variable t is isolated.

$$t = \frac{\hat{\mathbf{w}} \cdot \mathbf{D} - \hat{\mathbf{w}} \cdot \mathbf{S}}{\hat{\mathbf{w}} \cdot (\mathbf{C} - \mathbf{S})} \qquad (7.9)$$

Finally, Eq. 7.9 is substituted into Eq. 7.4 to determine the global coordinates of the intersection point $\mathbf{I_C}$.

$$\mathbf{I_C} = \mathbf{S} + \left(\frac{\hat{\mathbf{w}} \cdot \mathbf{D} - \hat{\mathbf{w}} \cdot \mathbf{S}}{\hat{\mathbf{w}} \cdot (\mathbf{C} - \mathbf{S})} \right)(\mathbf{C} - \mathbf{S}) \qquad (7.10)$$

The pixel row and column coordinate positions $(u_\mathrm{C}, v_\mathrm{C})$ of the projected sphere centres are determined by first translating the global coordinates of the intersection point $\mathbf{I_C}$ to a new origin at the detector centre, i.e. subtracting $\mathbf{D}$ from $\mathbf{I_C}$, then taking the dot product of the translated coordinates with the pixel row and pixel column vectors, respectively.

$$u_C = (\mathbf{I_C} - \mathbf{D}) \cdot \hat{\mathbf{u}} \qquad (7.11)$$

$$v_C = (\mathbf{I_C} - \mathbf{D}) \cdot \hat{\mathbf{v}} \qquad (7.12)$$

As an example, a reference object consists of M spheres. Each sphere is defined by its coordinate position in the global frame (x_m, y_m, z_m), where $m = 1, 2, \ldots, \mathrm{M}$. The reference object is imaged at N equally-spaced angles α_n of the stage to perform a full 360° revolution. The set of angles is given by Eq. 7.13.

$$\alpha_n = n \left(\frac{360°}{\mathrm{N}} \right) \qquad (7.13)$$

where $n = 0, 1, 2, \ldots, \mathrm{N} - 1$

For each value of α_n, the sphere centres are rotated accordingly about the axis of rotation, providing a new set of global coordinates $(x_{m,n}, y_{m,n}, z_{m,n})$. Analysis of each radiograph n produces a corresponding set of observed projected sphere centre pixel coordinates $(u_{m,n}, v_{m,n})_{\mathrm{obs}}$. Given the set of global sphere centre coordinate positions and the corresponding observed projection coordinates, two methods can be applied to solve for the CT geometrical parameters: analytical and minimization.

Analytical methods consist of solving discrete equations relating the sphere centre coordinates, the corresponding projected pixel coordinates, and the CT geometrical parameters. These methods typically rely on prior knowledge about one or more geometrical parameters, for example the alignment of the reference object in the global coordinate frame. The accuracy in the solved CT geometry is sensitive to errors in the values assigned to this a priori information. Separate procedures are often applied to acquire these initial parameters, e.g. by analysing the sinogram or employing reference instruments.

Minimization methods consist of comparing modelled and observed projection data. An initial set of geometrical parameters is used to construct a forward projection operator, producing a set of modelled pixel coordinates $(u_{m,n}, v_{m,n})_{\text{mod}}$ given an input set of sphere centre coordinates. Re-projection errors between modelled and observed pixel coordinates $du_{m,n}$ and $dv_{m,n}$ are evaluated, as shown in Eqs. 7.14 and 7.15.

$$du_{m,n} = (u_{\text{obs}} - u_{\text{mod}})_{m,n} \qquad (7.14)$$

$$dv_{m,n} = (v_{\text{obs}} - v_{\text{mod}})_{m,n} \qquad (7.15)$$

Optimization techniques are then applied to minimize the re-projection errors by iteratively modifying the geometrical parameters of the modelled forward projection operator. In a well-constructed minimization process, the geometrical parameters corresponding to the globally minimized re-projection error should correspond to the actual parameters of the test system. A common shortcoming in any minimization procedure is the presence of local minima in the cost function. To ensure that the minimization converges to a global minimum, one or more steps can be taken by the user. The initial geometrical parameters in the model can be estimated to within a reasonable value and to within reasonable constraints by applying, for example, separate analytical estimation steps. Alternatively, global minimization techniques (Floudas and Gounaris 2009) can be utilized.

The methods for measuring geometrical parameters described thus far are applicable for a given position of the sample rotation stage. The linear axes that control the position of the rotation axis are susceptible to the same 21 kinematic errors described for CMMs (Table 7.1). Reference instruments such as interferometers can be used to map the kinematic errors as a function of the position readout for each axis. Effects of load on the behaviour of the kinematic axes should be considered. Without a full error mapping of the CT kinematic axes, the geometrical estimation procedures based on the imaging of a reference object would have to be repeated for every new position of the sample rotation stage. Measured kinematic errors must be referenced to a 'home' position of the linear axes, where the 10 geometrical parameters are measured using the imaging method. Given a new kinematic position of the rotation stage, the CT geometrical parameters evaluated at the home position can be modified accordingly. Assuming the measured kinematic errors are repeatable, such a comprehensive mapping of the system geometry would eliminate the need to estimate geometrical parameters for every new position of the sample rotation stage.

While methods to estimate CT geometrical parameters have been developed, calibration of the CT geometrical parameters per the metrological requirements outlined in VIM has not been demonstrated. Research efforts should be focused on establishing traceability of the measured geometrical parameters by utilizing a traceable reference object. Furthermore, uncertainty in the calibrated parameters should be assessed; this topic is briefly discussed in the following section.

Uncertainty in the measured geometrical parameters

Several sources of error in the geometrical calibration procedure contribute to uncertainty in the measured parameters and are common to both analytical and minimization techniques. Some of the more significant sources are discussed below.

Error in the calibrated sphere centre coordinates: The coordinate position of the reference features, for example the sphere centres, serves as the traceable reference for measuring the CT geometrical parameters. The sphere centre coordinates can be measured, for example, by tactile CMM. Uncertainty in the measured coordinates is a result of several factors, including errors in the measuring instrument, the probing strategy applied to extract sphere centres (such as the number and location of measured surface points), and subsequent geometrical analysis of the measured data.

Error in the estimate of pixel coordinates of projected sphere centres: The pixel coordinates of the projected features must be determined by applying dedicated techniques to the radiographic images. Spheres are projected onto the detector as elliptical disks (in the special case that the sphere centre is coincident with the magnification axis, it is projected as a circular disk). The location of the projected sphere centre can be estimated, for example, as the weighted mean of the projected disk's intensity values or the centre of an ellipse fit to the projected sphere edges (Clackdoyle and Mennessier 2011). A more recent method for estimating the image coordinates of the projected sphere centre incorporates the fit ellipse center and supports (Fig. 7.15) into an analytical expression (Deng et al. 2015). Each method provides an estimate of the projected sphere centre coordinates and is subject to inherent errors. The size of the projected spheres can also affect the accuracy with which the projected centres are estimated, for example from digitization errors for small projected features. Error in the estimate is further exacerbated by noise in the detector and blurring due to the finite size of the X-ray source focal spot.

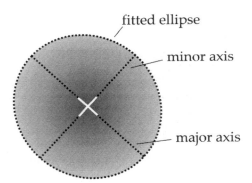

Fig. 7.15 The pixel coordinate of the projected sphere centre can be determined, for example, by fitting an ellipse to the edges of the projected sphere and extracting the ellipse centre. A more recent study also incorporates the fit ellipse supports in an analytical expression to estimate the coordinate of the projected sphere centre

Sample rotation stage errors: Several assumptions are made in the radiographic acquisition step of the geometrical calibration procedure. The sample stage is typically assumed to rotate ideally and that the angular displacement of the object between radiographs is known accurately. As with any kinematic system, rotary stages are prone to error motions and errors in the angular indexing (see Fig. 7.16). These inconsistencies will introduce errors unless they are measured, for example using standard procedures (ISO 2015) and compensated. In the absence of compensation, it is possible to incorporate the effects of these error motions into uncertainty in the sphere centre coordinates, e.g. as a function of rotation position α.

Focal spot drift: Spatial drift of the focal spot results in an inconsistency of the focal spot position between radiographs and therefore an inconsistency in the forward projection operator. In the case of repeatable drift, for example as a function of X-ray gun temperature and/or operation time, the focal spot position **S** can be modelled as a function of rotation angle α. Alternatively, a maximum drift value for the duration of the scan can be incorporated into the calculation of uncertainty in the measured geometrical parameters.

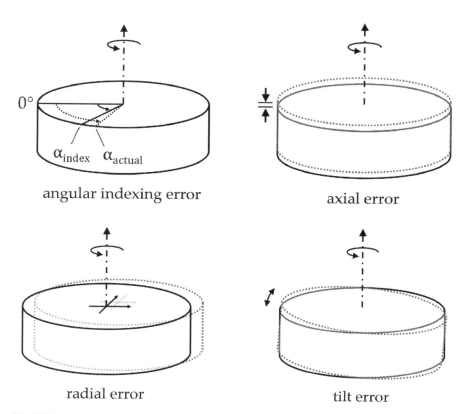

Fig. 7.16 Rotary stage error motions

Uncertainty in the analytically measured geometrical parameters can be evaluated utilizing the GUM method (see Sect. 7.5.1). Given that minimization methods do not utilize discrete equations, uncertainty in the solved geometrical parameters must be determined differently. A Monte Carlo approach (see Sect. 7.5.2) is a possible solution. In this technique, the minimization procedure is repeated, each time varying the input quantities (e.g. reference object coordinates, initial geometrical parameters, image coordinates of the projected sphere centres). The observed distribution of results can provide an indication for measurement uncertainty.

7.4.2 Calibration of the Imaging System

The imaging system is defined by the measurement of X-ray intensity at the detector. The signal emitted from typical diode sources—filament cathode and target anode—is characterized by a broad spectrum of photon energies as a result of bremsstrahlung emission and characteristic peaks (see Chap. 2). Energy-dispersive or photon-counting detectors are designed to discriminate incident photons by their energy. The ability to measure the spectral distribution of attenuated X-rays can be used to better inform tomographic reconstruction of the attenuating objects (Alvarez and Macovski 1976; Kanno et al. 2012). Alternatively, energy-integrating detectors measure the accumulated energy over the entire spectrum of incident X-rays. More detailed information on these detection technologies can be found elsewhere in the literature (see, for example, Gruner et al. 2001). Industrial CT systems most often employ energy-integrating detectors; the topics discussed in this section are therefore pertinent to this principle of X-ray detection (more details about X-ray detection can be found in Chaps. 2 and 3).

The intensity image recorded by each pixel on the detector is ideally proportional to the intensity incident upon the corresponding pixel area over the exposure time. The relationship between incident and recorded intensities can be generally referred to as the pixel response (other terms include detective efficiency or sensitivity). Ideally, detector response should be linear for all measured X-ray intensities. Detector response is shown to be approximately linear for lower intensity measurements (Williams and Shaddix 2007); however, linearity typically degrades for higher intensities. Additionally, response to incoming X-rays should ideally be uniform for all pixels in the detector. That is, the same X-ray signal incident on any given pixel should generate the same intensity output. Several factors contribute to non-linearity and spatial non-uniformity in the output of the detector. These factors include variations in the manufactured thickness of the scintillator, energy-dependent and geometry-dependent phenomena in the detector, and background signals. A more detailed description of these influence factors can be found in Barna et al. (1999) and Gruner et al. (2002). In this section, some of the more significant factors are introduced together with methods to characterize and correct them.

Zingers

Zingers are random events of emitted radiation, e.g. due to radioactive decay in the detector materials, that can occur at a frequency of approximately 0.8 Hz (Tate et al. 2005). In radiographic images, zingers manifest themselves as unpredictable spikes in pixel intensity. To detect zingers, two or more images are taken under nominally equivalent exposures. Differences in the intensity of the same pixel between images exceeding a certain threshold are identified as zingers. In Barna et al. (1999), an analytical equation is provided for defining a threshold value as a function of measured intensity to allow for statistical variations due to the Poisson distribution of emitted photons. According to Gruner et al. (2002), zingers can also be determined from a single image by identifying any pixel values that exceed the expected Poisson distribution for a localized intensity region. This method, however, is only efficient for radiographs without projections of sharp edges, for example, where a change in intensity between adjacent pixels could be expected to exceed Poisson statistics.

In the case that a zinger is identified, the intensity value of the corresponding pixel could either be substituted by the mean of neighbouring pixels or calculated from the lower value between the nominally identical images plus a statistically-determined offset to avoid statistical bias (Barna et al. 1999). While detection and removal of zingers for every radiographic image in a measurement can be impractical, it is strongly recommended for the images used in determining detector offset and flat-field correction(s). Uncorrected zingers in either of these two correction procedures will propagate to all radiographs to which the correction is applied.

Geometric image distortions

The intensities stored by the pixels in the radiograph should ideally correspond to a regular sampling grid across the detector area. As a result of imperfections in the various components of the detector, deviations from the regular grid can exist. Smoothly varying image distortions can be measured and corrected per the following method (Barna et al. 1999). An attenuating mask with regularly spaced holes (Fig. 7.17) is placed directly in front of the detector face. Each hole in the mask is assigned column and row indices $c = 1, 2, \ldots, N_C$ and $r = 1, 2, \ldots, N_R$, respectively.

Flood field illumination of the mask will provide a radiographic image consisting of a grid of bright spots. For each projected spot, the pixel column coordinates $u_{wm}(c, r)$ and pixel row coordinates $v_{wm}(c, r)$ corresponding to the weighted mean of intensity values are calculated. An ideal square lattice of equivalent grid count is least-squares fit to the set of weighted mean positions. Pixel column and row distortion vectors from the observed positions to the ideal positions are denoted $\Delta u(c, r)$ and $\Delta v(c, r)$, respectively, and are given by Eqs. 7.16 and 7.17.

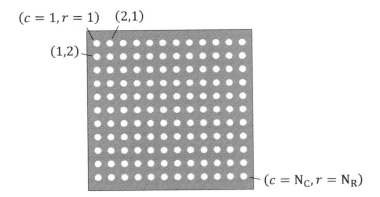

Fig. 7.17 Attenuating mask used for determining geometric image distortions

$$\Delta u(c, r) = u_{\text{fit}}(c, r) - u_{\text{wm}}(c, r) \tag{7.16}$$

$$\Delta v(c, r) = v_{\text{fit}}(c, r) - v_{\text{wm}}(c, r) \tag{7.17}$$

where $u_{\text{fit}}(c, r)$ and $v_{\text{fit}}(c, r)$ correspond to the coordinate positions of the ideal square lattice after least-squares fitting. Thus, the ideal lattice positions are given by the set of points $(u_{wm}(c, r) + \Delta u(c, r), v_{wm}(c, r) + \Delta v(c, r))$. A pixel-by-pixel distortion map is generated by interpolating from the observed distortions of the projected spot positions. In Barna et al. (1999), interpolation is achieved by constructing one-dimensional cubic splines in two steps. The first spline is constructed separately along each row r as a function of $u_{\text{wm}}(c, r)$ to interpolate along pixel columns i. The result is a set of interpolated $v(i, r)$ coordinates and interpolated distortion map $\{\Delta u(i, r), \Delta v(i, r)\}$. A second spline is constructed along each column i as a function of $v(i, r)$ to interpolate along pixel rows j. The result is a pixel-by-pixel distortion map $\{\Delta u(i, j), \Delta v(i, j)\}$ which can be applied to re-bin subsequent radiographic images to correct for image distortions. The re-binning of pixel intensities to the ideal pixel map is not trivial given that distortions are seldom (if ever) integer values. A detailed discussion on the considerations for effective re-binning is provided in Barna et al. (1999).

The thickness and material of the mask plate will depend on the X-ray energy used for imaging. For example, in Barna et al. (1999) a 50 μm tungsten foil plate was used for X-ray energies below 20 keV. The diameter of the holes should be slightly larger than the size of the detector pixels; 75 μm holes were used for a detector with 50 μm pixels. Spacing between holes should correspond to the expected scale of smoothly varying distortions, which depends on the type of image intensifier and the component used to transmit the measured signal from the scintillator to the CCD. For a 1024 × 1024 fibre-optically coupled detector with 50 μm pixels, distortions were expected to be smooth to within a few millimetres (Barna et al. 1999). For distortion measurement of that particular detector, hole spacing was 1 mm along both the horizontal and vertical directions. Small features with precise spacing can be made by photolithography and subsequently calibrated on a vision CMM. Calibrated feature positions can then be used to achieve traceable measurement and estimate uncertainty in geometrical distortions.

Detector offset

Detectors exhibit electronic noise (dark current) due to, e.g. thermal heating of the components, which results in an offset of the pixel intensities. Drifts in the detector offset can occur due to the temperature dependence; as a result, the pixel-by-pixel detector offset should be measured periodically (Hsieh 2015). Measurement of detector offset consists of acquiring several raw radiographic images DO_{raw} in the absence of X-ray illumination and calculating a mean offset image DO_{mean}. The notations used here follow the format provided in Williams and Shaddix (2007). For a set of N_{DO} raw images, the intensity of each pixel (u, v) in the mean offset radiographic image $DO_{mean}(u, v)$ is given by Eq. 7.18.

$$DO_{mean}(u, v) = \frac{1}{N_{DO}} \sum_{j=1}^{N_{DO}} DO_{raw}(j : u, v) \qquad (7.18)$$

where $j = 1, 2, \ldots, N_{DO}$ is the index of each raw image.

It is important to ensure that the exposure time used to acquire offset images is consistent with the exposure time used for the test radiographs I_{raw}. DO_{mean} can be subtracted pixel-by-pixel from test radiographs to provide an offset corrected test image I_{DO}, as shown by Eq. 7.19; this procedure is also known as background subtraction.

$$I_{DO}(u, v) = I_{raw}(u, v) - DO_{mean}(u, v) \qquad (7.19)$$

Gain correction

Gain correction is a procedure for measuring and correcting non-uniformity in the detector response. In the presence of uniform X-ray illumination, all detector pixels should ideally provide the same intensity output. Non-uniform response despite uniform incident X-rays results in variation of the pixel intensity outputs; a normalized gain correction map can be generated from the observed variations. The procedure for generating the gain correction map is herein provided following notation conventions similar to those found in Williams and Shaddix (2007). It is important to ensure that gain correction is performed using the same tube voltage, filtration (if any), and detector exposure as the test measurements.

Several raw radiographs GC_{raw} are acquired under flood field conditions, in which X-rays are generated in the absence of attenuators in the field of view. Raw radiographs are corrected for detector offset DO_{mean} and subsequently averaged into $\overline{GC}$, as in Eq. 7.20.

$$\overline{GC}(u, v) = \frac{1}{N_{GC}} \sum_{i=1}^{N_{GC}} \{GC_{raw}(i : u, v) - DO_{mean}(u, v)\} \qquad (7.20)$$

where $i = 1, 2, \ldots, N_{GC}$ is the index of the N_{GC} raw gain correction radiographs. The mean intensity of the image μ is given by Eq. 7.21.

$$\mu = \frac{1}{N_U \times N_V} \sum_{u=1}^{N_U} \sum_{v=1}^{N_V} \overline{GC}(u, v) \qquad (7.21)$$

where N_U and N_V are the number of pixel columns and rows, respectively, in the radiographic images. The mean gain correction image is normalized with respect to μ as follows.

$$GC(u, v) = \frac{\overline{GC}(u, v)}{\mu} \qquad (7.22)$$

Gain correction is applied by dividing each pixel (u, v) in offset-corrected test radiographs I_{DO} by the factor in the corresponding pixel location of the normalized gain correction image:

$$I_{GC}(u, v) = \frac{I_{DO}(u, v)}{GC(u, v)} \qquad (7.23)$$

Correction of non-linearity in pixel response can be performed as follows. Multiple gain correction maps are calculated at a series of emitted X-ray intensity levels along the expected range. Different X-ray intensities can be generated by varying the X-ray source current while keeping the acceleration voltage constant to maintain the same emitted spectrum (Kwan et al. 2006; Schmidgunst et al. 2007). Gain correction for the entire range of measured X-ray intensities is determined by interpolating between the various gain correction maps. The result is an analytical gain correction function for the entire range of intensities expected in a measurement (e.g. 0 to 65,535 for a 16-bit imaging system). In Seibert et al. (1998), a linear curve was used in the interpolation step, while a polynomial fit was applied in Kwan et al. (2006). In Schmidgunst et al. (2007), it was indicated that fitting the multiple intensity gain correction data with a piece-wise linear fit provides more effective interpolation than linear and polynomial fits.

The effectiveness of gain correction is based on the assumption that the incident X-ray intensity is equal for all pixels in the flood-field image. The Heel effect is a known factor that contributes to spatial inhomogeneity of the emitted spectra (Braun et al. 2010). Emission of X-ray photons at the source anode is not limited to the outer surface of the target. Electrons can penetrate into the target material, resulting in the emission of photons at a particular depth. Lower energy photons generated at a depth are attenuated by the target material as they propagate outward towards the aperture. This hardening of the beam at the anode is a function of the path length through the target, which varies with the beam exit angle. The result is an X-ray beam with a spatially inhomogeneous spectral profile. While energy-integrating detectors do not distinguish photons by their energy, accurate characterization and correction of several energy-dependent phenomena (e.g. beam hardening, scatter, and detector response) depend on accurate knowledge of the X-ray spectra throughout the generated beam.

The Heel effect, together with the energy-dependence of pixel response, can introduce inaccuracies in the generation of gain correction maps. A solution for addressing the effects of spectral variations was preliminarily investigated in Davidson et al. (2003), and Yu and Wang (2012). The procedure consists of generating gain correction maps in the presence of X-ray signals with varying levels of beam hardening, e.g. by applying various filters at the source aperture. Alternatively, detectors with well-known response characteristics can be used to provide reference measurements of the incident intensity as a function of narrow photon energy intervals for comparison with the test detector output. This method could also enable a path of traceability for detector intensity measurements (Haugh et al. 2012). Alternatively, a comprehensive model of the X-ray emission process can be developed, whereby the set of measured X-ray source characteristics (e.g. effective voltage and current, electron beam focusing behaviours, anode target material composition, and incidence geometry) can be used to calculate spectral and flux distributions across the beam profile. The implication of spatial inhomogeneity of the emitted beam on instrument calibration and on task-specific uncertainty is a topic deserving of further research.

Instrument drifts

The CT imaging system exhibits instabilities over time. These instabilities are overwhelmingly due to thermal effects but also include spatial fluctuations of the accelerated electron beam inside the X-ray source due to unstable magnetic focusing. In the presence of these temporal instabilities, periodic monitoring of the imaging system is recommended. For example, it might be useful to perform gain correction before and after test acquisition to determine fluctuations in gain correction maps and possibly interpolate over the scan time.

7.4.3 Calibration of the Tomographic Reconstruction Step

Tomographic reconstruction is an approximate solution to the inverse Radon transform (see Chap. 2). Given a set of 'ideal' radiographs, *i.e.* devoid of noise and acquired under ideal scanning conditions, the reconstruction algorithm introduces an assortment of inherent errors. These errors include cone-beam artifacts, digitization errors, and errors due to radiographic filtering prior to backprojection. While the general working principle is shared among the various algorithms available, it is not unreasonable to expect that the implementation of different algorithms to the same radiographic dataset will produce dissimilar volumes. Furthermore, differences in the reconstructed volume can be expected from the implementation of the same algorithm, albeit with variations in its parameters, e.g., choice of image filter (Ramp, Hanning, etc.) and method of voxel interpolation in the backprojection step. Irrespective of the reconstruction algorithm, its effects on measurements should be characterized and, if possible, reduced.

Tomographic reconstruction algorithms should be tested given a set of reference radiographic images. The type of algorithm and the specific parameters applied to reference images should be consistent with those used for the test data. Errors due to the tomographic reconstruction algorithm are typically dependent on the measured object(s). That is, characterizing the effects of the reconstruction algorithm with a dissimilar radiographic dataset might not capture the equivalent effects on the test dataset. For this purpose, it is critical that the characterization step be performed on a reference dataset similar to the experimental dataset.

7.5 Assessing Task-Specific CT Measurement Uncertainty

The assessment of task-specific measurement uncertainty requires input from the instrument calibration steps, spatial and material information about the measured object, environmental conditions during the measurements, and *preferably* repeat measurements for assessing statistical variations of measurement results. Commonly accepted methods for assessing measurement uncertainty are discussed below. It should be noted that, currently, only the comparator method has been confidently applied to assessing CT measurement uncertainty.

7.5.1 The GUM Method

The Guide to the Expression of Uncertainty in Measurement (GUM) is a standard document that outlines an analytical method to evaluate measurement uncertainty (BIPM 2008a). This method is often referred to by the acronymic form of the document in which it is described—the GUM method. The basis for the GUM method is that a measurement can be described by a model. In other words, solving the model is the mathematical equivalent of performing the measurement. The model consists of the function f that relates the measurand Y to the set of inputs X; that is,

$$Y = f(X) \tag{7.24}$$

where $X = [X_1, X_2, \ldots, X_N]$ and N is the total number of model inputs. In practical measurements, users only have an estimate x_i of each input X_i. The result of a measurement is therefore an estimate y for the measurand Y and is a function of multiple input estimates x_i. That is,

$$y = f(X_1 = x_1, X_2 = x_2, \ldots, X_N = x_N) \tag{7.25}$$

Uncertainty in the input estimates $u(x_i)$ will result in uncertainty of the estimated measurand $u(y)$. These uncertainties are categorized by the method with which they are evaluated. 'Type A' uncertainties are given by variations in the estimate of the input as a result of multiple observations. 'Type B' uncertainties are either provided in the instrument specifications or are determined from a priori knowledge about the statistical distribution of the input estimate. Assuming there is no correlation between inputs, uncertainty in the estimated measurand $u(y)$ is given by Eq. 7.26.

$$u^2(y) = \sum_{i=1}^{N} \left(\frac{\partial f}{\partial x_i} \right)^2 u^2(x_i) \tag{7.26}$$

where $\frac{\partial f}{\partial x_i}$ is the partial derivative of f with respect to X_i evaluated at $X_i = x_i$. In the case that some input estimates are correlated, covariance terms are added inside the square root (see Sect. 7.5.2 of the GUM). Equation 7.26 and the corresponding form with covariance terms express the *law of propagation of uncertainty*, which constitutes the analytical framework for evaluating uncertainty. A diagram for the application of the GUM method is shown in Fig. 7.18.

The law of propagation of uncertainty demands that the functional relationship f be continuously differentiable for all inputs, particularly for values of X_i in the vicinity of the estimates x_i. In the case that f does not satisfy this condition, the application of the GUM method to assessing uncertainty is limited. Currently, an analytical model for CT measurements is not available. As a result, the application of the GUM method by analytical means is not possible.

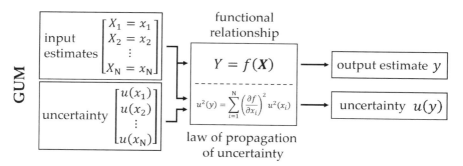

Fig. 7.18 Diagram (adapted from JCGM 101) illustrating the principle of uncertainty assessment by the GUM method. The law of propagation of uncertainty shown in this diagram is for input variables without covariance

7.5.2 Monte Carlo by Simulation

The Monte Carlo by simulation method (MCS) is a useful alternative to the GUM method for cases in which uncertainty propagation does not lend itself to the *law of propagation of uncertainty*. An example of this is when the functional relationship *f* is not continuously differentiable or it is too complex for evaluating the partial derivatives. The general principle for estimating uncertainty of measurement by the Monte Carlo method is as follows. The measurement model is repeatedly solved, each time varying the input quantities randomly within a specified distribution of values corresponding to their uncertainty. The observed distribution in measurement results from the repeated measurements is used to assess measurement uncertainty. Supplement 1 to the GUM (henceforth referred to as "JCGM 101") provides a thorough description of the Monte Carlo method as applied to a mathematical model (BIPM 2008b). Since simulation is the implementation of a mathematical model, principles of Monte Carlo can be applied to simulated measurements. The topics discussed in JCGM 101 are presented here.

The definition of a measurement model from Eq. 7.24 is recalled. The model output Y is a function of a set of N inputs X, where $X = [X_1, X_2, \ldots, X_N]$. Each input X_i is assigned a probability distribution function $g_{X_i}(\xi_i)$, where ξ_i is a variable representing the possible values of X_i. An example of a Gaussian probability distribution function is shown in Fig. 7.19. The shape and extent of each distribution function is determined from uncertainty in experimental measurements of the corresponding input estimate or from expert knowledge.

The measurement model is solved M times. For each iteration $r = 1, 2, \ldots, M$ of the simulated measurement, the quantity x_i assigned to each input X_i is randomly sampled from the corresponding distribution function $g_{X_i}(\xi_i)$. That is, for a given simulation run r, the set of input estimates x_r is given by Eq. 7.27.

$$x_r = [x_{1,r}, x_{2,r}, \ldots, x_{3,r}] \tag{7.27}$$

Fig. 7.19 In the Monte Carlo method, each input in the measurement model is assigned a probability distribution function $g_{X_i}(\xi_i)$ for the possible values ξ_i of the input estimate

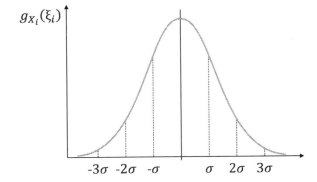

The corresponding output quantity y_r of the simulated measurement is therefore given by Eq. 7.28.

$$y_r = f(\mathbf{x}_r) \tag{7.28}$$

The set of M repeated simulations provides a set Y of output quantities,

$$Y = [y_1, y_2, \ldots, y_M] \tag{7.29}$$

which can also be presented by a distribution function $g_Y(\eta)$, where η is a variable describing the values of output quantity (Fig. 7.20).

The estimate $\tilde{y}$ of the measurement result is given by the mean of Y:

$$\tilde{y} = \frac{1}{M} \sum_{r=1}^{M} y_r \tag{7.30}$$

The standard uncertainty of the estimate $u(\tilde{y})$ is given by the standard deviation of Y (assuming $g_Y(\eta)$ is normally distributed):

$$u(\tilde{y}) = \sqrt{\frac{1}{M-1} \sum_{r=1}^{M} (y_r - \tilde{y})^2} \tag{7.31}$$

The number of simulations M should be large enough to ensure that a numerical tolerance for the estimated uncertainty is satisfied (see Sect. 7.9.2 in JCGM 101). In most cases, a value of M = 10^6 is expected to provide a 95% coverage interval.

Simulation of CT has progressed significantly in the past decade. However, more work is needed to completely model all physical phenomena. Furthermore, CT

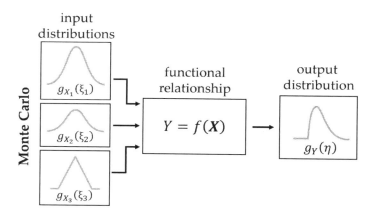

Fig. 7.20 Diagram (adapted from JCGM 101) illustrating the Monte Carlo method for assessing measurement uncertainty. In this example, the output Y is a function of three input variables

simulation is time-consuming. A dedicated solution similar to the virtual CMM (Trapet and Wäldele 1996) should be developed for handling the large number of recommended iterations in MCS. As a result, the application of MCS for assessing CT measurement uncertainty is limited.

7.5.3 Comparator Method

The comparator method, also known as the substitution method, is named after an instrument designed for transferring traceability from a reference object to a test object, most often gauge blocks. A typical gauge block comparator and corresponding diagram are shown in Fig. 7.21. Two styli contact the gauge block at its measuring faces. The position of each stylus is measured individually by way of a displacement measuring device, in this case differential transducers. Therefore, the output of the instrument is not an absolute length but rather the cumulative change in length as measured by the two transducers. Comparators are unique from other measuring instruments in that the absolute length is not provided by the instrument scale, but by the calibrated reference gauge block. The instrument output is used to measure differences between test and reference lengths; therefore, linearity of the instrument output is critical for accurate comparative measurements. Measurement uncertainty is calculated from repeated measurements of both the test and calibrated blocks, known differences in gauge block materials (as they pertain to elastic compression by anvil contact force, see Fig. 7.5), and environmental conditions (Beers and Tucker 1976; Doiron and Beers 1995).

The method for evaluating uncertainty in comparator measurements has been adapted for the evaluation of uncertainty by other instruments. This concept has been shown to be particularly useful for complex instruments, e.g. coordinate measuring systems, as there is no need for the comprehensive measurement model required in the GUM method. The international standard ISO 15530-3 describes the comparator method of uncertainty evaluation for coordinate measuring machines

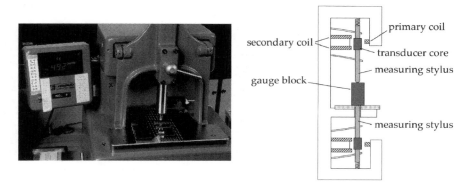

Fig. 7.21 Gauge block comparator and diagram (Images courtesy of NIST)

(ISO 2011). The VDI/VDE 2630 Part 2.1 (henceforth referred as the "guideline") is a guideline for assessing uncertainty of CT measurements by the comparator method, the contents of which are described below (VDI/VDE 2015).

Estimation of measurement uncertainty by the comparator method requires access to one or more calibrated reference objects similar to—if not exactly the same as—the test object(s). Criteria for similarity include shape, dimension, and material. The guideline emphasizes that, for objects that significantly attenuate the emitted intensity, differences in transmission lengths and attenuation coefficients between calibrated and test objects can introduce discrepancies not captured in the uncertainty assessment. The measurement procedure applied to the reference and test objects should be consistent, including object mounting, number and distribution of measuring points, scan settings, and ambient conditions. Any dissimilarity shall be taken into account in the uncertainty assessment.

Calibration of the reference object is often performed by tactile CMM. Some features are inaccessible by tactile CMM, for example inner features. In the absence of an alternative non-destructive calibration technique, the reference object can be physically segmented to make inner features accessible by tactile CMM or other CMS. Calibration of the reference object can be performed before or after CT measurement. If calibration is performed before, segmented reference objects shall be reassembled for CT measurements; changes in the material and dimensional integrity of the object due to segmentation should be considered in the estimation of calibration uncertainty.

The expanded uncertainty U_{MP} in measurement by the comparator method is given by Eq. 7.32.

$$U_{MP} = k \cdot \sqrt{u_{cal}^2 + u_{drift}^2 + u_p^2 + u_w^2 + u_b^2} \qquad (7.32)$$

where the contents of the radical correspond to the square of the standard uncertainty. Each term in Eq. 7.32 is explained in more detail below.

k	This coverage factor is selected based on the confidence interval with which the uncertainty is expressed. Confidence interval increases with coverage factor. The most common coverage factor of k = 2 corresponds to a 95% confidence interval
u_{cal}	The value u_{cal} corresponds to the standard uncertainty of the calibrated value. If the calibration certificate provides the expanded uncertainty U_{cal}, then the standard uncertainty u_{cal} can be calculated as follows: $u_{cal} = \frac{U_{cal}}{k_{cal}}$ where k_{cal} is the coverage factor with which the uncertainty in the calibrated value was expressed and which should be provided in calibration certificate
u_{drift}	Drifts in the dimensions of the reference object can occur over time as a result of, e.g., deformation from use, exposure to radiation, environmental conditions, and temporal instability of certain materials. Differences between successive calibrations can be used to estimate the expected drift at the time of the CT measurement. In the absence of multiple calibrations, expert knowledge with respect to, e.g., material behaviours and usage of the object, can be used to estimate this component

(continued)

(continued)

u_p	Multiple CT measurements of the reference object will expose statistical variations due to, e.g. measurement procedure, measurement strategy, and instabilities of the instrument. In the case that the reference object is similar to the test object, u_p is given by the standard deviation of N measurements of the reference object: $$u_p = \sqrt{\frac{1}{N-1} \cdot \sum_{i=1}^{N} (y_i - \bar{y})^2}$$ where y_i is the individual measurement result and $\bar{y}$ is the mean from N measurements. In the case that there are significant differences between reference and test object(s), separate procedures shall be undertaken to evaluate the contribution to uncertainty. To ensure a statistically significant estimate of u_p, a minimum of N = 20 measurements are recommended
u_w	This term estimates the contribution to uncertainty as a result of variations among test objects. At least two subcomponents u_{w1} and u_{w2} contribute to u_w in quadrature: $$u_w^2 = u_{w1}^2 + u_{w2}^2 + \cdots$$ u_{w1} is dedicated to variations in the production of multiple test objects. If uncertainty in only one test object is needed, then u_{w1} is negligible. Otherwise, u_{w1} can be determined from known manufacturing tolerances. Alternatively, the effects of production variations can be captured by performing repeat measurements on $j = 1, 2, \ldots, M$ test objects. In this case, a new u_p is calculated from the standard deviations $u_{p,j}$ of the repeat measurements on the test objects: $$u_p = \sqrt{\frac{1}{M} \sum_{j=1}^{M} u_{p,j}^2}$$ u_{w2} accounts for possible variations in the material composition among test objects, e.g. from the material supplier, and subsequent uncertainty in thermal expansion. This component is calculated as follows. $$u_{w2} = (t - 20°C) \cdot u_\alpha \cdot l$$ where t is the mean temperature during measurement, u_α is the standard uncertainty of the coefficient of thermal expansion for the object material, and l is the measured dimension. Subsequent components (u_{w3}, u_{w4}, etc....) are reserved for other task-specific uncertainty contributors not considered thus far
u_b	Bias can exist between the CT measurement of the reference object and its calibrated values. Bias is calculated as follows: $$b = \bar{y} - y_{cal}$$ where $\bar{y}$ is the mean of the N CT measurements on the reference object. It is recommended that bias be corrected in subsequent measurements of the test object(s). In this case, uncertainty in the calculation of the bias u_b must be considered. VDI/VDE 2630 Part 2.1 suggests that u_b should at least include effects due to uncertainty in the thermal expansion coefficient of the calibrated reference object: $$u_b = (t - 20°C) \cdot u_{\alpha b} \cdot l$$ where t is the mean temperature during the repeat CT measurements, $u_{\alpha b}$ is the standard uncertainty of the coefficient of thermal expansion for the reference object material, and l is the measured dimension

The application of the comparator method to CT measurements has been demonstrated by Müller et al. (2014). Several dimensions including plane-to-plane lengths (L_F and L_T), inner and outer diameter ($d_{1\text{-}3}$ and $D_{1\text{-}3}$, respectively), form (roundness $R_{1\text{-}2}$), and symmetry ($S_{1\text{-}2}$) were performed on a nickel-coated brass component from the manufacture of insulin pens. The component and the measured features are illustrated in Fig. 7.22.

Fig. 7.22 Various features were measured on a nickel-coated brass component for insulin pens. Uncertainty in their measurements was assessed using the comparator method. Figure reproduced with permission from Müller et al. (2014)

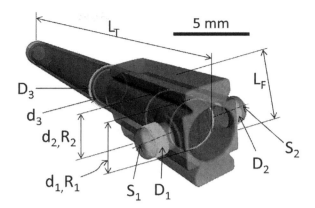

The results of the uncertainty calculation including correction for observed bias between CT measurements of the reference object and its CMM calibrated values are shown in Table 7.5. In this study, the u_{drift} component was neglected. Given that CT measurements can be relatively time consuming, Muller et al. estimated u_p from 9 repeat measurements as opposed to the recommended minimum of 20. The standard deviation in measurement results from the 9 measurements was subsequently multiplied by a safety factor of 1.2 to compensate for the reduced number of measurements.

Table 7.5 Results from the measurement of the various features on the insulin pen component and the corresponding uncertainty values

		Measured values (mm)		Uncertainty contribution (µm)				
		CMM	CT	u_{cal}	u_p	u_w	u_b	$U_{CT}(k=2)$
Measurand	d_1	3.3991	3.3850	0.90	1.60	0.01	0.44	3.8
	d_2	3.4020	3.3847	0.86	1.05	0.01	0.29	2.8
	d_3	3.4082	3.3914	0.87	0.60	0.01	0.17	2.1
	D_1	1.9077	1.9072	1.13	1.26	0.01	0.35	3.5
	D_2	1.9063	1.9106	0.50	1.54	0.01	0.43	3.4
	D_3	4.1174	4.1167	0.42	0.61	0.01	0.17	1.5
	L_F	6.2866	6.2843	1.34	1.06	0.02	0.29	3.5
	L_T	46.3990	46.3997	0.88	2.01	0.14	0.58	4.6
	R_1	0.0017	0.0561	0.63	6.33	0.00	1.76	13.2
	R_2	0.0010	0.0194	0.61	3.19	0.00	0.89	6.7
	S_1	0.0319	0.0421	1.58	0.56	0.00	0.15	3.4
	S_2	0.0487	0.0445	1.46	1.18	0.00	0.33	3.8

Table reproduced with permission from Müller et al. (2014)

7.6 Discussion

There have been—and continue to be—significant efforts in the research community to bring CT to measurement maturity. Industrial CT systems are powerful multi-purpose measuring instruments, allowing users to measure objects of various sizes, materials, and geometrical intricacies. However, the price of having such versatility in a measuring instrument is the complexity of ensuring accuracy for its measurements. It was shown in Sect. 7.4.1 that the ability to re-position the rotation stage makes geometrical calibration much more than an imaging procedure. Kinematic error motions must be combined with the set of geometrical parameters to generate a full geometrical error mapping of the system. It was shown in Sect. 7.4.2 that known inhomogeneity in emitted spectra and energy-dependent phenomena in the detector make calibration of the imaging system a non-trivial task.

Modelling of the entire CT measurement procedure, including physical phenomena such as X-ray scatter and beam hardening, is incomplete. Furthermore, the influence of tomographic reconstruction on measurement error is also deserving of more research. For these reasons, the GUM and Monte Carlo by simulation methods for assessing uncertainty are limited. The comparator method is the only method thus far to provide a confident estimate of CT measurement uncertainty.

Acknowledgements The author would like to acknowledge Dr. Claudiu Giusca at the National Physical Laboratory for his knowledgeable discussions on the structure of the chapter.

References

Alvarez RE, Macovski A (1976) Energy-selective reconstructions in X-ray computerized tomography. Phys Med Biol 21(5):733–744

Barna SL, Tate MW, Gruner SM, Eikenberry EF (1999) Calibration procedures for charge-coupled device x-ray detectors. Rev Sci Instrum 70(7):2927–2934

Beers JS, Tucker CD (1976) Intercomparison procedures for gage blocks using electromechanical comparators. National Bureau of Standards Interagency Report NBSIR, pp 76–979

BIPM (2006) The International system of units (SI), 8th edn. International Organisation for Standardisation, Geneva

BIPM JCGM 100 (2008a) Evaluation of measurement data—guide to the expression of uncertainty in measurement. International Organisation for Standardisation, Geneva

BIPM JCGM 101 (2008b) Evaluation of measurement data—supplement 1 to the "Guide to the expression of uncertainty in measurement"—propagation of distributions using a Monte Carlo method. International Organisation for Standardisation, Geneva

BIPM JCGM 200 (2012) International vocabulary of metrology—basic and general concepts and associated terms (VIM), 3rd edn. International Organisation for Standardisation, Geneva

Braun H, Kyriakou Y, Kachelrieß M, Kalender WA (2010) The influence of the heel effect in cone-beam computed tomography: artifacts in standard and novel geometries and their correction. Phys Med Biol 55:6005–6021

Clackdoyle R, Mennessier C (2011) Centers and centroids of the cone-beam projection of a ball. Phys Med Biol 56:7371–7391

Davidson DW, Fröjdh C, O'Shea V, Nilsson H-E, Rahman M (2003) Limitations to flat-field correction methods when using an X-ray spectrum. Nucl Instrum Methods Phys Res A 509:146–150

Deng L, Xi X, Li L, Han Y, Yan B (2015) A method to determine the detector locations of the cone-beam projection of the balls' centers. Phys Med Biol 60:9295–9311

Doiron T, Beers J (1995) The gauge block handbook. NIST Monograph 180

Ferrucci M, Leach RK, Giusca C, Carmignato S, Dewulf W (2015) Towards geometrical calibration of X-ray computed tomography systems—a review. Meas Sci Technol 26:092003. doi:10.1088/0957-0233/26/9/092003

Floudas CA, Gounaris CE (2009) A review of recent advances in global optimization. J Glob Optim 45:3–38

Gruner SM, Eikenberry EF, Tate MW (2001) Comparison of X-ray detectors. In: International tables for crystallography, vol F, pp 143–147

Gruner SM, Tate MW, Eikenberry EF (2002) Charge-coupled device area X-ray detectors. Rev Sci Instrum 73(8):2815–2843

Haugh MJ, Charest MR, Ross PW, Lee JJ, Schneider MB, Palmer NE, Teruya AT (2012) Calibration of X-ray imaging devices for accurate intensity measurement. Powder Diffr 27(2):79–86

Hermann G (2007) Geometric error correction in coordinate measurement. Acta Polytechnica Hungarica 4(1):47–61

Hsieh J (2015) Computed tomography: principles, design, artifacts, and recent advances, 3rd edn. SPIE Press, Bellingham

ISO 15530-3 (2011) Geometrical product specifications (GPS)—coordinate measuring machines (CMM): technique for determining the uncertainty of measurement Part 3 : Use of calibrated workpieces or measurement standards. International Organisation for Standardisation, Geneva

ISO 230-7 (2015) Test code for machine tools—Part 7: Geometric accuracy of axes of rotation. International Organisation for Standardisation, Geneva

Kanno I, Imamura R, Minami Y, Ohtaka M, Hashimoto M, Ara K, Onabe H (2012) Third-generation computed tomography with energy information of X-rays using a CdTe flat panel detector. Nucl Instrum Methods Phys Res A 695:268–271

Kwan ALC, Seibert JA, Boone JM (2006) An improved method for flat-field correction of flat panel x-ray detector. Med Phys 33(2):391–393

Lifton JJ, Malcolm AA, McBride JW (2015) On the uncertainty of surface determination in x-ray computed tomography for dimensional metrology. Meas Sci Technol 26:035003

Moroni G, Petrò S (2016) Impact of the threshold on the performance verification of computerized tomography scanners. Procedia CIRP 43:345–350

Muller P, Hiller J, Dai Y, Andreasen JL, Hansen HN, De Chiffre L (2014) Estimation of measurement uncertainties in X-ray computed tomography metrology using the substitution method. CIRP J Manufact Sci Technol 7:222–232

Muralikrishnan B, Sawyer DS, Blackburn CJ, Phillips SD, Borchardt BR, Estler WT (2009) ASME B89.4.19 performance evaluation tests and geometric misalignments in laser trackers. J Res Natl Inst Stand Technol 114:21–35

Muralikrishnan B, Ferrucci M, Sawyer DS, Gerner G, Lee V, Blackburn C, Phillips S, Petrov P, Yakovlev Y, Astrelin A, Milligan S, Palmateer J (2015) Volumetric performance evaluation of a laser scanner based on geometric error model. Prec Eng 40:139–150

Muralikrishnan B, Phillips S, Sawyer D (2016) Laser trackers for large-scale dimensional metrology: a review. Prec Eng 44:13–28

NBS (1975) The International Bureau of Weights and Measures 1875–1975, translation of the BIPM centennial volume, vol 420. NBS Special Publication, pp 77–85

Pereira PH (2012) Cartesian coordinate measuring machines. In: Hocken RJ, Pereira PH (eds) Coordinate measuring machines and systems, 2nd edn. CRC Press, Boca Raton, pp 61–67

Petrov P, Yakovlev Y, Grigorievsky V, Astrelin A, Sherstuk A (2011) Binary modulation rangefinder. US Patent US 7,973,912 B2

Santolaria J, Aguilar JJ (2010) Kinematic calibration of articulated arm coordinate measuring machines and robot arms using passive and active self-centering probes and multipose optimization algorithm based in point and length constraints. In: Lazinica A, Kawai H (eds) Robot manipulators new achievements. Intech, Vukovar

Santolaria J, Majarena AC, Samper D, Brau A, Velázquez J (2014) Articulated arm coordinate measuring machine calibration by laser tracker multilateration. Sci World J 2014:681853

Schmidgunst C, Ritter D, Lang E (2007) Calibration model of a dual gain flat panel detector for 2D and 3D X-ray imaging. Med Phys 34(9):3649–3664

Schwenke H, Knapp W, Haitjema H, Weckenmann A, Schmitt R, Delbressine F (2008) Geometric error measurement and compensation of machines—an update. CIRP Ann Manuf Technol 57:660–675

Seibert JA, Boone JM, Lindfors KK (1998) Flat-field correction technique for digital detectors. Proc SPIE 3336:348–354

Sładek J, Ostrowska K, Gąska A (2013) Modeling and identification of errors of coordinate measuring arms with the use of a metrological model. Measurement 46:667–679

Stolfi A, Thompson MK, Carli L, De Chiffre L (2016) Quantifying the contribution of post-processing in computed tomography measurement uncertainty. Procedia CIRP 43:297–302

Tate MW, Chamberlain D, Gruner SM (2005) Area X-ray detector based on a lens-coupled charge-coupled device. Rev Sci Instrum 76:081301

Trapet E, Wäldele F (1996) The virtual CMM concept. Advanced mathematical tools in metrology. World Scientific Publishing Company, pp 238–247

VDI/VDE 2630 Part 2.1 (2015) Computed tomography in dimensional measurement: determination of the uncertainty of measurement and the test process suitability of coordinate measurement systems with CT sensors. Verein Deutscher Ingenieure e.V., Dusseldorf

Weckenmann A, Hoffmann J (2012) Probing systems for coordinate measuring machines. In: Hocken RJ, Pereira PH (eds) Coordinate measuring machines and systems, 2nd edn. CRC Press, Boca Raton, pp 100–105

Williams TC, Shaddix CR (2007) Simultaneous correction of flat field and nonlinearity response of intensified charge-coupled devices. Rev Sci Instrum 78:123702

Yu Y, Wang J (2012) Beam hardening-respecting flat field correction of digital X-ray detectors. IEEE International Conference on Image Processing, pp 2085–2088

Chapter 8
Applications of CT for Non-destructive Testing and Materials Characterization

**Martine Wevers, Bart Nicolaï, Pieter Verboven, Rudy Swennen,
Staf Roels, Els Verstrynge, Stepan Lomov, Greet Kerckhofs, Bart Van
Meerbeek, Athina M. Mavridou, Lars Bergmans, Paul Lambrechts,
Jeroen Soete, Steven Claes and Hannes Claes**

Abstract Several important technological and economic trends are shaping the
research on non-destructive testing techniques. X-ray computed tomography is also
a NDT product of these ongoing developments and has become a very important tool
for doctors, material scientists, geologists, biologists, civil engineers, bio-engineers,
dentists, quality engineers, etc., all dealing with materials of which the fine internal
structure or the changes within the material are of outmost importance to understand
the behaviour of the material or to have insights in the processes going on. X-ray CT
with micron- and submicron resolution is now well accepted in those disciplines. In
the next paragraphs a wide variety of application fields will be addressed in order to
show how X-ray CT can be applied as an NDT technique for quality control, the
study of the material behaviour and its functional properties under specified
environmental conditions, and for production and material optimization.

Several important technological and economic trends are shaping the research on
non-destructive testing techniques. They include increasing computerization and

M. Wevers (✉) · S. Lomov
Department of Materials Engineering, KU Leuven, Leuven, Belgium
e-mail: martine.wevers@kuleuven.be

B. Nicolaï · P. Verboven
Department of Biosystems, KU Leuven, Leuven, Belgium

R. Swennen · J. Soete · S. Claes · H. Claes
Department of Earth and Environmental Sciences, KU Leuven, Leuven, Belgium

S. Roels · E. Verstrynge
Department of Civil Engineering, KU Leuven, Leuven, Belgium

G. Kerckhofs
Skeletal Biology and Engineering Research Center, KU Leuven, Leuven, Belgium

B. Van Meerbeek · A.M. Mavridou · L. Bergmans · P. Lambrechts
Department of Oral Health Sciences, KU Leuven, Leuven, Belgium

© Springer International Publishing AG 2018
S. Carmignato et al. (eds.), *Industrial X-Ray Computed Tomography*,
https://doi.org/10.1007/978-3-319-59573-3_8

automation, fusion of different NDT methods, new market expansions, use of the techniques as a process control tool, and a proliferation of merges, consolidations and joint ventures. Other important factors such as the development of new materials, the safety requirements in the civilian aerospace and the automotive sector, and awareness of plant maintenance have strongly increased the purchases of NDT equipment and services. Advanced computer technology is also enabling end users to quantitatively size flaws to better estimate the danger posed by such flaws. Rather than waiting to be forced to adopt non-destructive inspection techniques in order to meet safety standards, end users are willingly utilizing these more efficient methods as a cost-saving and quality improvement measure.

X-ray computed tomography is also a NDT product of these ongoing developments and has become a very important tool for doctors, material scientists, geologists, biologists, civil engineers, bio-engineers, dentists, etc., all dealing with materials of which the fine internal structure or the changes within the material are of outmost importance to understand the behaviour of the material or to have insights in the processes going on. X-ray CT is now well accepted in those disciplines as well as submicron or nano tomography facilities. In the next paragraphs a wide variety of application fields will be addressed in order to show how μ- and nano-CT can be applied as an NDT technique for quality control, the study of the material behaviour and its functional properties under specified environmental conditions, and for production and material optimization.

An important issue for modern microscopy is the three-dimensional (3D) information. Most users try to recognize the 3D structure inside an object from 2D micrographs. Most existing microscopes can visualize either the object surface or a transmission image through a thin section. Conclusions about the 3D structure can be obtained from the image of the surface or from a combination of several thin slices. In both ways, the information cannot be reliable. Additionally, it implies that the 3D internal object structure can only be investigated destructively. An additional significant aspect of modern microscopy is the quantitative interpretation of the images in terms of the microstructure of the object. Although most microscopes include or can be combined with powerful image processing systems, the interpretation of the image contrast is still the main problem. For instance, a 2D micrograph of the surface of an object does not allow deducing accurate morphological characteristics since this would require information about the third dimension. Moreover, the image contrast is not only generated by the morphology of the object but by other factors such as composition, etc. The interpretation can be improved by independently detecting several signals from the same object area, for example in the SEM by combining the secondary electron image with X-ray microanalysis. But even in that case the interpretation is still very cumbersome. On the other hand, reliable micro-morphological information could be easily obtained from a set of thin flat cross sections which reveal only density information, from which case accurate 2D and 3D numerical parameters of the internal microstructure could be calculated.

X-ray computed tomography is a technique which allows reconstructing the 3D internal structure of objects non-destructively without any prior preparation. For X-ray methods the contrast in the images is the result of a mixed combination of density and compositional information. Recently there has been a significant improvement in the development of X-ray microscopes using synchrotron sources but since these facilities are rather complicated and expensive and are not accessible for most researchers. The latest developments in NDT have therefore led to X-ray CT systems with nano- or submicron resolution which are nowadays often applied in many areas of materials research and development. Covering a resolution range of 500 nm up 50 μm or more, those systems play an important role in our long-standing efforts to unravel the internal structure, morphology, microstructural changes or damage development in materials because they provide high quality X-ray CT images from which quantitative data can be retrieved.

8.1 CT for Internal Quality Inspection of Biological Materials

This section mainly considers apple fruit as an example category of complex bio-logical materials with high water content and a cellular microstructure. Their visual, textural and nutritional quality has direct economic impact for fresh consumption. Fruit like other plant organs consist of different tissue types such as the epidermis (with cuticle), cortex parenchyma tissue, core tissue and vascular tissue; each with a different microstructural architecture. The cellular microstructure determines the physical properties of fruit. Fruit with many intercellular pores such as apple is in general softer and facilitates exchange of metabolic gasses with the environment. This is of particular relevance to commercial storage of apple where often controlled atmospheres are applied to preserve quality attributes. The gas composition of the atmosphere is critical and needs to be determined in advance by experimental means. Such experiments are expensive and often last several years to account for seasonal differences. CT can contribute to design and optimize storage conditions.

At the macroscopic level, internal defects, such as internal browning of apple and pear fruit, are often related to their microstructure and how it changes before or after harvest. The amount of loss due to internal disorders varies from year to year. Peak losses as high as 25% have been recorded of the total production in some years. Online sorting to remove fruit or vegetables with defects would allow increasing the amount of first class products to the market. To date there are, however, no reliable inline measurement technologies for internal quality defects commercially available. Non-destructive technologies (e.g., near infrared spectroscopy, nuclear magnetic resonance techniques or X-ray CT) have been introduced for quality evaluation of fresh produce (Nicolaï et al. 2014). Detection of several disorders, macroscopically, is possible to some extent with these techniques. X-ray CT is the most recent technique in this field, and was shown to be promising with respect to such quality assessment.

8.1.1 CT Systems for Biological Materials

Table 8.1 presents an overview of CT systems that have been used for imaging biological materials at different scales. Low energy systems are required. Gantry CT systems are well known devices with medical diagnostics purpose. The advantage of

M. Wevers et al.

Table 8.1 Typical CT systems for investigating biological materials

System	Gantry CT	Large field-of-view micro-CT	Small field-of-view micro-CT	Synchrotron nano-CT
Voltage range	70, 80, 100, 120 or 140 kV	25 kV to 160 kV	20 kV to 100 kV	6 kV to 250 kV
Maximum power	2×100 kW (dual source)	60 W	10 W	20 kW
Minimum spot size	700 μm	3 μm	<5 μm	<0.1 μm
Minimum voxel size	200 μm	1 μm	0.3 μm	0.2 μm
Detector	2×20 bit, 1000×64 (dual detector)	12 bit, 1024×1024	12 bit, 4000×2300	14 bit, 2048×2048
Maximum sample size diameter	78 cm	22 cm	2.7 cm	12 cm
Typical applications	Full body, multiple samples, fast scans	Organs	Tissue samples, microscopic resolution	Tissue samples, submicron resolution, phase contrast imaging

(continued)

Table 8.1 (continued)

System	Gantry CT	Large field-of-view micro-CT	Small field-of-view micro-CT	Synchrotron nano-CT
Example images for apple fruit				
	Cross section of disordered apples in a bin	Apple equatorial cross section (bar 1 cm)	Apple cortex tissue (bar 200 μm)	Apple cells and cell walls
Reference	Donis-González et al. (2014)	Herremans et al. (2015a)	Herremans et al. (2013)	Verboven et al. (2008)

medical CT scanners is that the source-detector assembly rotates at high speeds around the patient (or object) while it is translated through the gantry system, to acquire a so-called helical or helicoidal or spiral scan (see Sect. 1.2.1). On top of that, the system can be constructed with multiple source-detector pairs. In this way, full 3D images can be recorded within seconds (Hsieh 2009). This comes, however, at a high construction cost to accurately control moving mechanical and electronic parts (Donis-González et al. 2012). An advantage is that large samples fit in this system, which is at the expense of high resolutions. The systems have been used outside medical applications for example to develop image processing algorithms for the automatic classification of chestnuts internal quality (Donis-González et al. 2014).

A large field-of-view system (Table 8.1) is a full scale micro-CT device that could be self-assembled or bought from specialist suppliers. The source and detector are fixed, but the sample can rotate on a computer controlled rotating stage. Such systems could be particularly flexible, as the location of the sample and detector can be controlled, and numerous filters can be manually replaced. Relatively large samples can be scanned in the system. Moreover, testing stages and various holders that e.g. fixate samples, can be freely mounted in the measurement chamber. Such systems can be readily applied to scan individual organs such as fruit and explore the tissue structure and internal quality (Herremans et al. 2014; 2015a).

Small field-of-view micro-CT systems could be purchased as a desktop system, which implies sample dimensions are limited, in order to fit within the enclosed sample chamber. The system has a moveable detector and sample holder, which allows an optimal geometrical setting of the components. It has a large detector, enabling high spatial resolutions. Also obtaining offset and oversized scans is possible, by imaging a part of the sample, subsequently moving the sample or detector and performing another scan, and later merging the images. This kind of commercial systems often has limited flexibility towards changing components, but are generally user-friendly for microscopic 3D imaging of tissues (Herremans et al. 2013, 2015b).

Contrary to the previous systems, few large scale facilities produce X-rays by a different physical principle, and not by a conventional X-ray tube. In a synchrotron light-source electromagnetic synchrotron radiation is produced. Synchrotron X-ray radiation is almost parallel, monochromatic, highly brilliant and coherent. These characteristics have specific advantages in terms of the image quality when performing X-ray nano-CT. Due to the parallel beam, the reconstruction is more exact compared to cone or fan beam geometries. The high X-ray flux allows relatively fast imaging, at an excellent signal-to-noise ratio, and the use of optical magnification lenses enables to achieve high resolutions. The large coherence allows phase-contrast imaging that can be used for wall detection of cells in tissues (Verboven et al. 2008), which improves image processing. However, the limited accessibility to such large facilities is a drawback.

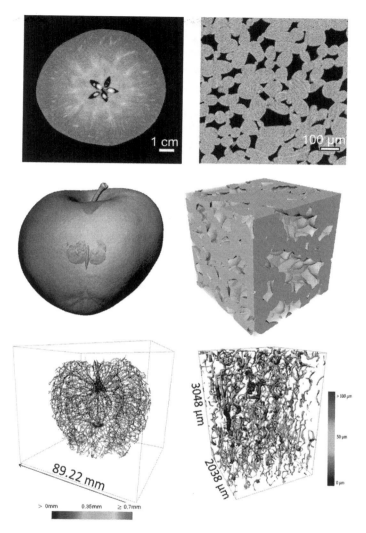

Fig. 8.1 X-ray CT slice and 3D visualization of an intact apple (*left*, pixel size 47 μm) showing core and vascular bundles and a cubical apple cortex tissue sample (*right*, pixel size 5 μm) showing cell clusters and intercellular spaces. On the bottom row, networks of vascular bundles (*left*) and of air spaces in the tissue (*right*) are depicted with respective color scales

8.1.2 Understanding Structure of Biological Materials

Figure 8.1 shows CT images of a whole apple fruit and a cubic fleshy cortex tissue sample, obtained by different CT systems at different pixel resolutions. Both vascular bundles and air spaces are important transport structures that can be resolved by CT imaging and dedicated processing. 3D image analysis has been a basis to better understand the development of these structures in growing fruit (Herremans et al. 2015a), explore differences between genotypes (Herremans et al. 2015b) and develop CAD models of fruit structures (Abera et al. 2014; Rogge et al. 2015). In a further step, CAD models can be used in computer simulations to quantify important processes in the structures such as fluid flow, gas exchange and tissue mechanics (Ho et al. 2011; Aregawi et al. 2013; Fanta et al. 2014; Ho et al. 2015).

Figure 8.2 demonstrates the ability of CT to detect different kinds of internal disorders in fruit non-destructively. Disorders can have different origins such as physical damage (e.g. bruising), physiological changes (e.g. browning) or diseases and infections. Internal disorders affect the porous tissue structure which alters the attenuation in the CT image (Lammertyn et al. 2003; Herremans et al. 2014).

This is evidenced by microstructure analysis as is demonstrated in Fig. 8.3: tissue breakdown starts with collapse of turgid cells due to membrane damage, followed by leaking of intercellular liquid into the pore spaces. This free water ultimately diffuses out of the tissue leaving only cavities and remainders of cell

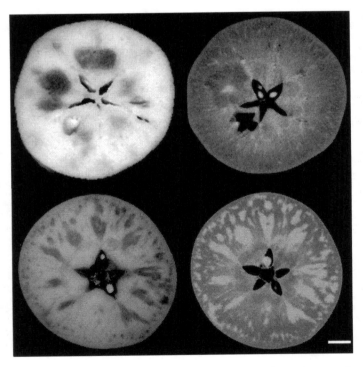

Fig. 8.2 Internal disorders of apple fruit: (*top*) browning disorder and cavity formation in 'Braeburn' apple which develops during long-term storage in suboptimal controlled atmosphere, (*bottom*) watercore disorder in 'Rebellón' apple. Visualization after cutting the fruit at the equator (*left*) and the corresponding CT image (*right*). The photograph is from approximately the same, however obviously not identical, position. Scale bar measures 10 mm, pixel size 50 μm

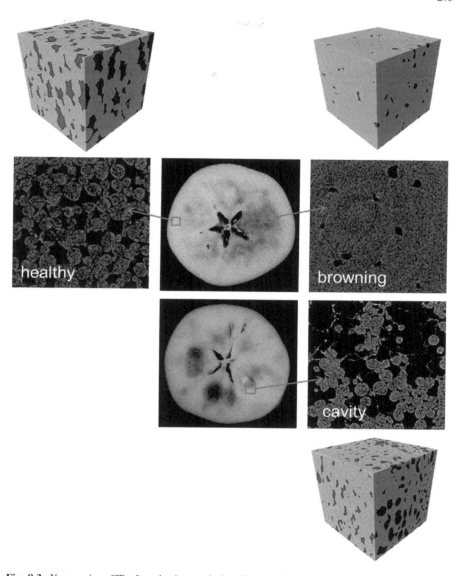

Fig. 8.3 X-ray micro-CT of apple tissues during disorder development. Healthy tissue has an intact cellular structure. Browning is induced by damage to cell membranes and subsequent leakage into intercellular spaces. At later stages the free water moves out of the fruit resulting in cavities in between remainders of the cell walls

walls. Different types of disorders pertain to the same behaviour as was evidenced in different fruit species for different kind of postharvest problems, including ripening (Herremans et al. 2013; Cantre et al. 2014).

8.1.3   Other Developments in CT Imaging of Biological Materials

Other developments in CT imaging of biological materials include the use of contrast agents and phase contrast imaging. The use of contrast agents in X-ray imaging is mainly known from clinical and biomedical applications (Lusic and Grinsta 2013). But not so much in other biology oriented disciplines. Contrast agents are applied to provide larger attenuation differences between different components of the tissue by which they enhance contrast. These contrast agents interact physically, chemically and/or biochemically with the target tissue. Thereto, contrast agents containing heavy elements (e.g., gold, bismuth or iodine) are mostly used for X-ray CT (Dhondt et al. 2010). A large variety of contrast agents exists, both natural and man-made. Recently, contrast agents are often nano-particle based. Typical examples for synthetic nanoparticles are micelles, liposomes, gold nanoparticles, silica or nanotubes (Shilo et al. 2012; Ahn et al. 2013; Hwang et al. 2014; Cole et al. 2015); and for natural nanoparticles lipoproteins, viruses or fer-retin (Cormode et al. 2010). Both types of nanoparticles can be specifically engineered for the case under study and can also be modified with targeting molecules to alter target sites. Although contrast agents have not yet been widely introduced in micro-CT scanning of plant materials or foods, they possibly have a good potential to extract microstructure information from samples which would otherwise be invisible. Unlike clinical applications, it can often be difficult to apply the contrast agent in the tissue in vivo, e.g., in the vascular system of plants or fruit (Drazeta et al. 2004; Hwang et al. 2014). Usually, contrast agents are applied ex vivo (Dhondt et al. 2010). For each tissue type, an appropriate technique will have to be devised to insert the contrast agent into the tissue, e.g., by diffusion via immersion of the sample in a solution of the contrast agent or via vacuum impregnation (Panarese et al. 2013).

Phase contrast imaging makes uses of differences in refraction index of different components in structure to aid visualization of small details in the microstructure. While this can be achieved without dedicated optical elements in synchrotron radiation X-ray imaging systems by means of so-called propagation-based imaging, micro-CT systems are now emerging that make use of gratings for imaging. Synchrotron phase-contrast imaging can be applied over a wider range of settings than grating based systems, that, however, will be more accessible. In biological material science, phase contrast imaging has already been successfully used for separating muscle, fat and connective tissue in animal soft tissues (Bech et al. 2009; Jensen et al. 2011; Miklos et al. 2014; Rousseau et al. 2015), and distinguishing single cells in dense plant tissues (Verboven et al. 2008; Lauridsen et al. 2014).

8.2   CT in Hydrocarbon Reservoir Characterization

With CT, 3D images of reservoir rocks can be obtained in a non-destructive way. Based on the acquired data, pores can be differentiated from solid material. Mineralogy of the rocks can be quantified, especially when applying dual energy CT. After segmentation, porosity can be quantified, not only as bulk parameter but after image segmentation individual pore shapes can be assessed in 3D. An intrinsic problem with CT is, however, that high resolution images with voxel sizes <1 μm^3 can only be acquired from very small samples which often are not representative for the reservoir rock under investigation. CT from larger samples, that can be considered as representative elementary volumes, are less detailed. Recent developments in multiple point geostatistics (MPS) allow working out an upscaling strategy integrating the detailed information acquired at high resolution, based on acquiring training images, with the representative lower resolution conditioning data. Data acquired by CT allow finally to better understand certain petrophysical properties of reservoir systems, such as permeability, tortuosity, acoustics, ... and are useful in monitoring fluid flow, such as wormhole formation and CO_2 sequestration.

Computed Tomography (CT), which allows to visualize the interior of the rock sample in a non-destructive way has been widely used in geosciences (Ketcham and Carlson 2001; Mees et al. 2003), especially in reservoir geology (Carlson 2006a, b) and building stone characterization (Dewanckele et al. 2012), for the characterization of volcanic rocks (e.g. Shea et al. 2010; Vonlathen et al. 2015), in paleontology (Lukeneder et al. 2014), and related geoscience research fields.

In reservoir studies the initial focus was on the visualization of internal structure, especially the distribution of the pores and the study of heterogeneities, ... (Ketcham and Carlson 2001; Ketcham 2005). Multi-phase flow experiments were already visualized in the late 1980s (Vinegar and Wellington 1986). With time a more quantitative approach was developed, not only in reconstructing the pore network, but also in deducing the mineralogy using dual energy approaches (Long et al. 2009; Remeysen and Swennen 2008).

In the following recent developments in reservoir studies using CT data, with a focus on porosity classification and linkage of pore shape to acoustic and other petrophysical measurements are reported on. Attention will also be given to recent developments in monitoring fluid flow and CO_2 sequestration.

8.2.1   Data Acquisition and Treatment

Artefact reduction (e.g. beam hardening and ring artifacts), spatial resolution and sample size play a crucial role when using CT images for detailed reservoir analysis. Because of its dependency on so many variables, resolution is commonly used to characterize the size of the reconstructed voxels (Van Marcke 2008; Cnudde and Boone 2013). Another drawback relates to the partial volume effect, since the borders of the scanned sample or a component like porosity in the sample may not coincide with the borders of the pixels of the detector. However, this effect is not always entirely negative in reservoir studies because objects smaller than the voxel size can still be inferred if the density difference with the surrounding components is large enough. For example micro-porosity and fractures smaller than the resolution can still be detected by CT (Ketcham 2006).

The key objective of CT is to create images of features of interest, which have sufficient contrast and are as free as possible of artifacts. Before applying any further processing steps it is possible to use image enhancement techniques (i.e. noise removal, application of filtering procedures or enhancing the image contrast) to make the images more suitable for analysis. In order to differentiate pores from matrix, an image segmentation needs to be performed. The purpose of this operation in reservoir studies is to separate solid phases such as rock constituents from pores. Common segmentation algorithms such as simple thresholding, edge detecting and active contours can be applied (see Wirjadi 2007 for an exhaustive survey of existing methods). However, applying a dual thresholding algorithm is often more appropriate. It uses two intervals of the histogram in order to determine the segmentation. Voxels corresponding to the first 'strong' threshold are classified as foreground voxels, while voxels selected by the second threshold are only considered foreground if they are connected to voxels already selected by the 'strong' threshold (Fig. 8.4). The advantages of this algorithm are the reduced sensitivity to residual noise in the dataset and the selection of less insulated foreground voxels.

The pores identified in the binary images form the basis for further calculations. Because the segmentation process in any 3D image analysis is important, special attention is paid, whenever possible, to the segmentation step by comparing the porosity value obtained from the dataset with helium porosity measurements. In Fig. 8.5, data from a number of 10 cm diameter cores scanned by medical CT together with 1.5 inch (3.8 cm) diameter plugs scanned by GE Phoenix Nanotom (with a 15.8 μm^3 voxel size resolution) of continental carbonates (the latter were added to obtain a sufficient number of data points) are shown. The r2 value of the regression curve is 0.91 which points to a good correlation between porosity values obtained by both approaches. Notice that the fitted curve does not pass through zero as it theoretically should. This can be explained by the existence of micro-porosity, which is not detectable due to the resolution of the medical and Nanotom CT scans. The intercept provides an indication of the amount of micro-porosity missed due to the resolution problem, at least when a homogeneous reservoir system is assessed.

Binarization of the images, i.e. identifying voxels as foreground and background in a two component system is the next step of pore space characterization. From this dataset global quantities such as the total porosity can directly be calculated as well as the volume of interconnected pores as far as they are visually connected.

In most cases the acquired 3D-dataset is saved either as volume data or as a stack of 2D slices. Both result in the generation of massive datasets that need specialized software for correct visualization. Several commercial as well as open source software packages exist such as Avizo, VG studio, Paraview, Osirix, CTAn,

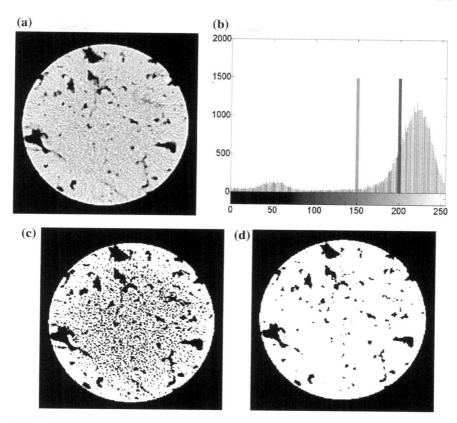

Fig. 8.4 Dual thresholding: **a** Original slice; **b** Histogram of the attenuation coefficients; the strong threshold is indicated in *red*; the weak threshold is indicated in *green*; **c** Rock selection using only the strong threshold; **d** Resulting slice using both the strong and weak threshold (Claes et al. 2016)

Morpho+, Matlab etc. It is clear that a lot of computational power is required to perform this task.

In order to achieve a correct assessment of the pore shapes, it is necessary to split complex pores into elementary pore bodies by determining the pore throats in the 3D dataset. Pore throats are those areas where the pore diameter reaches a local minimum. The identification of these pore throats is performed using a so called 'watershed transformation' approach (Beucher and Meyer 1992). The watershed algorithm has many applications in image processing by simulating the flooding from a set of labelled regions in the 3D image. These regions are expanded according to the distance map until the watershed lines are reached. Hence, the process can be seen as a progressive immersion of a landscape. Three steps are characteristic for this algorithm: (1) calculating the distance map of the binary images, (2) determining the local maxima in this new dataset, (3) calculating the

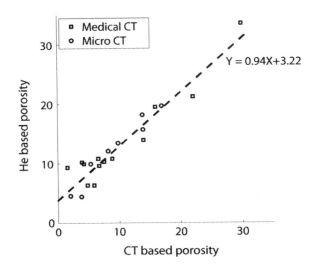

Fig. 8.5 Correspondence between He porosity and CT based porosity measurements

dividing lines (watershed lines) between different pore bodies (defined by the local maxima) based on a 26-connectivity in 3D (Meyer 1994).

8.2.2 Hydrocarbon Reservoir (Analogue) Characterization

The high intensity, natural collimation, monochromaticity and energy tunability of synchrotron X-ray sources could potentially be used to provide CT images of improved quality. The advantages of these systems would be that images could be produced more rapidly with better spatial resolution and reduced beam artifacts. In addition images, in some cases, could be acquired with elemental sensitivity. The high intensity, natural collimation, monochromaticity and energy tunability of synchrotron X-ray sources could potentially be used to provide CT images of improved quality. Main aim in reservoir studies is either to differentiate rock from porosity, allowing to reconstruct the pore network, or to differentiate pore blocking minerals. Here, the reconstructed and analyzed CT images provide information on the X-ray attenuation of the composing minerals, which is mainly depending on their intrinsic density and atomic number, at resolutions depending on sample size, focal spot size of the X-ray source as well as the resolution of the X-ray detector. This will be discussed first together with pore surface shape, Representative Elementary Volume (REV) and upscaling. A particularly additional application of CT reconstruction, addressed subsequently, is linking fluid flow to petrophysical rock properties. In this way CT can be used to evaluate reservoir quality (Carlson 2006a, b). The approach up to now is mostly applied for conventional reservoirs (Berg et al. 2014; Andrew et al. 2015), but recently also unconventional reservoirs, like shales, are studied (Akbarabadi et al. 2014; Fogden et al. 2015).

8.2.2.1 Porosity and Mineralogy Characterization

Due to recent optimization of X-ray sources and detectors, flexible, multifunctional, high-resolution X-ray CT scanners have been developed allowing 3D quantification at (sub)micron scale, which is often required for the study of pore structure and for the determination of matrix characteristics. Notice that in reservoir geology 1 μm is often considered as the percolation threshold in fluid flow, but microporosity <1 μm might still be of importance in some reservoir studies (e.g. shale gas). Because of the large numerical data-sets generated by CT, also quantitative conclusions can be drawn (Remeysen and Swennen 2008).

In addition, the generated datasets can be used as basis for fluid-flow analysis (Blunt et al. 2013; Youssef et al. 2013). These models provide important information about parameters such as porosity, hydraulic conductivity and diffusivity (Van Marcke 2008). This results in a realistic characterization of reservoir rocks concerning their productivity. Moreover the CT datasets form an excellent starting point to perform quantitative analysis with regard to the compositional structure of the sample. The results of these analyses include data on the volumes of different fractions, as well as surface and other morphological parameters (Brabant et al. 2011).

X-ray computed tomography has been used in sandstone reservoir characterization. Mineralogy can be deduced, such as differentiating quartz, zircon and for example rutile inclusions, as well as calcite and ankerite cements (Long et al. 2009). However, one should be aware that X-ray attenuation coefficients are a function of both atomic number and material density. Based on dual energy CT, the latter parameters can be deduced. However, differentiating feldspar from quartz grains remains less straightforward, certainly if the latter are partially altered. In addition identifying detrital and authigenic clay minerals is most of the time impossible, due to the small grain size of the phases and due to the partial volume effect. Nevertheless average grain size, sphericity and packing can be inferred certainly from high resolution CT images. Furthermore, several key pore structure parameters, including porosity, pore size distribution, pore connectivity, surface area, hydraulic radius and aspect ratio can be quantified. More detailed pore shape characterization highlights reservoir heterogeneity and allows to calculate main pore networks and quantify disconnected pore ganglia (Schmitt et al. 2016). But an intrinsic problem with high resolution CT images relates to the size of the samples which always is very small, questioning the representativeness of the acquired data. In contrast, if a larger sample is scanned, then some details will fall below the resolution. Selecting an appropriate resolution is function of the research question to be addressed and should be a compromise between the pore size and the representative element volume for the specific property or process of interest (Peng et al. 2012; Cnudde et al. 2011). Often a combination of moderate resolution CT images on larger samples and high resolution (eventually synchrotron CT) needs to be carried out. Several of the acquired parameters can easily be compared with petrophysical measurement, such as helium porosimetry, mercury intrusion porosimetry (MIP), nuclear magnetic resonance (NMR) relaxometry, etc. In addition pore shape characterization helps interpreting petrophysical properties of reservoir rocks, such as permeability, surface area, fractal dimension of the surface area, and complex electrical properties. With regard to synchrotron CT, it should be stressed that it allows submicron resolution with a monochromatic beam of tunable energy (Nico et al. 2010), but these facilities are not easily accessible for daily use for most researchers.

With regard to porosity classification in carbonates the most frequently applied system in carbonate rocks by petroleum geologists is the classification introduced by Choquette and Pray (1970). This classification mainly relates to the sedimentological characteristics of the samples and hence is closely linked to the depositional environment, diagenetic history, fracturing and origin of the carbonates studied. More petrophysically based porosity classifications were developed by Archie (1952) and Lucia (1995). These classifications aim to establish a link between porosity types and fluid flow properties. However, the disadvantage of all these classifications is the subjectivity of describing the abundance and shape of the pores and hence not making a correct and objective interpretation of the porosity parameter. More recently, Lønøy (2006) introduced a useful classification based on pore characteristics and sizes. The latter author is also one of the first to add in his classification a measure addressing the porosity distribution. Samples were divided between uniform and patchy distribution, however, all the proposed parameters are still descriptive and relative to each other. The rock is described by information about the porosity type, rather than by definition of the pore itself e.g. intercrystal or interparticle porosity. Hence the problem remains of correctly assessing the porosity distribution based on parameters deduced from a 2D (thin) section of the sample. In order to acquire sufficient and correct information about the porosity distributions, the step towards 3D datasets is an absolute necessity. In fact, porosity is a 3D parameter. In order to link it with permeability values and other rock parameters, such as acoustic measurements and dynamic shear moduli, an understanding of the pore connectivity in 3D is essential (Weger et al. 2009). This data acquisition can be done by using X-ray computed tomography.

Despite the fact that shape is one of the most fundamental properties of an object, it remains very difficult to characterize and quantify it objectively. In geosciences most shape descriptors are used to describe the external shape of particles and rock fragments (e.g. Goudie et al. 2003; Evans and Benn 2014). For example shape parameters have been largely used to describe and predict the hydraulic behaviour of sediment grains (Janke 1966; Dobkins Jr. and Folk 1970) as well as to analyze the size and shape of volcanologic particles (Riley 2003; Ersoy et al. 2008). Shape descriptors are powerful tools, which allow to describe shape in an objective manner. Different approaches of describing shapes have been suggested such as parametrizations (Klette 1996), analyzing invariant features (Sharp et al. 2002) and Fourier transforms (Vranic and Saupe 2001).

A shape descriptor needs to express the broad and medium scale aspects of the morphology of the considered sample. In reservoir studies the focus lays on the description of form as a 3D characteristic of the pore. In order to quantify the shape of the object, it is necessary to define the dimensions of the object. In literature

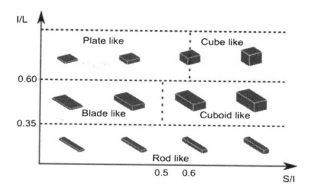

Fig. 8.6 Overview of the proposed pore shapes using the ratios of the longest (*L*), intermediate (*I*) and shortest (*S*) pore dimension, based on a projected ellipsoid within the assigned pore (Claes et al. 2016)

standard practice is first to calculate the longest dimension L, followed by the intermediate axes I being the longest dimension perpendicular to L, and finally determining the smallest axes S, which is perpendicular to both L and I. Based on the ratio of I/L and S/I, it is possible to define five shape classes: i.e. rod, blade, plate, cuboid and cube like in Fig. 8.6 Thus rock typing based on pore shape parameters becomes quantitative. An additional advantage of this classification is that the data provides information about the orientation of pores. This allows assessing the anisotropy of the porosity parameter.

The 3D classification (Claes et al. 2016) allows to better reflect the influence of the porosity parameter on other petrophysical parameters. An example of the difference between the pore networks is illustrated in Fig. 8.7. Sample A is dominated by sub-horizontal orientated pores, which are plate or cuboid like shaped. Sample B

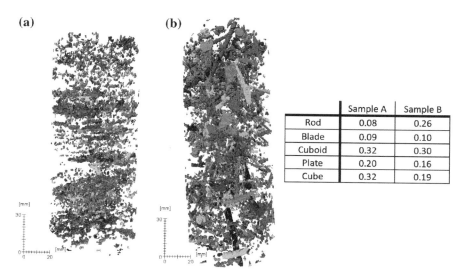

	Sample A	Sample B
Rod	0.08	0.26
Blade	0.09	0.10
Cuboid	0.32	0.30
Plate	0.20	0.16
Cube	0.32	0.19

Fig. 8.7 3D volume rendering of the porosity network in continental carbonates: **a** sample dominated by sub-horizontal orientated pores and **b** sample with a heterogeneous pore network. The volume fractions of the respective shapes as defined in Fig. 8.6 are also given

on the contrary has a much more heterogeneous pore network. Here, moldic rod like shaped pores are much more abundant. The pore shape distribution will have an important effect on the petrophysical characteristics of the sample such as permeability and acoustic response (see below).

8.2.2.2 Pore Surface Shape

Quantification of random surfaces of shapes is not straightforward. In the past different techniques for describing the morphological characteristics have been developed such as: spherical harmonics (Brechbühler et al. 1995), hyperquadrics (Hanson 1988) and superquadrics (Terzopoulos and Metaxas 1990). In numerous geological research topics, the surface area plays an important role in establishing accurate calculations and simulations. One of these fields is the modeling of fluid-rock interactions. Brosse et al. (2005) reported that the reactive surface area is a crucial parameter to determine the accuracy of the model. In most studies a Brunauer-Emmett-Teller (BET) surface area is calculated in order to estimate the surface area (Brunauer et al. 1938). However the steric aspects are not taken into account in most models resulting in a full sphere to be considered as reaction surface. Moreover the non-spherical character and roughness of the reaction surface cannot be quantified (Peysson 2012; Elkhoury et al. 2013).

In permeability simulations the fluid-fluid and fluid-solid interfacial areas are important parameters to consider when determining conductivity (Mostaghimi et al. 2013; Bultreys et al. 2015). In recent algorithms the contact surface areas are measured for each pore individually. Moreover in multi-phase flow simulations the contact surfaces can become complex, resulting in overestimations of the surface area due to digitalization effects.

In Fig. 8.8 examples are provided to describe surfaces using spherical harmonic functions. These functions are the 3D extension of Fourier series and are especially well suited for modelling shapes (see workflow proposed by Chung et al. 2008).

In order to assess the influence of digitalization on surface area calculations, spheres with different radii are generated and the surface area is calculated and compared with the analytical value of the surface area. In Fig. 8.9 three different methods are compared: simple voxel face counting, the marching cube algorithm and the surfaces generated using spherical harmonic approximation. A negative error value is observed if the surface area is underestimated. Simple voxel face counting results in a large overestimation of the surface area of around 50%. The error observed when using the marching cube algorithm is around 8% even for fine digitalized spheres, which is also observed by Dalla et al. (2002). The error of the surface area calculated using spherical harmonics varies between −6% for the smallest sphere and 0.1% for the largest sphere. For a sphere with a radius of 10 pixels the error becomes smaller than 1% and remains this small for larger spheres.

An additional advantage of using spherical harmonic generated surfaces is that curvature can be calculated at every point of the surface. By its definition curvature is undefined at non-differentiable features such as sharp edges and corners. The

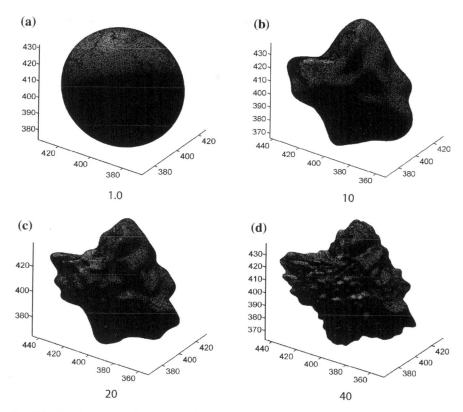

Fig. 8.8 Spherical harmonic representation of the pore surface. The parameter indicates the order of the parameterization

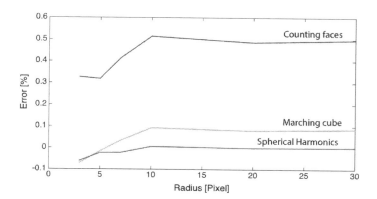

Fig. 8.9 Error of the computed surface area of a sphere versus the number of pixels used to digitize the radius

curvature calculation map proves to be an excellent starting point to define the roughness of the surface on outcrop models as illustrated by Lai et al. (2014). This approach can also be used on smaller spatial scales.

8.2.2.3 Representative Elementary Volume

Heterogeneity in grain size, bio- and allochem composition and pore distribution is a general characteristic especially of carbonate reservoir rocks and relates to their geological history, i.e. sedimentary origin, burial history and diagenesis. These processes have an influence on the complexity of the pore network. However, when the scale of observation becomes large enough, the material can be viewed as a homogeneous continuum containing enough representative information of the heterogeneity. In order to address this problem, the concept of the Representative Elementary Volume (i.e. the smallest value that can be taken as a representation for the entire sample area/volume that does not respond to small changes in volume or location) was first introduced by Bear (1972).

The representative elementary volume (REV) is an important parameter in several research fields including reservoir rock typing. Clausnitzer and Hopmans (1999) determined that the side length of the REV cube should be approximately 5.15 times the diameter of randomly packed spheres. Baveye et al. (2002) used the same workflow as Clausnitzer and Hopmans (1999) in soil samples to estimate the size of the REV, but did not succeed to establish a rule of thumb concerning the size of the REV. Razavi et al. (2007) estimated the size of the REV in sandstones with elongated particles to be between 5 and 11 times the longest dimension of the particles.

Based on CT reservoir studies, two different methods for calculating the REV can be put forward. The first method is based on the original definition of Bear (1972) and uses the chi-square statistical test to objectify the method. The second method uses applications of geostatistical techniques such as variogram models to assess the size of the REV. Both methods have a different approach and generate complementary information. The statistical method is more general and allows to deduce 95% confidence intervals of the size of the REV. The second method provides more information about the spatial distribution of the studied parameter and the critical direction (anisotropy) of the REV. Hence the combination of both methods allows to thoroughly analyze the size of the REV (Claes et al. 2016).

8.2.2.4 Multiple Point Geostatistics

Multiple point geostatistics (MPS) have in recent years gained an important place for modelling geological structures on different spatial scales (from μm to m). For example in the field of groundwater flow modelling they allow to build models on a decametre scale (Mariethoz et al. 2015). Okabe and Blunt (2004) used this technique to model the pore-space of siliciclastic samples on a μm scale. These samples have a more homogeneous pore size and shape distribution compared to carbonate rocks and hence are easier to model. Based on MPS artificial "numerical rock" samples can be generated which can be used for example to investigate the influence of rock heterogeneity (rock typing) on permeability, and thus on fluid flow. Moreover the technique permits to bridge the gap between different datasets with different resolutions, and to quantify the effect of using training images to jump from one resolution scale to another. It also allows developing an upscaling strategy (Soete et al., submitted).

MPS enables to model heterogeneity by directly inferring the patterns from Training Images (TI). The general concept is to use as TI the most detailed dataset, which has the best resolution and to derive condition data from the dataset with the lowest resolution.

Figure 8.10 provides a schematic outline of the workflow. The image on the left-hand side is a detailed image obtained using micro-CT. Hence this dataset, with voxel resolution of $4 \times 4 \times 4$ μm^3, has detailed information about small porosity present in this specific lithology type. The image on the right hand side originates from a medical CT dataset of the same sample, which typically incorporates a much larger sample volume (notice the difference in scale in Fig. 8.10). This dataset provides information about the pore network on a larger spatial scale and will be

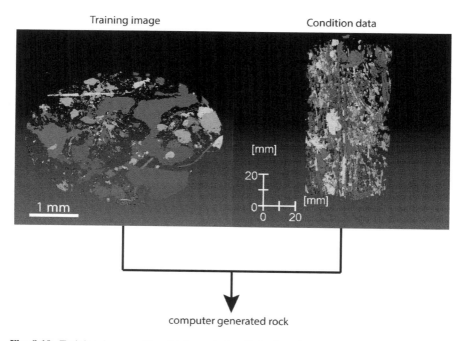

Fig. 8.10 Training image with a high resolution (*left picture*) and condition data with a low resolution (*right picture*) from the input of the Multiple Point Workflow (Images are color coded to depict connectivity, i.e. connected pore volumes have the same color)

used as condition data in the simulations. Thus the TI, which is representative (= REV) for the matrix of the medical CT, will simulate the matrix micro-porosity, which was below the resolution of the medical CT. This approach allows retaining the connectivity of the porosity network on a larger spatial scale, resulting in more reliable permeability simulations.

8.2.2.5 μCT and Rock Petrophysics

In their study on acoustic velocities of continental carbonates, Soete et al. (2015) were able to explain certain outliers, based on a μCT study of related pore shapes. In Fig. 8.11 porosity is plotted versus V_p graph, with indication of the regression line of the entire dataset. Samples dominated by rod- and/or cuboid-like pore shapes (>55 vol.%, i.e. Al08, AL09 and BU11) increase compressional-wave velocities. These samples plot above the regression line. Flattened pore shapes, i.e. plate- and blade-like pores, lower the acoustic wave propagation in travertines (>50 vol.%, i.e. samples Al18 and SU18). Equidimensional cubic pore shapes do not significantly increase or decrease compressional-wave propagation and thus plot close to the regression line. The volume shares of the different pore shapes exert further control on acoustic velocities, additionally to the influence of pore types.

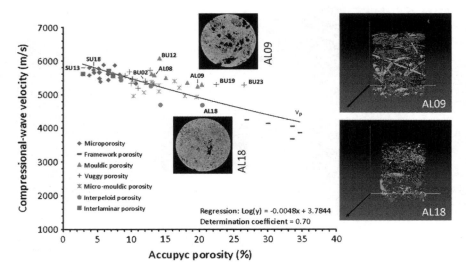

Fig. 8.11 P-wave velocity versus porosity for the 7 dominant pore types in the continental carbonate samples. The exponential trend line through the data is given as a reference. Velocity scatter at equal porosity is caused by the presence of different pore types and shapes. Slices and volume rendered pore networks, extracted from μCT scans are given. 2D slices perpendicular to the long axis of sample Al09 and sample Al18. Dominance of cuboid and rod-like pore shapes are observed in sample Al09. Plate-like, cube and blade-like pores dominate in sample Al18. The labelling, conducted on pore networks assign colors as a measure of connectivity

More specifically, micro-CT scans (Fig. 8.11) of sample Al09 and Al18 which possess rather similar porosity (respectively 19.7 and 20.2%), and which yield quiet different V_p-values (respectively 5253 and 4703 m/s), revealed that their pore networks are significantly different. The 2D μCT slices show a dominance of large (millimetre-sized) oval shaped pores in sample Al09, compared to a dominance of more uniformly distributed oblate (micro-) pores (10 s to 100 s of micrometres in size) in Al18. Hence, the porosity distribution is significantly different for both samples, patchy versus uniform respectively, implying that body waves will propagate in a different manner (Anselmetti and Eberli 1993; Brigaud et al. 2010; Verwer et al. 2008). Acoustic velocities in continental carbonates show a first-order dependency on porosity, with an inverse linear relation. Second-order velocity deviations correlate with pore size and shape complexity. Small and complex pores are associated with negative acoustic velocity deviations. Large, simple and stiff pores result in increased velocities.

Soete et al. (submitted) simulated permeability based on μCT reconstructed pore networks (Fig. 8.12) and calculated the total porosity (Ø) and the connected porosity in the z direction ($Ø_{c,z}$) on carbonate reservoir systems, more specifically on continental carbonate samples.

The connected porosity was always lower than the total porosity. For some samples the lowering was limited, which meant that the majority of recorded pore objects in the μCT scans contributed to the connectivity. Other samples displayed large differences between total and connected porosity, as for example in simulation

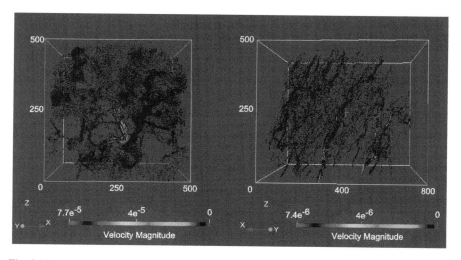

Fig. 8.12 Examples of 3D streamline velocity distributions within continental carbonate samples, with preferential flow in the z-direction. The velocity magnitude is given in non-dimensional lattice units

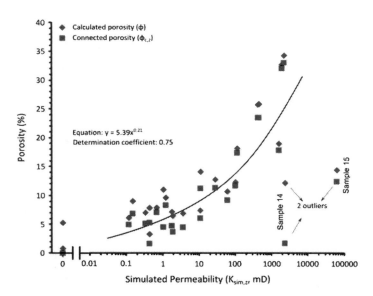

Fig. 8.13 Cross plot of the calculated total and connected μCT porosity (respectively blue and *red symbols*) versus the permeability

14 (12.4 vs. 1.4%). The latter was indicative for the pore network heterogeneity and for the presence of isolated porosity in the direction of the simulation and at the resolution of the acquired CT images. In Fig. 8.13 the simulated permeability along the z-axis ($K_{sim,z}$) of the samples is plotted against Ø and $Ø_{c,z}$. Porosity was in this case plotted on the vertical axis of the graph to clearly show the lowering from Ø to $Ø_{c,z}$. A power-law relationship is observed between $K_{sim,z}$ and $Ø_{c,z}$, with a determination coefficient of 0.75. This implies that good permeability estimates were obtained from μCT porosity calculations at 16 μm resolution. However, it has to be kept in mind that pores <16 μm resolution could influence flow properties in the considered carbonates, but were not taken into account in the calculated permeability estimations. Notice, however, that two outliers with high permeabilities, but relatively low porosities occur.

Soete et al. (submitted) also calculated the tortuosity (τ) and plotted it against permeability (Fig. 8.14). The observed power-law relation between $τ_z$ and $K_{sim,z}$ is consistent with findings of studies in volcanic rocks, where increasing tortuosity results in lower permeabilities (Bouvet de Maisonneuve et al. 2009; Degruyter et al. 2010a; 2010b). The tortuosity of the pore network allowed to better understand the discrepancies between porosity and permeability. The two outliers from the porosity—simulated permeability regression line (sample 14 and 15 in Fig. 8.13) were also highlighted in the tortuosity—simulated permeability cross plot (Fig. 8.14). Both samples are part of a reservoir system dominated with rod shaped moldic porosity where $Ø_{c,z}$ consists of only a few of these vertical rod-shaped molds. Despite the limited porosity, both samples are characterized by a low

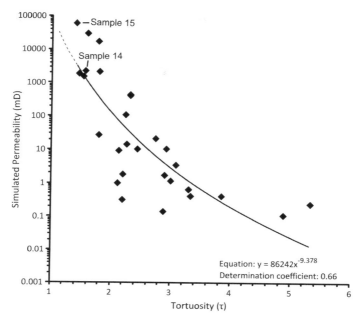

Fig. 8.14 Tortuosity—simulated permeability cross plot. A moderate power-law relationship is observed between both parameters

tortuosity, indicative for the straight, high velocity flow paths that the rod-shaped molds provide through the sample.

8.2.2.6 CT in Fluid Flow Characterization

Fluid flow simulations can start from the petrophysical rock properties deduced from static CT reconstruction. However, fluid flow characterization from tomography is greatly improved by adding data acquired from flow analysis experiments (e.g. Coles et al. 1994, 1998; Dvorkin et al. 2009; Karpyn et al. 2009). There are some practical challenges for fluid flow analysis experiments, like the space needed to couple the fluid devices and lines, and the possibility to make the fluid flow lines able to rotate with the sample. Here devices with a rotating source counter these problems. The main challenges, however, are related to image quality. Therefore, fluid processes should be slow or stable while scanning and sufficient exposure time is needed in order to obtain good signal-to-noise ratio (Cnudde and Boone 2013). Most experiments are thus limited to slow fluid flow, in turn limiting the applicability of the technique. New iterative CT reconstruction algorithms, however, enable fluid flow experiments with faster dynamics (Van Eyndhoven et al. 2015; Kazantsev et al. 2016). By separating stationary (solid matter) from dynamic regions (fluid flow), the redundancy of information in the temporal domain can be employed to improve spatial resolution (Kazantsev et al. 2016). The attenuation curve of a particular voxel in the dynamic region can be modeled by a piecewise constant function over time, in accordance with the actual advancing fluid/air boundary. In this way reconstruction is possible from substantially fewer projections per rotation without image quality loss, increasing the temporal resolution, enabling faster non-reacting fluid flow experiments (Van Eyndhoven et al. 2015; Kazantsev et al. 2016).

Fluid flow experiments lead to substantial advancements for pore network reconstruction of conventional reservoir rocks or along fractures (Alajmi and Grader 2009; Karpyn et al. 2009). However for complex pore networks, such as in carbonates (Soete et al. 2015), with a large range in pore sizes it is difficult to gather the experimental data needed for simulations. Furthermore this poses great computational challenges. This challenge can be partly countered by upscaling. Bultreys et al. (2015) used MPS for upscaling based on a dual pore network model. Furthermore by imaging pore-scale displacement processes during fluid injection, the quantifiable non-wetting phase cluster size redistribution (e.g. Georgiadis et al. 2013) allowed to study the fundamental mechanisms like invasion percolation for drainage, which can be validated by experimental data.

Computer tomography is also used in research related to CO_2 sequestration. Here, the understanding the fundamental mechanical and chemical processes involved is not restricted to the reconstruction of realistic pore geometries from nano-scale to reservoir scale, but also relates to fluid flow simulations and reactive transport modeling. Applying Computational Fluid Dynamics solvers on CT generated 3D data allows verifying existing sequestration models and studying dynamic precipitation and dissolution processes involved. CT scanning can thus be deployed in many subfields of carbon storage. High Resolution X-ray CT (HRXCT) has been used to visualize the difference in residual trapping of supercritical CO_2 (Chaudhary et al. 2013; Altman et al. 2014). Furthermore, Aminzadeh et al. (2013) and Altman et al. (2014) used CT to demonstrate how adding nanoparticles to the in situ brine increases the CO_2 saturation behind the front and deceases gravity segregation. Time-lapse CT imaging, fluid simulations and reactive transport modeling are applied to study environmental changes in the pore domain. Visualizing tracer displacement through rocks or simulating pore plugging due to calcite cementation helps reconstructing flow paths and its alteration due to CO_2 injection (Fourar et al. 2005; Nico et al. 2010; Mehmani et al. 2012; Altman et al. 2014; Van Stappen et al. 2014). Lastly, synchrotron tomography offers the advantage of being equipped with a coherent, high flux, tunable monochromatic X-ray source allowing phase contrast imaging in multi-phase systems and chemically sensitive imaging. The latter allows predicting CO_2 distributions throughout the porous network (Nico et al. 2010). As stated by Altman et al. (2014), the integration of molecular dynamics simulations, pore-scale experiments, pore-scale simulations, in which computer tomography has been of major importance, and the study of natural analog sites provides useful insight in the efficacy of capillary, solubility, dissolution, and mineral trapping for geological CO_2 storage.

8.2.3 Conclusion

Computed tomography (CT) is a powerful tool in hydrocarbon reservoir characterization. It not only allows to quantify bulk porosity, but based on the acquired data pore shape characteristics between reservoir units can be quantified. It allows to use a pore classification in 3D, which is a necessity for permeability simulations, after image segmentation, as well as linkage of pore shape characteristics and its linkage to petrophysical measurements such as acoustics, tortuosity, etc. Pore network reconstruction is essential for understanding fluid flow (inclusive CO_2 sequestration), which is an essential component of reservoir analysis. However, an important question that can be raised based on the acquired CT data, is whether the data are Representative for the Elementary Volume. Furthermore, upscaling becomes possible thanks to Multiple Point geoStatistical analysis, using training image derived from high resolution CT-images and less detailed CT images as conditional data. The integration of all these approaches especially with regard to multiphase fluid flow experiments with simulations, at this moment has mainly computational limitations. Future advancements in algorithm development and computer technology, however, will increase its applicability.

8.3 CT to Characterize the Behaviour of Materials

For the discussion on the use of X-ray CT for building materials research the focus will be on the behaviour of porous, often brittle building materials such as brick, mortar, stone and concrete. CT is applied to visualize and characterize the material's inner structure, to analyze the mechanical behaviour under loading, and to characterize the moisture transport in building materials.

8.3.1 Visualization and Characterization

Numerous studies on building materials benefit from a visualization and characterization of the material's inner structure. Unfortunately, building materials typically have pore radii ranging from nanometres to millimetres, which makes current CT facilities still inadequate to detect the smallest pores. But, as regards the visualization of micro-scale phenomena CT has already proven to be of interest, as indicated by various studies on the visualization of (larger) pores, cracks (Roels et al. 2003) and fibres.

In the example presented in Fig. 8.15, X-ray micro-CT was applied to visualize the fibre orientation in steel fibre reinforced self-compacting concrete (SFRSCC). In Andries et al. 2015, two types of SFRSCC mixtures were investigated in order to determine the influence of the flow distance on fibre distribution and orientation.

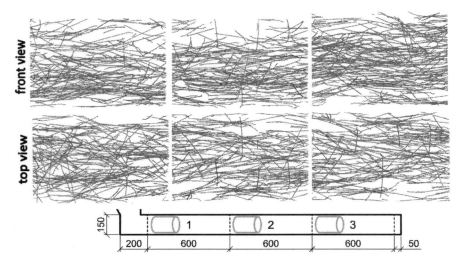

Fig. 8.15 Front and top view of three core samples obtained by micro-CT (Andries 2015)

Beam elements were cast and after curing, standardized prisms of 150×150 600 mm^3 were cut from them. The prisms were tested under three-point bending load to determine the post-cracking behaviour in function of the flow length. Additionally, the spatial distribution and orientation of the fibres were measured by means of manual counting and image analysis. The results were compared to a 2D (Fig. 8.15) and 3D representation obtained by micro-CT on core samples with a diameter of 100 mm. For the micro-CT, a resolution of 81 µm was achieved on a Nikon Metrology XT H450 system, with applied X-ray source parameters being 360 kV and 0.205 mA. The orientation coefficients determined with different measurement methods were in good agreement with each other. The results indicate that, independent from the viscosity of the mixture, preferential fibre orientation is induced by shear flow of the concrete.

8.3.2 Mechanical Behaviour

When focusing on the mechanical behaviour of brittle construction materials, micro-CT is applied to analyse the material's inner structure and fracture propagation at micro scale. This can be done by applying in situ loading during the test program (inside or outside the X-ray chamber), by accelerating the corrosion of reinforcement bars in reinforced concrete by inducing an electric current, etc. The aim is to visualise the initiation and propagation of micro cracking and pore collapse during subsequent loading stages, to relate the damage nucleation zones to anomalies in the material's micro structure, such as pores and inclusions, and to relate the micro mechanical behaviour to the constitutive relations observed at the macro scale (Verstrynge et al. 2016).

As a first example, the failure mechanisms of brick masonry samples with cement and lime mortars were investigated under compressive loading using micro-CT combined with an in situ mechanical press (Hendrickx et al. 2010). Tests were performed on brick-mortar-brick cylinders of 29×48 mm^2 (diameter $\times$ height). Four types of mortar were selected, and one of them had two different curing procedures. For each of the mortars, 7 cylindrical specimens were tested. Four of them were pressed in a hydraulic press to estimate the compressive strength (outside the X-ray chamber). Three were pressed inside the X-ray chamber and loaded up to fixed percentages of their estimated strength, at which point intermediate scans were performed. The applied X-ray source parameters were: 100 kV and 0.35 mA. The magnification was 3.8 and the pixel size 45.6 micron. Micro-CT allowed to visualise the spatial distribution of propagating cracks (Fig. 8.16). Unfortunately, the resolution was not high enough to visualise the crack initiation, which is a micro scale phenomenon. Only the larger meso and macro cracks were visible, which allowed to make a rough discrimination between different failure mechanisms.

A second example focuses on the influence of cementation degree and moisture on the mechanical behaviour of ferruginous sandstone (Verstrynge et al. 2014a, b). Micro-CT analysis was part of a larger test program which focused at different scales: micro (sandstone inner structure), meso (sandstone blocks) and macro (masonry) (Verstrynge et al. 2014a, b). Three types of ferruginous sandstone were included: low and high quality Diestian ferruginous sandstone and Brusselian ferruginous sandstone, all originating from Belgium and applied as building stone in numerous historical monuments. At micro level, compressive tests under

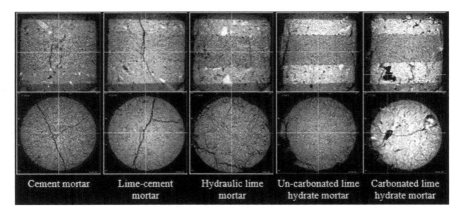

Fig. 8.16 Vertical and horizontal section for crack visualization after failure of the specimen (Hendrickx et al. 2010)

sustained loading were performed on dry and saturated sandstone samples (diameter 50 mm, height 60 mm). For micro-CT scanning, small cylindrical sandstone samples (diameter ±10 mm, height 14 mm to 15 mm) were scanned at different stages of a compression test. A resolution of 7.8 µm was achieved on a SkyScan 1172 system, with applied X-ray source parameters being 100 kV and 0.1 mA. The applied X-ray scanning system did not have an in situ loading stage. To enable stress application during scanning, a small transparent compression cell was fabricated in polyetherimide (PEI). This allowed the application of a constant compressive deformation while scanning.

As a result of the micro-CT tests, horizontal slices were obtained which show the inner structure of the ferruginous sandstone samples during the stepwise compression tests. These micro-CT images can be compared with the results of the stress-strain curves (Fig. 8.17). It was observed that crack initiation in saturated samples could be visualized at an earlier stage during the compression tests due to their decreased internal friction and less brittle behaviour. In general, the scan resolution was not sufficient to observe the very initiation of micro cracks and the behaviour of the clay minerals. However, crack propagation, quartz gains (0.1 mm to 1 mm), coagulated clay fragments and relatively large pores could clearly be distinguished.

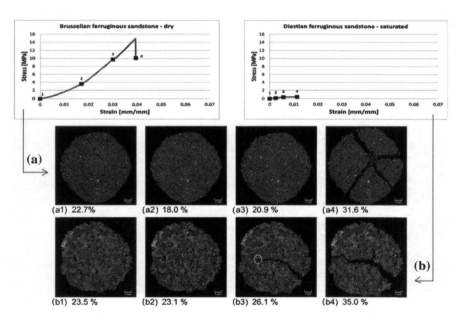

Fig. 8.17 Results of micro-CT on stepwise loaded ferruginous sandstone: stress-strain graph of dry, Brusselian **a** and saturated, low-quality Diestian **b** sample. Horizontal slices with indication of 2D porosity (Verstrynge et al. 2014a, b)

8.3.3 Moisture Transport

Moisture transfer in building materials is a key factor for the durability and sustainability of built structures. To quantify this moisture transfer in porous materials, the microfocus X-ray projection method can be applied (Roels and Carmeliet 2006). In this method, which is based on the attenuation of materials, the moisture content w (kg/m^3) at each position is quantified by logarithmically subtracting the reference image of a dry reference sample from an image taken during moisture transfer (also called the 'wet' image):

$$w = - \frac{\rho_w}{\mu_w} \left(\ln(I_{wet}) - \ln(I_{dry}) \right) \tag{8.1}$$

with ρ_w the density of the liquid phase (1000 kg/m^3 for water), μ_w (–) the attenuation coefficient of water (or in general liquid) and I_{dry} and I_{wet} the attenuated intensity for a dry and a wet sample, respectively.

A first example demonstrates the applicability of the method for analyzing liquid water uptake processes in a fractured brick sample (Roels et al. 2003). The study was performed on the Tomohawk AEA system, with the Philips HOMX 161 X-ray tube, and the source parameters being 115 kV and 0.04 mA. As shown in Fig. 8.18a, the moisture profile in the brick matrix and the waterfront in the fracture are clearly visible. Both show to be in good agreement with the simulated water uptake. In this way, the results obtained based on the X-ray projection method serve as a validation of the simulation model.

A second example (Fig. 8.18b) deals with the detection of interstitial condensation in wall assemblies with interior insulation and with the study of the working

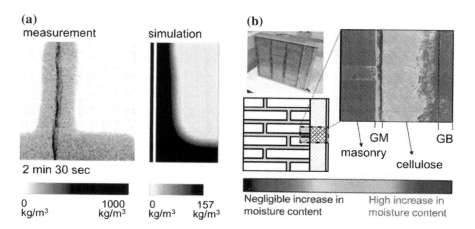

Fig. 8.18 **a** Measured (*left*) and simulated (*right*) moisture front in a naturally fractured brick during a water uptake experiment (Roels et al. 2003), **b** moisture distribution in a wall assembly (*GM* glue mortar, *GB* gypsum board)

mechanism of those interior insulation systems. In the study, 3 cm thick test walls positioned in a hot box-cold box were first exposed to winter conditions (Vereecken Roels 2014). After 6 weeks, the test walls were taken out of the hot box–cold box and the attenuated intensity (I_{wet}) was measured and compared to the reference image taken before the start of the experiment (I_{dry}). Also this study was performed on the Tomohawk AEA system, with the Philips HOMX 161 X-ray tube. Because of (1) a potential modification of the X-ray spectrum during the experiment, which lasted for weeks, (2) a potential deviation of one-dimensional moisture transport due to the extremely thin test walls (only 3 cm) and (3) rather small changes in attenuation induced by vapour transport, the study was limited to a qualitative analysis. As shown in Fig. 8.18b, this resulted in sufficient information on the working mechanism of the insulation system. For the wall with cellulose insulation, moisture induced by interstitial condensation was partially stored in the glue mortar and partially redistributed inwards by the cellulose insulation.

Remark that the examples shown in this chapter are restricted to X-ray radiography; though if the CT scans can be obtained sufficiently fast in comparison to the moisture flow in the material, the projection method can be applied for a three-dimensional quantification as well.

8.3.4 Conclusion

CT is shown to be a valuable non-destructive technique for the exploration of porous building materials. A number of typical studies that make use of a static visualisation and characterisation of the inner structure were mentioned and the value of CT in the analysis of the mechanical and hygric behaviour of building materials was briefly illustrated.

8.4 CT Research on Composites, Foams and Fibrous Materials

Fibre reinforced composites, and fibre assemblies in general, have a distinct internal structure, which can be random, especially in short fibre composites, or periodic, as in textiles or textile composites. The mechanical behaviour of a fibre reinforced composite is largely defined by its structure, in combination with mechanical properties of the fibres and the matrix. Micro-CT is an efficient tool for characterization of the fibrous internal structure, as well as for creating models for prediction of the mechanical properties of composites. Apart from the fibrous structures, cellular solids (e.g. foams, balsa wood) are important structural materials, which are used as a core for composite lightweight sandwich structures.

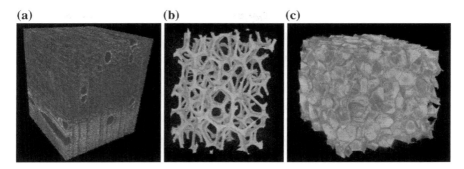

Fig. 8.19 Micro-CT images of cellular materials: (**a**) balsa wood (natural honeycomb), (**b**) open-cell foam, (**c**) closed-cell foam (Shishkina 2014)

8.4.1 CT of Cellular Materials (Foams)

Figure 8.19 shows examples of micro-CT images of cellular materials. A cellular solid can be defined as a material made of "interconnected network of solid struts or plates, which form the edges and faces of cells" (Gibson and Ashby 1997). The term 'cell wall' relates to foams as a general term which comprises cell faces, edges and vertices together.

Following (Pinto et al. 2013), a foam characterization based on its micro-CT im-age is done in the following steps:

Step 1 Image binarization. The image contrast is enhanced and a median filter is applied to reduce the noise of the image and preserve edges, then a con-volve filter is applied to obtain an image where the edges are revealed.

Step 2 Cells identification. This is an interactive step. The user selects/validates the cells with no binarization/border defects which later on will be measured.

Step 3 Measuring. Automatic measurement of the size and the anisotropy ratio for each selected cell.

Step 4 Calculations of the cell distribution parameters, as the homogeneity of the cell size distribution, asymmetry coefficient and the average three-dimensional cell size.

Figure 8.20 shows an example of such a characterization (Shishkina 2014)

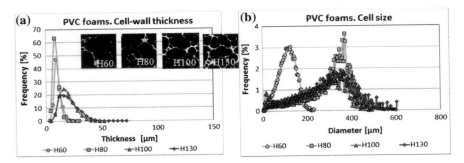

Fig. 8.20 Distributions of local cell-wall thickness **a** and cell diameter **b** in PVC foams; Hxx designates the average foam density of xx •10-3 g/cm^3 (*inset* sections of a micro-CT images of the foams) (Shishkina 2014)

8.4.2 CT of Random Fibre Composites

Random fibre composites are reinforced by discontinuous fibres, which are characterised by their orientation distribution (or orientation tensor), length distribution and waviness. Processing of micro-CT images of random fibre composites aims to measure these characteristics, or to reconstruct the detailed geometry of the fibres.

The most of the published work and software tools deal with assemblies of straight fibres, which is the most practically relevant case. Methods based on direct identification of the individual fibres orientation can utilize the methods well-developed for physical sample sectioning (Mlekusch 1999)and for measuring the elliptical fibre cross-sections on the slices (Thi et al. 2015). If the fibrous reinforcement has planar fibre arrangements, 2D image analysis algorithms to identify fibre-like objects can be used (Graupner et al. 2014; Fliegener et al. 2014).

For a 3D fibre arrangements a general-purpose micro-CT analysis software VGStudio MAX has an add-on module which allows analysis of composite components non-destructively by use of computed tomography data. With the Fibre Composite Material Analysis Module the following parameters can be calculated: local and global fibre orientation, local and global fibre concentration, deviation from predefined reference orientation, local fibre orientation in a plane projection, and other statistical parameters such as fibre distribution. This calculation is based on the analysis of the local variation of the grey scale image field, for example, using Hessian matrix or Hough transformation.

The most advanced tools, as *FibreScout* (Weissenbock et al. 2014) (see Fig. 8.21), combine segmentation of the micro-CT image down to individual fibres, identification of the fibre orientation and length with classification of the fibres and fibre regions. The latter, in combination with the visualization of the fibres metadata, allows visual inspection of the positioning of fibre groups and the calculation of fibre parameters statistics, which can be further used in modelling of the composite mechanical properties.

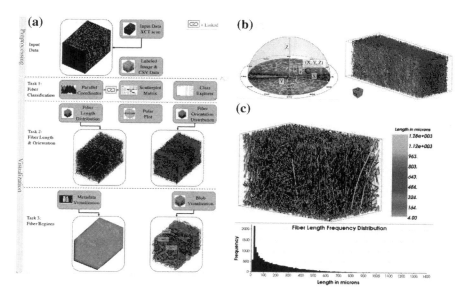

Fig. 8.21 Analysis of a random fibre composite with FibreScout (Weissenbock et al. 2014): **a** general analysis scheme; **b** fibre orientations; **c** fibre length

Analysis of micro-CT images of composites reinforced with long discontinuous wavy fibres, as well as non-woven textile materials, face the problem of identification of the shape of the wavy fibres. This can be done using the expression (Abdin et al. 2014)

$$r(s) = A\left(\mathbf{r}_1 \sin\left(n_1 \frac{\pi s}{L} + \psi_1\right) + \mathbf{r}_2 \sin\left(n_2 \frac{\pi s}{L} + \psi_2\right)\right), \qquad (8.2)$$

where $r(s)$ is the radial position in relation to a certain axis, s the coordinate along the curved fiber axis, and other parameters are randomly generated amplitude, period and phase of two harmonics describing the fibre shape. Identification of the parameters is done in (Abdin et al. 2014) based on the segmentation of the micro-CT image with *Mimics* software. It allows creation of random instances of the fibrous assembly with the identified parameters, which realistically represents the micro-CT observations, as shown in Fig. 8.22.

8.4.3 CT of Textiles and Textile Composites

The last decade saw a rapid growth of µCT applications for textiles and textile composites (Desplentere et al. 2005; Blacklock et al. 2012; Harjkova et al. 2014; Naouar et al. 2014; Pazmino et al. 2014; Barburski et al. 2015). This growth was in

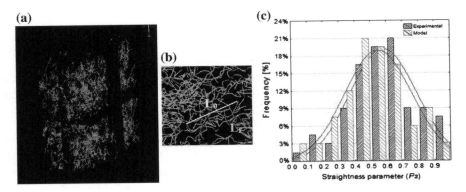

Fig. 8.22 A model of internal geometry of random wavy steel fibre reinforced composite: **a** micro-CT image, processed with Mimics; **b** identification of individual fibres, fibre chord length L_0 and fibre length L_f; **c** comparison of fibre straightness (= $L0/Lf$, see **b**) as measured on the micro-CT image and simulated with the model

line with the fast development of the μCT hardware. The X-ray computer tomography apparatus, from desktops to synchrotron-based instruments, allow the registration of 3D images with submicron and micron resolution, with a contrast sufficient to distinguish organic and carbon-based materials (like polymer matrices and carbon fibres) and the image post-processing is effective in eliminating the imaging artefacts. Figure 8.23 depicts a gallery of micro-CT images different textiles and textile composites.

Micro-CT images are used as virtual counterpart of physical cross-sectioning of textile structures, as a tool to produce slices and measure features of the textile internal structure (Desplentere et al. 2005; Harjkova et al. 2014), including its changes after the textile deformation (Naouar et al. 2014; Pazmino et al. 2014; Barburski et al. 2015). The precision of micro-CT registration allows characterization of statistical parameters of variability of the yarn placement (Blacklock et al. 2012; Bale et al. 2012; Vanaerschot et al. 2016).

Beautiful 3D images, however, are not used in full as the virtual representation of the internal structure of textiles and textile composites, because the voxel structure of the image lacks directionality, which is the paramount local characteristic of fibre reinforced composites and fibrous assemblies. Directionality defines the local axes of anisotropy, which, in turn, lead to anisotropic properties of the dry or impregnated fibrous bundle or a ply (like the local permeability and the local stiffness tensor). Another important use of the directionality lies in the fact that it (or rather the local anisotropy) provides an additional image feature for segmentation, together with the grey scale values of the image voxels. The segmentation or thresholding based on the grey scale only is prone to uncertainties, which are made worse by similarity of the density and chemical composition of carbon or organic or cellulose fibres and organic matrices. Addition of the second feature—anisotropy—allows the use of powerful segmentation methods based on cluster analysis, bringing a decisive improvement of the segmentation precision. Moreover, the

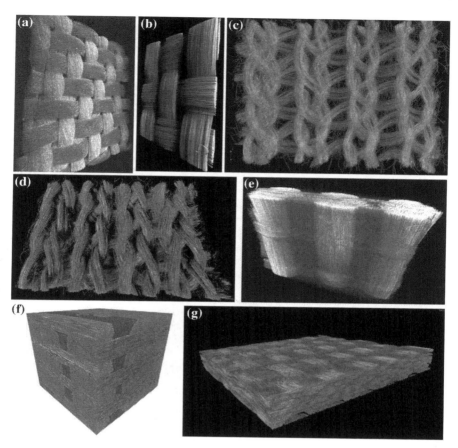

Fig. 8.23 Micro-CT images of textiles and textile composites: **a** flax; and **b** glass fibre woven fabrics; **c** glass; and **d** steel fibre knitted fabrics; **e** glass and **f** carbon fibre 3D woven fabrics; **g** alumina fibre woven laminate

identification of the local directionality makes the identification of yarns of different directions (warp vs weft in woven fabrics, braiding vs axial yarns in braids etc.) and the analysis of the local fibre misalignment possible.

The VoxTex software (KU Leuven) employs methods of 3D image processing, which use in full the local directionality information, retrieved using the analysis of the local structure tensor (Straumit et al. 2015). The processing results in a voxel 3D array (the voxel dimensions can be the same as in the initial μCT image or larger), with each voxel carrying information on (1) material type (matrix; yarn/ply, with identification of the yarn/ply in the reinforcement architecture; void) and (2) fibre direction for fibrous yarns/plies. The knowledge of the material phase volume and the known characterization of the textile structure allows as-signing to the voxels (3) the fibre volume fraction. With this basic voxel model built, it can be further

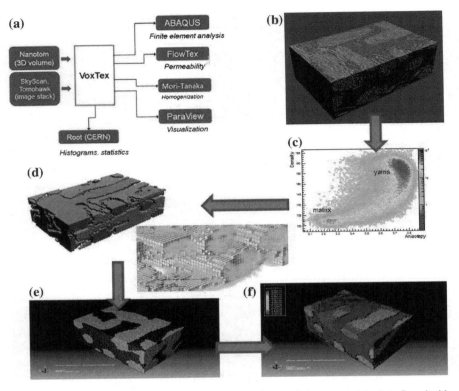

Fig. 8.24 VoxTex software: sources of the input and use of the output (**a**); data flow inside VoxTex: micro-CT image (**b**), segmentation (**c**), the segmented image and voxel directions (**d**), the FE model (**e**) and the solution (**f**)

used for different type of the material analysis, as shown in Fig. 8.24. The image segmentation gives direct values of the material porosity and the positioning of voids. The segmented voxel model is visualized via Para-View. The information of the direction of the fibres is processed to yield histograms of the fibre misalignment angles and the deviations of the yarns from the nominal directions.

After segmentation into voxels representing inter-yarn space (matrix and voids) and yarn volumes the voxel model is ready to be transferred to a CFD software to calculate the homogenized permeability solving Navier-Stokes or Stokes equations. Intra-yarn/ply permeability can be taken into account by assigning local permeability to voxels, belonging to yarn/plies volumes, using the homogenization formulae (Gebart 1992; Endruweit et al. 2014) and the information on the fibre volume fraction and the fibre direction in the voxel. The Brinkmann equation is solved in this case. VoxTex is integrated with FlowTex solver of KU Leuven (Verleye et al. 2008) for such calculations. In (Straumit et al. 2016) the possibility of correct (within the experimental scatter) calculation of a textile reinforcement permeability based on X-ray micro-computed tomography registration of the textile internal architecture is demonstrated on an example of carbon fibre non-crimp fabrics.

Calculation of mechanical properties of the composite can proceed in two routes. Considering a representative volume of the composite and assigning the mechanical properties to voxels, according to the voxel fibre volume fraction and fibre direction or to properties of the matrix, a homogenization routine (iso-strain, Mori-Tanaka …) can be employed to calculate homogenized stiffness of the composite. These capabilities are integrated in VoxTex. The voxel model with as-signed material properties per voxel is ready to be transferred to FE software. VoxTex produces an ABAQUS input file for the voxel model, or invokes its own FE solver. Using a continuum damage model for unidirectional fibre reinforced material, the calculation can include the damage initiation and growth and simulate a reduction of the load-carrying ability of the textile composite with damage progression.

8.5 CT to Optimize the Next Generation Biomaterials

Instead of classical treatments for large bone trauma's, which focus on the reconstruction of bone using for example an Ilizarov fixator (Fabry et al. 1988), bone tissue engineering (TE) emphasizes on regeneration of the tissue (Langer and Vacanti 1993) using. The combination of open porous biomaterials, also called bone TE scaffolds, with osteogenic cells and/or growth factors can, after a process of cell seeding and/or bioreactor culture, be implanted in large bone defects for healing (Janssen et al. 2006; Timmins et al. 2007; Bakker et al. 2008; Cancedda et al. 2007; Kruyt et al. 2008; Salgado et al. 2004). For the healing to be successful, a bone TE scaffold needs to fulfil certain requirements, such as biocompatibility, favourable cell-material interactions, optimal pore structure and pore dimensions, proper surface macro- and microstructure, suitable chemical composition and dissolution/degradation behaviour, mechanical properties equivalent to bone, etc. (Place et al. 2009; Barradas et al. 2011; Habraken et al. 2016; Habibovic and de Groot 2007). Ideally, the scaffold meets all these requirements, but in most cases compromises have to be made between the different constraints. Thus, as most of these characteristics are coupled, scaffold design, production, characterization and translation to clinical applications remain a challenging task (Place et al. 2009; Barradas et al. 2011). Additionally, existing bone TE strategies still suffer from a limited reproducibility of results and lack of robustness against external perturbations owing to, amongst others, a large variability in material properties due to inconsistencies in the production processes. Reducing the material variability is imperative in order to improve the predictability of the outcome of current TE strategies, hereby avoiding a trial-and-error approach and enabling a more reliable translation to the clinics.

Therefore, suitable characterization for screening and quality control of biomaterials is necessary to provide input for the design and fabrication of optimized TE scaffolds, aiming at an increase in repeatability and robustness avoiding trial-and-error. Micro-CT can provide a solution as it is a non-destructive 3D imaging technique from which an extensive set of morphological parameters can be determined. Micro-CT has been frequently reported in the literature as a relevant tool for the characterization of the 3D morphology of scaffolds (van Lenthe et al. 2007; Peyrin et al. 2007; Jungreuthmayer et al. 2009), and bone formation within (Guldberg et al. 2008; Papadimitropoulos et al. 2007; Chai et al. 2012; Jones et al. 2004; Jones et al. 2007). In what follows, some examples are provided to show the added value of micro-CT compared to standard characterization techniques, as well as to highlight the potency of contrast-enhanced micro-CT (CE-CT) for the evaluation of soft tissue formation in the biomaterials.

8.5.1 Materials Characterization

8.5.1.1 Morphological Analysis

Apart from allowing non-destructive visualization of the internal structure of porous biomaterials, micro-CT image analysis also allows to quantify the morphological properties of these materials. In a recent study, it was shown that by combining micro-CT morphological characterization and empirical modelling as an innovative and robust screening approach, the material-specific variability for cell-based TE scaffold design could be reduced (Kerckhofs et al. 2016). Via micro-CT characterization a quantitative scaffold library of morphological and compositional properties was built of six CE approved CaP-based scaffolds (CopiOs®, BioOss™, Integra Mozaik™, chronOS Vivify, MBCP™ and ReproBone™—Figs. 8.25 and 8.26), and of their bone forming capacity (cf. paragraph 8.5.3) and in vivo scaffold degradation when combined with human periosteal derived cells (hPDCs) was assessed. The empirical model, based on the micro-CT based scaffold library, allowed identification of the construct characteristics driving optimized bone formation, i.e. (a) the percentage of β-TCP and dibasic calcium phosphate, (b) the concavity of the CaP structure, (c) the average CaP structure thickness and (d) the seeded cell amount (taking into account the seeding efficiency). Additionally, the model allowed to quantitatively predict the bone forming response of different hPDC-CaP scaffold combinations, thus providing input for a more robust design of optimized constructs and avoiding trial-and-error.

8.5.1.2 Surface Properties

It has been shown that also the surface roughness and topology of biomaterials can have an important effect on the cell proliferation and differentiation, and as a result the osteoinductive capacity of the scaffold (Salgado et al. 2004; Habibovic et al. 2006; Habibovic and de Groot 2007; Mustafa et al. 2001; Galli et al. 2005).

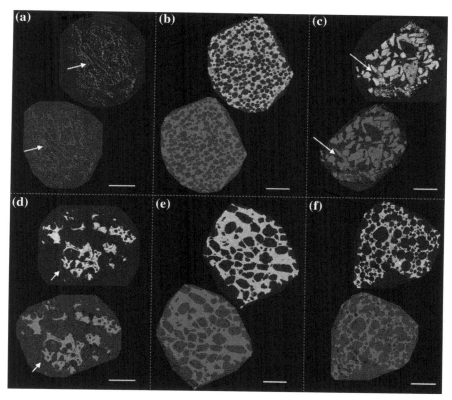

Fig. 8.25 ReproBone™ scaffold, along with their corresponding segmented images (*bottom*). In the segmented images, the ROI drawn around the scaffold can also be depicted. The *white arrows* indicate the collagen network inside the scaffold. Scale bar = 1 mm

However, commercially available profiling systems fail when determining the surface roughness of porous materials. High-resolution micro-CT imaging has provided a solution for this, as it was shown to allow quantification of the surface roughness of 3D porous structures in a non-destructive manner (Kerckhofs et al. 2012). More in detail, 2D cross-sectional micro-CT images of the porous structures have been used to extract the profile lines of the biomaterial surface (Fig. 8.27), both at the outer surface as well as inside the structure. These profile lines were then used to calculate the surface roughness parameters using a MatLab tool. Comparing the roughness parameters of flat substrates determined both with commercially available (optical and contact) profiling systems and the micro-CT-based roughness measurement protocol showed that micro-CT can be applied accurately and in a robust manner for surface roughness quantification of 3D complex porous materials with a micro-scale roughness. This tool will allow to determine the influence of the surface topology on the cell behaviour and the bone forming capacity of the scaffold.

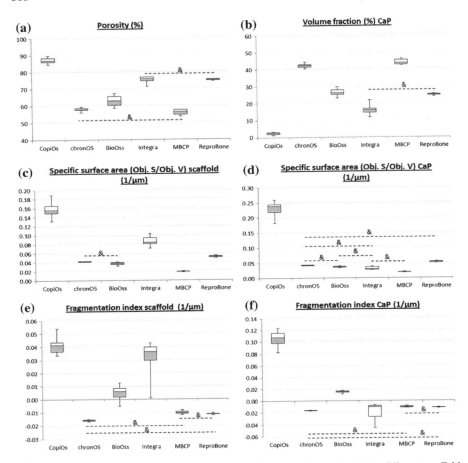

Fig. 8.26 (Kerckhofs et al. 2016). Micro-CT based structural analysis of the different scaffold types ($n = 12$): **a** porosity of the full scaffold, **b** volume fraction of the CaP structure, **c** fragmentation index of the full scaffold, **d** fragmentation index of the CaP structure, **e** specific surface area of the full scaffold and **f** specific surface area of the CaP structure. All the results are expressed in a box plot. '&' indicates non-significant differences ($p > 0.05$)

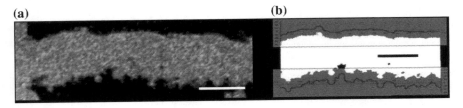

Fig. 8.27 Kerckhofs et al. (2012). A typical high resolution (voxel size = 1.5 μm) 2D micro-CT cross-sectional image of a single strut of a porous Ti6Al4 V biomaterial and (**b**) a binarized section of (**a**) with the corresponding profile lines. Scale bars = 200 μm

8.5.2 *In Vitro Evaluation*

In order to further advance the field of bone TE, another important aspect is to identify the most optimal bioreactor process parameters. To be able to do this, amongst others it is important to visualize and quantify the neo-tissue growth (i.e. cells and extracellular matrix) into the 3D biomaterial. Most currently used standard techniques, such as histological sectioning and Live/Dead assay, only allow assessing the neo-tissue distribution in two dimensions, with a limited depth resolution or with a loss in information. Also here, micro-CT could play an important role. However, when using desktop micro-CT in standard absorption mode, without the use of a contrast agent, it has not been possible to visualize non-mineralized tissues in 3D scaffolds. Contrast-enhanced CT provides a valuable solution, as it allows visualizing in 3D engineered neo-tissue formation in TE scaffolds after bioreactor culture (Sonnaert et al. 2015). The use of contrast agents to stain the neo-tissue in combination with micro-CT image analysis provides 3D quantitative read-outs regarding the amount of neo-tissue and its distribution throughout the TE scaffold during bioreactor based cell culture (Fig. 8.28), and thus allows optimizing the bioprocess parameters. The development of robust tools and methodologies such as CE-CT to assess important and potentially critical quality characteristics of TE constructs (i.e. biomaterials with cells/matrix), such as neo-tissue quantity and homogeneity, can facilitate the gradual transformation of 'TE constructs' in well characterized 'TE products'.

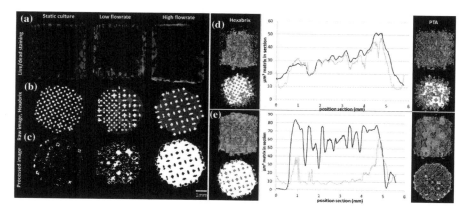

Fig. 8.28 a Live/dead staining of constructs cultured in different conditions for 21 days, **b** corresponding 2D grey-scale CE-CT cross-sections using Hexabrix®, **c** the corresponding binarized and processed cross-sections, **d** CE-CT results for a construct cultured at 4 ml/min for 14 days **e** CE-CT results for a construct cultured at 0.04 ml/min for 28 days; from *left* to *right*: Hexabrix 3D reconstruction and 2D cross-section, representative longitudinal neo-tissue distribution throughout the complete TE construct i.e. volume of neo-tissue per slice, with a slice thickness of 3.75 μm, in function of the scaffold height (*black* Hexabrix, *grey* PTA), PTA 3D reconstruction and 2D cross-section

8.5.3 In Vivo Evaluation

In order to evaluate the potential of a biomaterial, typically it will be implanted in a rodent model and the tissue formation inside is evaluated. However, like in organs, tissues formed in biomaterials after in vivo implantation have a spatial heterogeneity. Thus, measurements made in 2D, like histological sectioning, only partially reveal the degree of change induced during development. Innovations in imaging techniques are fundamental to fully understand the complex 3D events during tissue formation. As mentioned earlier, micro-CT has been frequently to quantify bone formation within a biomaterial (Guldberg et al. 2008; Papadimitropoulos et al. 2007; Chai et al. 2012; Jones et al. 2004; Jones et al. 2007; Kerckhofs et al. 2016). However, soft tissue contrast is inherently poor. As shown before, a recent shift in micro-CT imaging focuses on the use of X-ray opaque contrast agents for visualizing soft tissues, such as neotissue (Sonnaert et al. 2015), cartilage (Kerckhofs et al. 2013, 2014; Bansal et al. 2011; Xie et al. 2010), blood vessels (Lusic and Grinstaff 2013; Fei et al. 2010) and connective tissues (Wong et al. 2012; Metscher 2009). For the other soft tissues, however, there are currently no tissue-specific X-ray opaque contrast agents. Using trial-and-error, several groups have assessed different chemicals to stain these tissues with varying degrees of success, but without in-depth validation or sufficient knowledge of the tissue-specific binding mechanisms (Metscher 2009; Pauwels et al. 2013). Additionally, most of those contrast agents are invasive or toxic (Buytaert et al. 2014), preventing subsequent histological processing. Recently, novel contrast agents (metal-substituted polyoxometalates—POMs) have been developed to visualize the adipocytes and blood vessel network in the bone marrow compartment (Fig. 8.29) in a non-invasive manner allowing subsequent immunological staining

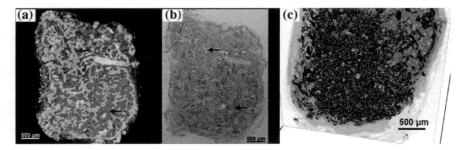

Fig. 8.29 A typical **a** cross-sectional CE-CT image (using POM staining—light grey is bone, darker grey is bone marrow and the black arrows indicate the blood vessels and adipocytes) of a TE explant (CaP-containing scaffold coated with a growth factor and seeded with hPDCs after 6 weeks of ectopic implantation in a nude mouse model) and **b** the corresponding histological section (Haematoxylin and Eosin staining—*pink* is bone, *blue* is bone marrow, *white* areas are blood vessels or fat cells). **c** A 3D representation of the blood vessel network (in *red*) in the bone marrow of the explants. In greyscale, a cross-section through the explants is shown (grey-scales are inverted compared to (**a**))—courtesy of Dr. Erwan Plougonven—ULg)

(Kerckhofs et al. 2016). These contrast agents, combined with micro-CT imaging, will allow visualizing and quantifying simultaneously multiple tissues with only one imaging modality in a faster, less invasive and more quantitative manner than histology. As a result, using CE-CT, a more in-depth and 3D quantitative screening of a TE construct can be performed.

8.5.4 Conclusion

To conclude, micro-CT is a promising imaging technique not only for the morphological characterization of biomaterials for bone tissue engineering, but also for the characterization of their surface topology, and for the formation of bone and other soft tissues within the biomaterial after in vivo implantation in an animal model. Hence, it is the ideal tool to advance and optimize existing TE therapies and to facilitate a more reliable translation to the clinics.

8.6 CT to Optimize Dental Restorations and to Evaluate Dental Implants

Within the dental profession and oral health research community, there exists an ongoing interest to develop restorations with improved mechanical properties and that are easy to place. Even though it is often stipulated that clinical research is the ultimate test for dental restoratives, laboratory tests are an essential tool in understanding and improving fundamental issues. In this aspect, the advent of micro-computed tomography has increased the possibilities of testing dental materials tremendously. It offers a non-destructive 3D-evaluation of surface structures as well as internal structures on a microscopic level; this cannot be achieved with imaging techniques in vivo, due to the high radiation doses and the limitations inherent to the dimensions of the scanned object. Due to recent improvements in the resolution of commercially available desktop scanners (Kampschulte et al. 2016), as well as in image processing techniques, the number of applications is considerably increasing. An overview of several applications will be discussed here, together with their possible advantages and pitfalls.

8.6.1 Finite-Element Analysis

Stress and strain within the restored tooth complex are of major concern, since they are related to cracks and fractures within the restoration, the remaining tooth tissue, or the interface between both substrates, ultimately leading to clinical failure

(Ferracane 2008; Schneider et al. 2010; Zhang et al. 2013). Due to the anatomical shape and layered structure of teeth, the stress and tensor fields very complex and does not allow simplification; hence, finite element analysis is often employed to visualize and interpret the stress distributions. Geometry acquisition and modification of FE models have been enormously facilitated by implementation of micro-CT images (Verdonschot et al. 2001; Magne 2007; Kato and Ohno 2008; Rodrigues et al. 2009). Micro-CT images of natural teeth and restorations can be segmented, based on image density thresholds. Filtering and smoothing of the images is often required to reduce noise and scanning artifacts and to optimize the meshing procedure (Magne 2007; Rodrigues et al. 2009). Using micro-CT based models, the effect of different material's properties, such as shrinkage of resin composites during the polymerization (Kuijs et al. 2003) and elasticity of composites and ceramics (Della Bona et al. 2013; Duan and Griggs 2015), as well as their geometric designs (Rodrigues et al. 2012), has been evaluated in various studies.

8.6.2 Quantification of Voids, Gaps and Microleakage

Gap formation and marginal adaptation at the interface between tooth and restoration is of great importance, because it may result in recurrent caries and pulpal complications (Roulet 1994). Air bubbles, voids and other flaws within the bulk of the restoration or within the adhesive cement, at the other hand, may act as critical flaws, were cracking and fracturing is initiated. Various methods have been used to evaluate marginal adaptation and microleakage (Roulet 1994; Heintze 2007; Heintze and Zimmerli 2011). In vivo, these evaluations are severely limited by the access to the surface, which is basically restricted to the occlusal, vestibular and lingual surfaces. Evaluation then occurs through visual inspection, whether with aid of an explorer or not (Hickel et al. 2007, 2010a, b), or indirectly by silicone impressions and resin replicas (Contrepois et al. 2013). In most in vitro studies, the specimen is sectioned and either gaps or (micro-)leakage of a colored dye is quantified on the retrieved slices (Neves et al. 2014). Justified criticisms have been made that the number of measurements that can be retrieved this way is limited, and the outcome is highly dependent on the position and number of sections that can be made (Neves et al. 2014), since microleakage is proven to be highly uniform (Sun et al. 2009). Moreover, this technique inevitably destructs the sample, which has two main disadvantages: firstly, the procedure cannot be repeated; secondly, the sectioning itself can cause specimen processing artifacts induced during the sectioning, which can be mistaken for originally present flaws.

When micro-CT is used to study microleakage, the tracer needs to be radiopaque. Silver nitrate has been used as a tracer in various studies, with varying degrees of success (Neves et al. 2014; Han et al. 2015; Jacker-Guhr et al. 2015; Kim and Park 2014; Eden et al. 2008). However, other authors assumed that microleakage could be predicted by calculating interfacial gaps, without prior use of a tracer (Sun et al. 2009). Microleakage protocols were often combined with mechanical loading protocols, to determine their influence on the microleakage patterns. Based on the numerous different applications, micro-CT has been proven a versatile tool in the evaluation of internal and marginal adaptation.

However, the use of micro-CT also has some disadvantages: the spatial resolution obtained is not only determined by the specifications of the device, but also by the dimensions of the sample. In most studies, the used voxel size varies between 10 μm and 20 μm, which is a much lower accuracy than can be obtained with SEM or TEM. The contrast resolution is also dependent on the density and the linear attenuation of the used materials; most adhesives are rather translucent, approximating the values of air, making it impossible to distinguish on the scans (Carrera et al. 2015). On the other hand, severe beam hardening, scattering and diffraction induced noise can occur with materials with more radiopaque restorative materials, principally metals and zirconia-based ceramics (Tan et al. 2011). Low contrast between different substances and noise and artifacts, may render automatic thresholding very difficult, if not impossible, due to a high amount of overlap of the substances' peaks on the histograms. The researcher is than often reliant on manual segmentation methods (Borba et al. 2011), which are inevitably susceptible to interpretive bias. Beam hardening can be avoided by using a monochromatic beam, such as in synchrotron micro-CT (De Santis et al. 2005; however, the availability of this equipment is limited. When using a polychromatic beam, a filter—usually alumina or copper—can be used to reduce beam hardening effects.

8.6.3 Image Correlation and Image Registration

As previously mentioned, stress and strain originating from polymerization stress and mechanical loading have been widely studied. The non-destructive nature of micro-CT allows a sample containing resin composite, to be scanned before and after polymerization (Hirata et al. 2014; Chiang et al. 2010; Cho et al. 2011; Zeiger et al. 2009), or before and after mechanical loading (Kim and Park 2014). This is a very interesting asset to verify shrinkage or loading effects in an experimental setting. The benefit of using a 3D technique is that not only the volumetric shrinkage can be calculated, but the location of the shrinkage can be verified (Hirata et al. 2014), which is possibly even more important. Images before and after the challenge can then be aligned—in case the sample has moved between scans—and can be compared to each other. In image registration, an algorithm is used to automatically align both images. After alignment, images can be subtracted (Hirata et al. 2014) or tensor fields representing the occurred deformation can be calculated (Chiang et al. 2010; Cho et al. 2011; Van Ende et al. 2013).

8.6.4 Conclusion

Continuous improvement of image resolution and image processing techniques, have resulted in an increase of useful applications of micro-CT to optimize dental restorations. The most important assets are related to the ability to evaluate the sample internally, on a microscopic level in a non-destructive manner.

Borba M., Cesar P.F., Griggs J.A., Della Bona Á. Adaptation of all-ceramic fixed partial dentures.

8.7 CT for Research on Dental Hard Tissues

In recent years, micro-CT is useful in a wide variety of applications in dentistry, especially when analyzing radiopaque structures (Swain and Xue 2009; Zou et al. 2011; Davis et al. 2013). In particular, micro-CT imaging allows 2D and 3D observation of the internal and external tooth morphology and pathology, whereas quantitative and qualitative analysis of density or mineral concentration distribution (mineral content) is feasible (Davis et al. 2013). Some examples of micro-CT applications on dental hard tissues research are given in the following paragraphs.

8.7.1 Visualization of the Morphology and Analysis of Tooth Structure

Micro-CT is an excellent tool for the analysis of tooth structure, as it helps to differentiate between enamel, calculus, dentin, cementum and pulp (Fig. 8.30). Furthermore, by 3D reconstruction and analysis of the obtained 2D sections of the tooth, important information can be drawn on the pulp cavity and the root canal morphology (Fig. 8.31). The detailed information of the root canal morphology is essential for the dentists, in order to provide a successful endodontic treatment. The reconstructed sections can be visualized in all orientation planes and thus, an insight into normal and complex morphologic root canal variations (ramifications, isthmuses, C-shape configuration etc.) can be achieved. The root and root canal curvature, together with root canal diameter and configuration can be visualized and measured (Peters et al. 2000; Oi et al. 2004).

Furthermore, by using the 3D data it is also possible to show the tissue loss due to endodontic instrumentation (increase of the root canal volume). The post-treatment volume can be subtracted from the pre-treatment volume, to yield the tooth structure removed. In this way different shaping and cleaning techniques of the root canal can be compared and correlated, in order to evaluate the effectiveness and the invasive nature of each technique (Bergmans et al. 2001, 2002a, 2003).

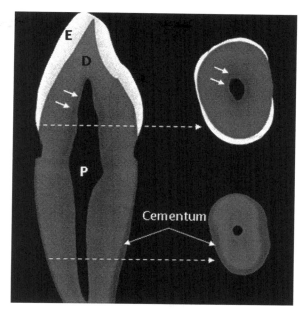

Fig. 8.30 Single slices of a tooth. Enamel (*white*) has a higher mineral concentration than dentin (*grey*). Observe the difference of the mineral composition in the coronal part of the pericanalar dentin (*arrows*) and that of Cementum (*C*), Enamel (*E*), Dentin (*D*), Pulp (*P*)

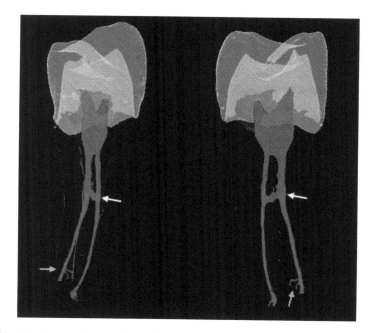

Fig. 8.31 3D volume rendering of a tooth. Enamel (*white*), dentin (*grey*) and root canal (*red*) can be segmented separately in order to highlight the tooth morphology and the root canal complexity. Observe the interradicular anastomoses (*white arrows*) and periapical delta (*green arrows*)

8.7.2 *Visualization and Analysis of Tooth Resorption*

Micro-CT is extremely useful in understanding complex phenomena in dentistry, such as tooth resorption (Bergmans et al. 2002b; Luso and Luder 2012; Gunst et al. 2013; Mavridou et al. 2016a). There are different types of tooth resorption and the most complex one is external cervical resorption. By using CT scanners with advanced technical properties at a micro-CT resolution range, it is possible to segment enamel, dentin, pulp and bone into different parts, based on pixel grey values or mineral density (Mavridou et al. 2016a). 2D histological images match perfectly with the microtomographical images (Fig. 8.32). However, by using 3D images important structural aspects can be derived, which could not be evaluated only by using conventional histomorphometry. In particular, the high spatial resolution and image quality makes it possible to assess: the initiation point of the resorption (portal of entry), the Heithersay resorption channels inside dentin and enamel, the pulp reaction through calcification, the pericanalar resorption resistant Sheet (PRRS) which surrounds the root canal, the extent of bone-like tissue formation inside the tooth and the interconnections with the external root surface (Fig. 8.33). The 3D thickness of the PRRS can be measured by selecting a region of interest and by using the structural analysis routines of specialized software (Mavridou et al. 2016b). Accurate models representative of this condition can be used to get a better insight on the evolving phenomena and would significantly help dental researchers and clinicians to understand the complexity of this condition.

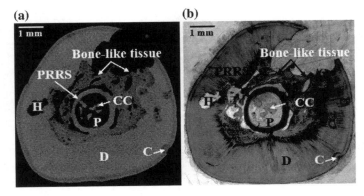

Fig. 8.32 **a** 2D tomographical trasaxial slice of a tooth match the **b** histological image. Enamel (*E*) dentin (*D*), pulp (*P*), cementum (*C*), pericanalar resorption resistant sheet (*PRRS*), Heithersay resorption channel (*H*), intracanal calcifications (*CC*) (Mavridou et al. 2016a)

Fig. 8.33 3D models of tooth with external cervical resorption showing the complexity of external cervical tooth resorption. External tooth surface morphology can be correlated with the internal tooth structure. Enamel (*E*) Dentin (*D*) Pulp (*P*) (Mavridou et al. 2016b)

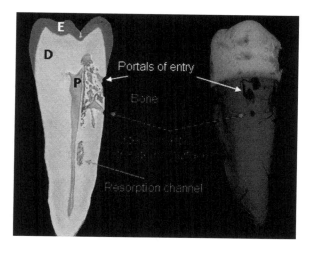

8.7.3 Visualization and Analysis of Tooth Caries

Micro-CT has attracted the interest in caries research due to its non-destructive nature and the possibility to measure the mineral concentration in dental hard tissues (Huang et al. 2010; Neves et al. 2010; Taylor et al. 2010).

Early detection of dental caries and precise caries diagnostic methods are important in order to identify lesions before cavitation while the enamel surface layer is still intact and to determine whether preventive care (remineralization of the damaged tissue) or operative care (removal of the carious tissue) is necessary.

By using micro-CT, demineralized tissue can be discerned from sound dental tissue (Fig. 8.34). In addition, it is possible to investigate the de-/remineralization of the tooth structure, which is associated with the initiation, progression and reversal of carious lesions. In this way, micro-CT can be useful to validate the outcome of diagnostic tests for caries extension assessment. Furthermore by using the 3D data, the volume of the tissue removed during caries excavation can be investigated. In this way different caries excavation techniques can be evaluated in terms of effectiveness and minimal-invasiveness potential (Neves et al. 2011).

Despite the high applicability of micro-CT in mineral density determination, even more promising results can be obtained if further optimization of the scanning and reconstruction process is achieved. This optimization is believed to be achieved by suitable calibration of the mineral density, beam filtration during scanning and beam hardening correction during reconstruction (Zou et al. 2009, 2011).

Fig. 8.34 Enamel
demineralization (*black
arrows*), intact surface layer
(*green arrow*) and dentin
demineralization (*white
arrows*). Pulp reaction with
calcification (*red arrow*).
Calculus (*yellow arrow*)

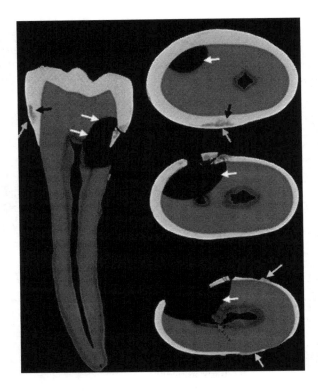

8.7.4 Visualization and Analysis of Microcracks

Novel non-destructive methods can visualize the development of microcracks in a
tooth. Some typical examples of visualisation of such microcracks can be seen in
the root-dentin (Shemesh 2016; Versiani et al. 2015) or in the crown of the tooth
(Fig. 8.35). In the first case the microcracks were induced by root canal instrumentation and filling techniques, whereas in the second case by restoration or
occlusion forces. The use of micro-CT is considered essential as they can help the
clinician to select the less invasive instrumentation and obturation techniques and
understand the etiology of their formation.

8.7.5 Visualization of Tooth Development Anomalies:
Enamel Invagination and Evagination

Dens invagination occurs in different clinical and radiographic forms.
Cone-beam-CT helps to define the type and extent of the invagination and the
relation to the pulp tissue (Neves 2013). Dens evaginatus is an uncommon anomaly
exhibited by protrusion of a tubercle from occlusal surfaces of posterior teeth, and

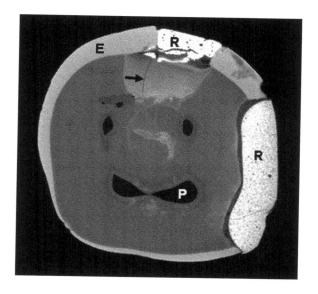

Fig. 8.35 Example of microcrack formation in the crown of a tooth. Microcrack (*black arrow*), enamel (*E*), dentin (*D*), pulp (*P*), restoration (*R*)

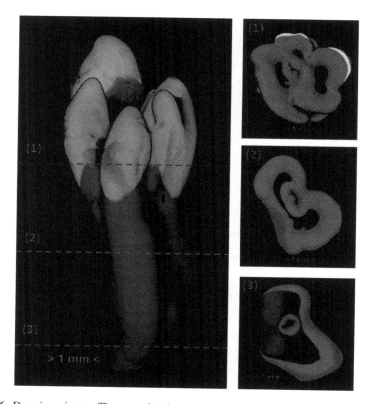

Fig. 8.36 Dens invaginatus. The enamel (*white*) and dentin (*grey*) invagination toward the pulp creates three dimensional complex clinical and endodontic problems. Invagination (*red arrow*)

lingual surfaces of anterior teeth. It occurs primarily in Asian people (6.3%). Micro-CT reveals the complexity of these anomalities and helps to develop new treatment strategies (Fig.8.36).

8.8 Conclusion

The search for the hierarchical structure of complex material systems is one example but the use of advanced image analysis for the determination of global and local strains during in situ loading (mechanical or thermal) of materials in the nano- and μCT is another. This information is a prerequisite to study and unravel by this new method the mechanical behaviour and failures but also to use the input for the modelling.

The research results presented on fruit, dental composites, building materials, radioactive waste, open porous structures and biomaterials - be it for quality control, material behaviour or optimization - underline the need and advantages of this non-destructive technique for a broad science community in which scientists also have the opportunity to learn from one another.

Acknowledgements The authors acknowledge financial support of the Flanders Agency for Innovation and Entrepreneurship (SBO project TomFood 120033), the Hercules foundation, the European Union's Seventh Framework Programmes (FP7 project PicknPack 311987 and FP7 project QUICOM 314562 and ITN Network INTERAQCT) and the FWO Research Foundation—Flanders.

References

Abdin Y, Lomov SV, Jain A, Van Lenthe H, Verpoest I (2014) Geometrical characterization and micro-structural modelling of short steel fiber composites. Compos A 67:171–180

Abera MK, Verboven P, Herremans E et al (2014) 3D virtual pome fruit tissue generation based on cell growth modeling. Food Bioprocess Technol 7:542–555. doi:10.1007/s11947-013-1127-3

Ahn S, Jung SY, Lee SJ (2013) Gold nanoparticle contrast agents in advanced X-ray imaging technologies. Molecules 18:5858–5890. doi:10.3390/molecules18055858

Akbarabadi M, Saraji S, Piri M (2014) High-resolution three-dimensional characterization of pore networks in shale reservoir rocks. In: Unconventional Resources Technology Conference (URTeC), p 1870621

Alajmi AF, Grader A (2009) Influence of fracture tip on fluid flow displacements. J Porous Media 12(5):435–447

Altman SJ, Aminzadeh B, Balhoff MT, Bennett PC, Bryant SL, Cardenas MB, Chaudhary K, Cygan RT, Deng W, Dewers T, DiCarlo DA, Eichhubl P, Hesse MA, Huh C, Matteo EN, Mehmani Y, Tenney CM, Yoon H (2014) Chemical and hydrodynamic mechanisms for long-term geological carbon storage. J Phys Chem C 118:15103–15113

Aminzadeh B, Chung DH, Bryant SL, Huh C, DiCarlo DA (2013) CO_2 leakage prevention by introducing engineered nanoparticles to the in-situ brine. Energy Procedia 37:5290–5297

Andrew M, Bijeljic B, Blunt M (2015) Reservoir condition pore-scale imaging of multiple fluid phases using X-ray microtomography. J Visualized Exp 96:52440. doi:10.3791/52440

Andries J, Van Itterbeeck P, Vandewalle L, Van Geysel A (2015) Influence of concrete flow on spatial distribution and orientation of fibres in steel fibre reinforced self-compacting concrete. Paper presented at the fib symposium, Copenhagen, Denmark

Anselmetti FS, Eberli GP (1993) Controls on sonic velocity in carbonates. Pure Appl Geophys 141 (2–4):287–323

Archie GE (1952) Classification of carbonate reservoir rocks and petrophysical considerations. AAPG Bull 36(2):278–298

Aregawi W, Defraeye T, Verboven P et al (2013) Modeling of coupled water transport and large deformation during dehydration of apple tissue. Food Bioprocess Technol 6:1963–1978. doi:10.1007/s11947-012-0862-1

Bakker A et al (2008) Quantitative screening of engineered implants in a long bone defect model in Rabbits. Tissue Eng Part C 14(3):251–260

Bale H, Blacklock M, Begley MR, Marshall DB, Cox BN, Ritchie RO (2012) Characterizing three-dimensional textile ceramic composites using synchrotron X-ray micro-computed-tomography. J Am Ceram Soc 95(1):392–402. doi:10.1111/j.1551-2916.2011.04802.x

Bansal PN et al (2011) Cationic contrast agents improve quantification of glycosaminoglycan (GAG) content by contrast enhanced CT imaging of cartilage. J Orthop Res 29(5):704–709

Barburski M, Straumit I, Zhang X, Wevers M, Lomov SV (2015) Micro-CT analysis of internal structure of sheared textile composite reinforcement. Compos A 73:45–54. doi:10.1016/j.compositesa.2015.03.008

Barradas AM et al (2011) Osteoinductive biomaterials: current knowledge of properties, experimental models and biological mechanisms. Eur Cell Mater 21:407–429 (Discussion 429)

Baveye P, Rogasik H, Wendroth O, Onasch I, Crawford JW (2002) Effect of sampling volume on the measurement of soil physical properties: simulation with X-ray tomography data. Meas Sci Technol 13(5):775–784

Bear J (1972) Dynamics of fluids in porous media. Courier Corporation

Bech M, Jensen TH, Feidenhans R et al (2009) Soft-tissue phase-contrast tomography with an X-ray tube source. Phys Med Biol 54:2747–53. doi:10.1088/0031-9155/54/9/010

Berg S, Armstrong R, Ott H, Georgiadis A, Klapp SA, Schwing A (2014) Multiphase flow in porous rock imaged under dynamic flow conditions with fast X-ray computed micro-tomography. Petrophysics 55(4):304–312

Bergmans L, Van Cleynenbreugel J, Beullens M, Wevers M, Van Meerbeek B, Lambrechts P (2002a) Smooth flexible versus active tapered shaft design using NiTi rotary instruments. Int Endod J 35(10):820–828

Bergmans L, Van Cleynenbreugel J, Beullens M, Wevers M, Van Meerbeek B, Lambrechts P (2003) Progressive versus constant tapered shaft design using NiTi rotary instruments. Int Endod J 36(4):288–295

Bergmans L, Van Cleynenbreugel J, Verbeken E, Wevers M, Van Meerbeek B, Lambrechts P (2002b) Cervical external root resorption in vital teeth. J Clin Periodontol 29(6):580–585

Bergmans L, Van Cleynenbreugel J, Wevers M, Lambrechts P (2001) A methodology for quantitative evaluation of root canal instrumentation using microcomputed tomography. Int Endod J 34(5):390–398

Beucher S, Meyer F (1992) The morphological approach to segmentation: the watershed transformation. Optical Engineering-New York-Marcel Dekker Incorporated 34:433–433

Blacklock M, Bale H, Begley M, Cox B (2012) Generating virtual textile compo-site specimens using statistical data from micro-computed tomography: 1D tow representations for the Binary Model. J Mech Phys Solids 60(3):451–470. doi:10.1016/j.jmps.2011.11.010

Blunt MJ, Bijeljic B, Dong H, Gharbi O, Iglauer S, Mostaghimi P, Paluszny A, Pentland C (2013) Pore-scale imaging and modelling. Adv Water Resour 51:197–216

Bouvet de Maisonneuve C, Bachmann O, Burgisser A (2009) Characterization of juvenile pyroclasts from the Kos Plateau Tuff (Aegean Arc): Insights into the eruptive dynamics of a large rhyolitic eruption. Bull Volc 71(6):643–658

Borba M, Cesar PF, Griggs JA, Della Bona Á (2011) Adaptation of all-ceramic fixed partial dentures. Dent Mater 27:1119–1126. doi:10.1016/j.dental.2011.08.004

Brabant L, Vlassenbroeck J, De Witte Y, Cnudde V, Boone MN, Dewanckele J, Van Hoorebeke L (2011) Three-dimensional analysis of high-resolution X-ray computed tomography data with Morpho+. Microsc Microanal 17:252–263

Brechbühler C, Gerig G, Kübler O (1995) Parametrization of closed surfaces for 3-D shape description. Comput Vis Image Underst 61(2):154–170

Brigaud B, Vincent B, Durlet C, Deconinck JF, Blanc P, Trouiller A (2010) Acoustic properties of ancient shallow-marine carbonates: effects of depositional environments and diagenetic processes (Middle Jurassic, Paris Basin, France). J Sed Res 80(9):791–807

Brosse E, Magnier C, Vincent B (2005) Modelling fluid-rock interaction induced by the percolation of CO_2-enriched solutions in core samples: the role of reactive surface area. Oil Gas Sci Technol 60(2):287–305

Brunauer S, Emmett PH, Teller E (1938) Adsorption of gases in multimolecular layers. J Am Chem Soc 60(1):309–319

Bultreys T, Van Hoorebeke L, Cnudde V (2015) Multi-scale, micro-computed tomography-based pore network models to simulate drainage in heterogeneous rocks. Adv Water Resour 78:36–49

Buytaert J et al (2014) Volume shrinkage of bone, brain and muscle tissue in sample preparation for micro-CT and light sheet fluorescence microscopy (LSFM). Microsc Microanal 20(4):1208–1217

Cancedda R, Giannoni P, Mastrogiacomo M (2007) A tissue engineering approach to bone repair in large animal models and in clinical practice. Biomaterials 28(29):4240–4250

Cantre D, Herremans E, Verboven P et al (2014) Characterization of the 3-D microstructure of mango (*Mangifera indica* L. cv. Carabao) during ripening using X-ray computed microtomography. Innov Food Sci Emerg Technol 24:28–39. doi:10.1016/j.ifset.2013.12.008

Carlson W (2006a) Three-dimensional imaging of earth and planetary materials. Earth Planet Sci Lett 249:133–147

Carlson W (2006b) Three-dimensional imaging of earth and planetary materials. Earth Planet Sci Lett 249:133–147

Carrera C, Lan C, Escobar-Sanabria D, Li Y, Rudney J, Aparicio C et al (2015) The use of micro-CT with image segmentation to quantify leakage in dental restorations. Dent Mater 2015:1–9. doi:10.1016/j.dental.2015.01.002

Chai YC et al (2012) Ectopic bone formation by 3D porous calcium phosphate-Ti6Al4V hybrids produced by perfusion electrodeposition. Biomaterials 33(16):4044–4058

Chaudhary K, Cardenas MB, Wolfe WW, Maisano JA, Ketcham RA, Bennett PC (2013) Pore-scale trapping of supercritical CO_2 and the role of grain wettability and shape. Geophys Res Lett 40:3878–3882

Choquette P, Pray L (1970) Geologic nomenclature and classification of porosity in sedimentary carbonates. AAPG Bull 54(2):207–250

Chiang Y-C, Rösch P, Dabanoglu A, Lin C-P, Hickel R, Kunzelmann K-H (2010) Polymerization composite shrinkage evaluation with 3D deformation analysis from microCT images. Dent Mater 26:223–231. doi:10.1016/j.dental.2009.09.013

Cho E, Sadr A, Inai N, Tagami J (2011) Evaluation of resin composite polymerization by three dimensional micro-CT imaging and nanoindentation. Dent Mater 27:1070–1078. doi:10.1016/j.dental.2011.07.008

Chung MK, Dalton KM, Davidson RJ (2008) Tensor-based cortical surface morphometry via weighted spherical harmonic representation. IEEE Trans Med Imaging 27(8):1143–1151

Claes S, Soete J, Cnudde V, Swennen R (2016) A 3 dimensional classification for mathematical pore shape description in complex carbonates. Mathematical Geosciences, (accepted)

Clausnitzer V, Hopmans JW (1999) Determination of phase-volume fractions from tomographic measurements in two-phase systems. Adv Water Resour 22(6):577–584

Cnudde V, Boone M, Dewanckele J, Dierick M, Van Hoorebeke L, Jacobs P (2011) 3D characterization of sandstone by means of X-ray computed tomography. Geosphere 7:54–61. doi:10.1130/GES00563.1

Cnudde V, Boone MN (2013) High-resolution X-ray computed tomography in geosciences: a review of the current technology and applications. Earth Sci Rev 12:1–17. doi:10.1016/j.earscirev.2013.04.003

Cole LE, Ross RD, Tilley JM et al (2015) Gold nanoparticles as contrast agents in X-ray imaging and computed tomography. Nanomedicine (Lond) 10:321–341. doi:10.2217/nnm.14.171

Coles ME, Spanne P, Muegge EL, Jones KW (1994) Computed microtomography of reservoir core samples. In: Proceedings of international symposium of the society of core analysts

Coles ME, Hazlett RD, Spanne P, Soll WE, Muegge EL, Jones KW (1998) Pore level imaging of fluid transport using synchrotron X-ray microtomography. J Pet Sci Eng 19:55–63

Contrepois M, Soenen A, Bartala M, Laviole O (2013) Marginal adaptation of ceramic crowns: a systematic review. J Prosthet Dent 110. doi:10.1016/j.prosdent.2013.08.003

Cormode DP, Jarzyna PA, Mulder WJM, Fayad ZA (2010) Modified natural nanoparticles as contrast agents for medical imaging. Adv Drug Deliv Rev 62:329–338. doi:10.1016/j.addr.2009.11.005

Dalla E, Hilpert M, Miller CT (2002) Computation of the interfacial area for two-fluid porous medium systems. J Contam Hydrol 56(1–2):25–48

Davis GR, Evershed AN, Mills D (2013) Quantitative high contrast X-ray microtomography for dental research. J Dent 41(5):475–482

De Santis R, Mollica F, Prisco D, Rengo S, Ambrosio L, Nicolais L (2005) A 3D analysis of mechanically stressed dentin-adhesive-composite interfaces using X-ray micro-CT. Biomaterials 26:257–270. doi:10.1016/j.biomaterials.2004.02.024

Degruyter W, Bachmann O, Burgisser A (2010a) Controls on magma permeability in the volcanic conduit during the climactic phase of the Kos Plateau Tuff eruption (Aegean Arc). Bull Volc 72(1):63–74

Degruyter W, Burgisser A, Bachmann O, Malaspinas O (2010b) Synchrotron X-ray microtomography and lattice Boltzmann simulations of gas flow through volcanic pumices. Geosphere 6(5):470–481

Della Bona Á, Borba M, Benetti P, Duan Y, Griggs JA (2013) Three-dimensional finite element modelling of all-ceramic restorations based on micro-CT. J Dent 41:412–419. doi:10.1016/j.jdent.2013.02.014

Desplentere F, Lomov SV, Woerdeman DL, Verpoest I, Wevers M, Bogdanovich A (2005) Micro-CT characterization of variability in 3D textile architecture. Compos Sci Technol 65:1920–1930

Desrues J, Viggiani G, Besuelle P (2010) Advances in X-ray tomography for geomaterials. ISTE, London, pp 143–148

Dewanckele J, De Kock T, Boone MA, Cnudde V, Brabant L, Boone MN, Fronteau G, Van Hoorebeke L, Jacobs P (2012) 4D imaging and quantification of pore structure modifications inside natural building stones by means of high resolution X-ray CT. Sci Total Environ 416:436–448. doi:10.1016/j.Scitoenv.2011.11.018

Dhondt S, Vanhaeren H, Van Loo D et al (2010) Plant structure visualization by high-resolution X-ray computed tomography. Trends Plant Sci 15:419–422. doi:10.1016/j.tplants.2010.05.002

Dobkins JE Jr, Folk RL (1970) Shape development on Tahiti-nui. J Sediment Res 40(4):1167–1203

Donis-González IR, Guyer DE, Pease A (2012) Application of Response Surface Methodology to systematically optimize image quality in computer tomography: A case study using fresh chestnuts (Castanea spp.). Comput Electron Agric 87:94–107. doi:10.1016/j.compag.2012.04.006

Donis-González IR, Guyer DE, Fulbright DW, Pease A (2014) Postharvest noninvasive assessment of fresh chestnut (Castanea spp.) internal decay using computer tomography images. Postharvest Biol Technol 94:14–25. doi: 10.1016/j.postharvbio.2014.02.016

Drazeta L, Lang A, Hall AJ et al (2004) Causes and effects of changes in xylem functionality in apple fruit. Ann Bot 93:275–282. doi:10.1093/aob/mch040

Duan Y, Griggs JA (2015) Effect of elasticity on stress distribution in CAD/CAM dental crowns: glass ceramic vs. polymer-matrix composite. J Dent 43:742–749. doi:10.1016/j.jdent.2015.01.008

Dvorkin J, Derzhi N, Qian F, Nur A, Nur B, Grader A, Baldwin C, Tono H, Diaz E (2009) From micro to reservoir scale: permeability from digital experiments. Leading Edge 28(12):1446–1453

Eden E, Topaloglu-Ak A, Cuijpers V, Frencken J (2008) Micro-CT for measuring marginal leakage of class II resin composite restorations in primary molars prepared in vivo. Am J Dent 21:393–397

Elkhoury JE, Ameli P, Detwiler RL (2013) Dissolution and deformation in fractured carbonates caused by flow of CO_2-rich brine under reservoir conditions. Int J Greenhouse Gas Control 16: S203–S215

Elliott JC, Bromage TG, Anderson P, Davis G, Dover SD (1989) Application of X-ray microtomography to the study of dental hard tissues. In: Fearnhead RW, Fearnhead DW (eds) Tooth enamel, 5th edn. Florence Publishers, pp 429–433

Elliott JC, Boakes R, Dover SD, Bowen DK (1988) Biological applications of microtomography. In: Sayre D, Howells M, Kirz J, Rarback H (eds) X-ray Microscopy, 2nd edn., Springer, Heidelberg, pp 349–355

Endruweit A, Zeng X, Long AC (2014) Multiscale modeling of combined deterministic and stochastic fabric non-uniformity for realistic resin injection simulation. Adv Manuf Polym Compos Sci 1:3–15. doi:10.1179/2055035914Y.0000000002

Ersoy O, Aydar E, Gourgaud A, Bayhan H (2008) Quantitative analysis on volcanic ash surfaces: application of extended depth-of-field (focus) algorithm for light and scanning electron microscopy and 3D reconstruction. Micron 39(2):128–136

Evans D, Benn D (2014) A practical guide to the study of glacial sediments, Routledge

Fabry G et al (1988) Treatment of congenital pseudarthrosis with the Ilizarov technique. J Pediatr Orthop 8(1):67–70

Fanta SW, Abera MK, Aregawi W et al (2014) Microscale modeling of coupled water transport and mechanical deformation of fruit tissue during dehydration. J Food Eng 124:86–96. doi:10. 1016/j.jfoodeng.2013.10.007

Fei J et al (2010) Imaging and quantitative assessment of long bone vascularization in the adult rat using microcomputed tomography. Anat Rec (Hoboken) 293(2):215–224

Ferracane JL (2008) Buonocore lecture. Placing dental composites–a stressful experience. Oper Dent 33:247–257. doi:10.2341/07-BL2

Fliegener S, Luke M, Gumbsch P (2014) 3D microstructure modeling of long fiber reinforced thermoplastics. Compos Sci Technol 104:136–145. doi:10.1016/j.compscitech.2014.09.009

Fogden A, Latham S, McKay T, Marathe R, Turner ML, Kingston A, Senden T (2015) Micro-CT analysis of pores and organics in unconventionals using novel contrast strategies. In: Unconventional resources technology conference (URTeC), p 1922195

Fourar M, Konan G, Fichen C (2005) Tracer tests for various carbonate cores using X-ray CT. SCA paper, pp 1–12

Galli C et al (2005) Comparison of human mandibular osteoblasts grown on two commercially available titanium implant surfaces. J Periodontol 76(3):364–372

Gebart BR (1992) Permeability of unidirectional reinforcements for RTM. J Compos Mater 26 (8):1100–1133

Georgiadis A, Berg S, Makurat A, Maitland G, Ott H (2013) Pore-scale micro-computed-tomography imaging: Nonwetting-phase cluster-size distribution during drainage and imbibitions. Phys Rev E 88(033002):1–9. http://dx.doi.org/10.1103/PhysRevE.88.033002

Gibson LJ, Ashby MF (1997) Cellular solids: structure and properties. Cambridge University Press, Cambridge

Goudie AS, Anderson M, Burt T, Lewin J, Richards K, Whalley BWorsley P (eds) (2003) Geomorphological techniques. 2nd ed. Routledge, London, 709 pages

Graupner N, Beckmann F, Wilde F, Muessig J (2014) Using synchroton radiation-based micro-computer tomography (SR Î¼-CT) for the measurement of fibre orientations in cellulose fibre-reinforced polylactide (PLA) compo-sites. J Mater Sci 49(1):450–460. doi:10.1007/ s10853-013-7724-8

Guldberg RE et al (2008) 3D imaging of tissue integration with porous biomaterials. Biomaterials 29(28):3757–3761

Gunst V, Mavridou A, Huybrechts B, Van Gorp G, Bergmans L, Lambrechts P (2013) External cervical resorption: an analysis using cone beam and microfocus computed tomography and scanning electron microscopy. Int Endod J 46(9):877–887

Habibovic P, de Groot K (2007) Osteoinductive biomaterials–properties and relevance in bone repair. J Tissue Eng Regen Med 1(1):25–32

Habibovic P et al (2006) Osteoinduction by biomaterials–physicochemical and structural influences. J Biomed Mater Res A 77(4):747–762

Habraken W et al (2016) Calcium phosphates in biomedical applications: materials for the future? Mater Today 19(2):69–87

Han S-H, Sadr A, Tagami J, Park S-H, Non-destructive evaluation of an internal adaptation of resin composite restoration with swept-source optical coherence tomography and micro-CT. Dent Mater 2015:1–7. doi:10.1016/j.dental.2015.10.009

Hanson AJ (1988) Hyperquadrics: Smoothly deformable shapes with convex polyhedral bounds. Comput Vis Graph Image Process 44(2):191–210

Harjkova G, Barburski M, Lomov SV, Kononova O, Verpoest I (2014) Weft knit-ted loop geometry measured with X-ray micro-computer tomography. Textile Res J 84:500–512. doi: http://trj.sagepub.com/content/early/2013/12/06/0040517513503730

Heintze SD, Zimmerli B (2011) Relevance of in vitro tests of adhesive and composite dental materials. A review in 3 parts. Part 3: in vitro tests of adhesive systems. Schweiz Monatsschr Zahnmed 121:1024–40. doi:smfz-2011-11-01 [pii]

Heintze SD (2007) Laboratory tests on marginal quality and bond strength. J Adhes Dent 9:77–106

Hendrickx R, Buyninckx K, Schueremans L et al (2010) Observation of the failure mechanism of brick masonry doublets with cement and lime mortars by microfocus X-ray computed tomography. In: Paper presented at the 8th International Masonry Conference, Dresden, The Netherlands

Herremans E, Melado-Herreros A, Defraeye T et al (2014) Comparison of X-ray CT and MRI of watercore disorder of different apple cultivars. Postharvest Biol Technol 87:42–50. doi:10.1016/j.postharvbio.2013.08.008

Herremans E, Verboven P, Bongaers E et al (2013) Characterisation of "Braeburn" browning disorder by means of X-ray micro-CT. Postharvest Biol Technol 75:114–124. doi:10.1016/j.postharvbio.2012.08.008

Herremans E, Verboven P, Hertog ML et al (2015) Spatial development of transport structures in apple (Malus × domestica Borkh.) fruit. Front Plant Sci. doi:10.3389/fpls.2015.00679

Herremans E, Verboven P, Verlinden BE et al (2015b) Automatic analysis of the 3-D microstructure of fruit parenchyma tissue using X-ray micro-CT explains differences in aeration. BMC Plant Biol 15:264. doi:10.1186/s12870-015-0650-y

Hickel R, Peschke A, Tyas M, Mjör I, Bayne S, Peters M et al (2010a) FDI World Dental Federation—clinical criteria for the evaluation of direct and indirect restorations. Update and clinical examples. J Adhes Dent 12:259–272. doi:10.3290/j.jad.a19262

Hickel R, Peschke A, Tyas M, Mjör I, Bayne S, Peters M et al (2010b) FDI World Dental Federation: clinical criteria for the evaluation of direct and indirect restorations-update and clinical examples. Clin Oral Investig 14:349–366. doi:10.1007/s00784-010-0432-8

Hickel R, Roulet J, Bayne S, Heintze S, Mjör I, Peters M et al (2007) Recommendations for conducting controlled clinical studies of dental restorative materials. Science Committee Project 2/98–FDI World Dental Federation study design (Part I) and criteria for evaluation (Part II) of direct and indirect restorations includi. J Adhes Dent 9(1):121–147

Hirata R, Clozza E, Giannini M, Farrokhmanesh E, Janal M, Tovar N et al (2014) Shrinkage assessment of low shrinkage composites using micro-computed tomography. J Biomed Mater Res B Appl Biomater 2014:1–9. doi:10.1002/jbm.b.33258

Ho QT, Rogge S, Verboven P et al (2015) Stochastic modelling for virtual engineering of controlled atmosphere storage of fruit. J Food Eng. doi:10.1016/j.jfoodeng.2015.07.003

Ho QT, Verboven P, Verlinden BE et al (2011) A three-dimensional multiscale model for gas exchange in fruit. Plant Physiol 155:1158–1168. doi:10.1104/pp.110.169391

Hsieh J (2009) Computed tomography. principle, design, artifacts, and recent advances, 2nd edn. SPIE and Wiley, Washington, USA

Huang TT, He LH, Darendeliler MA, Swain MV (2010) Correlation of mineral density and elastic modulus of natural enamel white spot lesions using X-ray microtomography and nanoindentation. Acta Biomater 6(12):4553–4559

Hwang BG, Ahn S, Lee SJ (2014) Use of gold nanoparticles to detect water uptake in vascular plants. PLoS One 9:e114902. doi:10.1371/journal.pone.0114902

Jacker-Guhr S, Ibarra G, Oppermann LS, Lührs AK, Rahman A, Geurtsen W (2015) Evaluation of microleakage in class V composite restorations using dye penetration and micro-CT. Clin Oral Investig 2015:1–10. doi:10.1007/s00784-015-1676-0

Janke NC (1966) Effect of shape upon the settling velocity of regular convex geometric particles. J Sed Res 36(2):370–376

Janssen FW et al (2006) A perfusion bioreactor system capable of producing clinically relevant volumes of tissue-engineered bone: In vivo bone formation showing proof of concept. Biomaterials 27(3):315–323

Jensen TH, Böttiger A, Bech M et al (2011) X-ray phase-contrast tomography of porcine fat and rind. Meat Sci 88:379–383. doi:10.1016/j.meatsci.2011.01.013

Jones AC et al (2004) Analysis of 3D bone ingrowth into polymer scaffolds via micro-computed tomography imaging. Biomaterials 25(20):4947–4954

Jones AC et al (2007) Assessment of bone ingrowth into porous biomaterials using MICRO-CT. Biomaterials 28(15):2491–2504

Jungreuthmayer C et al (2009) A comparative study of shear stresses in collagen-glycosaminoglycan and calcium phosphate scaffolds in bone tissue-engineering bioreactors. Tissue Eng Part A 15(5):1141–1149

Kampschulte M, Langheinirch AC, Sender J, Litzlbauer HD, Schwab JD, Martels G et al (2016) Nano-computed tomography : technique and applications nanocomputertomografie. Technik und Applikationen 2016:146–54

Karpyn ZT, Alajmi A, Radaelli F, Halleck PM, Grader AS (2009) X-ray CT and hydraulic evidence for a relationship between fracture conductivity and adjacent matrix porosity. Eng Geol 103(3–4):139–145

Kato A, Ohno N (2008) Construction of three-dimensional tooth model by micro-computed tomography and application for data sharing. Clin Oral Investig 13:43–46. doi:10.1007/s00784-008-0198-4

Kazantsev D, Van Eyndhoven G, Lionheart WRB, Withers PJ, Dobson KJ, McDonald SA, Atwood R, Lee PD (2016) Employing temporal self-similarity across the entire time domain in computed tomography reconstruction. Phil Trans R Soc A 373:20140389. http://dxdoi.org/10.1098/rsta.2014.0389

Kerckhofs G et al (2013) Contrast-enhanced nanofocus computed tomography images the cartilage subtissue architecture in three dimensions. Eur Cell Mater 25:179–189

Kerckhofs G et al (2014) Contrast-enhanced nanofocus x-ray computed tomography allows virtual three-dimensional histopathology and morphometric analysis of osteoarthritis in small animal models. Cartilage 5(1):55–65

Kerckhofs G et al (2016) Combining microCT-based characterization with empirical modelling as a robust screening approach for the design of optimized CaP-containing scaffolds for progenitor cell-mediated bone formation. Acta Biomater

Kerckhofs G et al (2012) High-resolution microfocus X-ray computed tomography for 3D surface roughness measurements of additive manufactured porous materials. Adv Eng Mater 2012

Kerckhofs G et al (2016) Simultaneous 3D visualization and quantification of the bone marrow adiposity and vascularity using novel contrast agents for contrast-enhanced computed tomography, in IBMS2016March. Brugge, Belgium

Ketcham RA (2005) Computational methods for quantitative analysis of three-dimensional features in geological specimens. Geosphere 1:32–41

Ketcham RA (2006) Accurate three-dimensional measurements of features in geological materials from X-ray computed tomography data. In: Desrues J, Viggiani G, Besuelle, J, (eds) Advances in X-ray Tomography for Geomaterials, ISTE, London, 143–148

Ketcham RA, Carlson WD (2001) Acquisition, optimization and interpretation of X-ray computed tomographic imagery: applications to the geosciences. Comput Geosci 27:381–400

Kim HJ, Park SH (2014) Measurement of the internal adaptation of resin composites using micro-CT and its correlation with polymerization shrinkage. Oper Dent 39:E57–E70. doi:10.2341/12-378-L

Klette R (1996) A Parametrization of digital planes by least-squares fits and generalizations. Graphical Models Image Process 58(3):295–300

Kruyt M et al (2008) Analysis of the dynamics of bone formation, effect of cell seeding density, and potential of allogeneic cells in cell-based bone tissue engineering in goats. Tissue Eng Part A 14(6):1081–1088

Kuijs RH, Fennis WMM, Kreulen CM, Barink M, Verdonschot N (2003) Does layering minimize shrinkage stresses in composite restorations ? J Dent Res 82:967–971

Lai P, Samson C, Bose P (2014) Surface roughness of rock faces through the curvature of triangulated meshes. Comput Geosci 70:229–237

Lammertyn J, Dresselaers T, Van Hecke P et al (2003) MRI and x-ray CT study of spatial distribution of core breakdown in "Conference" pears. Magn Reson Imaging 21:805–815. doi:10.1016/S0730-725X(03)00105-X

Langer R, Vacanti JP (1993) Tissue engineering. Science 260(5110):920–926

Lauridsen T, Glavina K, Colmer TD et al (2014) Visualisation by high resolution synchrotron X-ray phase contrast micro-tomography of gas films on submerged superhydrophobic leaves. J Struct Biol 188:61–70. doi:10.1016/j.jsb.2014.08.003

Long H, Swennen R, Foubert A, Dierick M, Jacobs P (2009) 3D quantification of mineral components and porosity distribution in Westphalian C sandstone by microfocus X-ray computed tomography. Sed Geol 220:116–125. doi:10.1016/j.sedgeo.2009.07.003

Lønøy A (2006) Making sense of carbonate pore systems. AAPG Bull 90(9):1381–1405

Lucia FJ (1995) Rock-Fabric: petrophysical classification of carbonate pore space for reservoir characterization. AAPG Bull 9(9):1275–1300

Lukeneder A, Lukeneder S, Gusenbauer C (2014) Computed tomography and laser scanning of fossil cephalopods (Triassic and Cretaceous) Denisia 32, zugleich Kataloge des oberosterre-ichischen Landesmuseums Neue Serie 157:81–92

Lusic H, Grinsta MW (2013) X-ray-computed tomography contrast agents

Lusic H, Grinstaff MW (2013) X-ray-computed tomography contrast agents. Chem Rev 113(3):1641–1666

Luso S, Luder HU (2012) Resorption pattern and radiographic diagnosis of invasive cervical resorption. A correlative microCT, scanning electron and light microscopic evaluation of a case series. Schweizer Monatsschrift für Zahnmedizin 122:914–930

Magne P (2007) Efficient 3D finite element analysis of dental restorative procedures using micro-CT data. Dent Mater 23:539–548. doi:10.1016/j.dental.2006.03.013

Mariethoz G, Straubhaar J, Renard P, Chugunova T, Biver P (2015) Environmental modelling and software constraining distance-based multipoint simulations to proportions and trends. Environ Model Softw 72:184–197

Mavridou AM, Pyka G, Kerckhofs G, Wevers M, Bergmans L, Gunst V, Huybrechts B, Schepers E, Hauben E, Lambrechts P (2016a) A novel multimodular methodology to investigate external cervical tooth resorption. Int Endod J 49(3):287–300

Mavridou AM, Pyka G, Wevers M, Lambrechts P (2016b) Applying Nano-CT technology in endodontology: understanding external cervical root resorption. Paper presented in European Society of Endodontics conference, Barcelona 16–19 September 2015. Int Endod J 49(1):41. doi:10.1111/iej.12496

Mees F, Swennen R, Van Geet M, Jacobs P (2003) Applications of X-ray computed tomography in geology and related domains: Introductory Paper Geol Soc Lond Spec 215:1–6

Mehmani Y, Sun T, Balhoff MT, Eichhubl P, Bryant S (2012) Multiblock pore-scale modeling and upscaling of reactive transport: application to carbon sequestration. Transp Porous Med 95:305–326

Metscher BD (2009) MicroCT for comparative morphology: simple staining methods allow high-contrast 3D imaging of diverse non-mineralized animal tissues. BMC Physiol 9:11

Meyer F (1994) Topographic distance and watershed lines. Sig Process 38(1):113–125

Miklos R, Nielsen MS, Einarsdóttir H et al (2014) Novel X-ray phase-contrast tomography method for quantitative studies of heat induced structural changes in meat. Meat Sci 100C:217–221. doi:10.1016/j.meatsci.2014.10.009

Mlekusch B (1999) Fibre orientation in short-fibre-reinforced thermoplastics II. Quantitative measurements by image analysis. Compos Sci Technol 59:547–560

Mostaghimi P, Blunt MJ, Bijeljic B (2013) Computations of absolute permeability on micro-CT images. Math Geosci 45(1):103–125

Mustafa K et al (2001) Determining optimal surface roughness of TiO$_2$ blasted titanium implant material for attachment, proliferation and differentiation of cells derived from human mandibular alveolar bone. Clin Oral Implant Res 12(5):515–525

Naouar N, Vidal-Sallã E, Schneider J, Maire E, Boisse P (2014) Meso-scale FE analyses of textile composite reinforcement deformation based on X-ray computed tomography. Compos Struct 116:165–176. doi:http://dx.doi.org/10.1016/j.compstruct.2014.04.026

Neves AA, Jaecques S, Van Ende A, Cardoso MV, Coutinho E, Lührs A-K et al (2014) 3D-microleakage assessment of adhesive interfaces: exploratory findings by μCT. Dent Mater 30:799–807. doi:10.1016/j.dental.2014.05.003

Neves FS (2013) Radicular dens invaginatus in a mandibular premolar: cone-beam computed tomography findings of a rare anomaly. Oral Radiol. 29(1):70

Neves FS, Pontual, AD, Campos PSF, Frazao MAG, de Almeida SM, Ramos-Perez FMD

Neves AA, Coutinho E, De Munck J, Lambrechts P, Van Meerbeek B (2011) Does DIAGNOdent provide a reliable caries-removal endpoint? J Dent 39(5):351–360

Neves AA, Coutinho E, Vivan Cardoso M, Jaecques SV, Van Meerbeek B (2010) Micro-CT based quantitative evaluation of caries excavation. Dent Mater 26(6):579–588

Nico PS, Ajo-Franklin JB, Benson SM, McDowell A, Silin DB, Tomutsa L, Wu Y (2010) Synchrotron X-ray micro-tomography and geological CO$_2$ sequestration. In advances in computed tomography for geomaterials, GeoX 2010 (eds Alshibli KA, Reed AH), Wiley, Hoboken, NJ, USA

Nicolaï BM, Defraeye T, De Ketelaere B et al (2014) Nondestructive measurement of fruit and vegetable quality. Annu Rev Food Sci Technol 5:285–312. doi:10.1146/annurev-food-030713-092410

Oi T, Saka H, Ide Y (2004) Three-dimensional observation of pulp cavities in the maxillary first premolar tooth using micro-CT. Int Endod J 37(1):46–51

Okabe H, Blunt MJ (2004) Prediction of permeability for porous media reconstructed using multiple-point statistics. Phys Rev E 70(6):66135

Panarese V, Dejmek P, Rocculi P, Gómez Galindo F (2013) Microscopic studies providing insight into the mechanisms of mass transfer in vacuum impregnation. Innov Food Sci Emerg Technol 18:169–176. doi:10.1016/j.ifset.2013.01.008

Papadimitropoulos A et al (2007) Kinetics of in vivo bone deposition by bone marrow stromal cells within a resorbable porous calcium phosphate scaffold: An X-ray computed microtomography study. Biotechnol Bioeng 98(1):271–281

Pauwels E et al (2013) An exploratory study of contrast agents for soft tissue visualization by means of high resolution X-ray computed tomography imaging. J Microsc 250(1):21–31

Pazmino J, Carvelli V, Lomov SV (2014) Micro-CT analysis of the internal deformed geometry of a non-crimp 3D orthogonal weave e-glass composite reinforcement. Compos B 65:147–157. doi:10.1016/j.compositesb.2013.11.024

Peng S, Hu Q, Dultz S, Zhang M (2012) Using X-ray computed tomography in pore structure characterization for a Berea sandstone: resolution effect. J Hydrol 472–473:254–261

Peters O, Laib A, Ruegsegger P, Barbakow F (2000) Three-dimensional analysis of root canal geometry by high-resolutin computed tomography. J Dent Res 79(6):1405–1409

Peyrin F et al (2007) SEM and 3D synchrotron radiation micro-tomography in the study of bioceramic scaffolds for tissue-engineering applications. Biotechnol Bioeng 97(3):638–648

Peysson Y (2012) Permeability alteration induced by drying of brines in porous media. Eur Phys J Appl Phys 60(2):24206

Pinto J, Solorzano E, Rodriguez-Perez MA, de Saja JA (2013) Characterization of the cellular structure based on user-interactive image analysis procedures. J Cell Plast 49(6):555–575. doi:10.1177/0021955x13503847

Place ES, Evans ND, Stevens MM (2009) Complexity in biomaterials for tissue engineering. Nat Mater 8(6):457–470

Razavi MR, Muhunthan B, Al Hattamleh O (2007) Representative elementary volume analysis of sands using X-ray computed tomography. Geotech Test J 30(3):212–219

Remeysen K, Swennen R (2008) Application of microfocus computed tomography in carbonate reservoir sedimentology: possibilities and limitations. Mar Pet Geol 25:486–499. doi:10.1016/j.marpetgeo.2007.07.008

Riley CM (2003) Quantitative shape measurements of distal volcanic ash. J Geophys Res 108 (B10):1–15

Rodrigues FP, Li J, Silikas N, Ballester RY, Watts DC (2009) Sequential software processing of micro-XCT dental-images for 3D-FE analysis. Dent Mater 25:47–55. doi:10.1016/j.dental.2009.02.007

Rodrigues FP, Silikas N, Watts DC, Ballester RY (2012) Finite element analysis of bonded model Class I "restorations" after shrinkage. Dent Mater 28:123–132. doi:10.1016/j.dental.2011.10.001

Roels S, Carmeliet J (2006) Analysis of moisture flow in poreus materials using microfocus X-ray radiography. Int J Heat Mass Transf 49:4762–4772

Roels S, Vandersteen K, Carmeliet J (2003) Measuring and simulating moisture uptake in a fractured porous medium. Adv Water Resour 26(3):237–246

Rogge S, Defraeye T, Herremans E et al (2015) A 3D contour based geometrical model generator for complex-shaped horticultural products. J Food Eng 157:24–32. doi:10.1016/j.jfoodeng.2015.02.006

Roulet JF (1994) Marginal integrity: clinical significance. J Dent 22:S9–S12

Rousseau D, Widiez T, Di Tommaso S et al (2015) Fast virtual histology using X-ray in-line phase tomography: application to the 3D anatomy of maize developing seeds. Plant Methods 11:55. doi:10.1186/s13007-015-0098-y

Salgado AJ, Coutinho OP, Reis RL (2004) Bone tissue engineering: State of the art and future trends. Macromol Biosci 4(8):743–765

Schmitt M, Halisch M, Müller C, Fernandes CP (2016) Classification and quantification of pore shapes in sandstone reservoir rocks with 3-D X-ray micro-computed tomography. Solid Earth 7:285–300. doi:10.5194/se-7-285-2016

Schneider LFJ, Cavalcante LM, Silikas N (2010) Shrinkage stresses generated during resin-composite applications: a Review. J Dent Biomech. doi:10.4061/2010/131630

Sharp G, Lee S, Wehe D (2002) ICP registration using invariant features. IEEE Pattern Anal Mach Intell 24(1):90–102

Shea T, Houghton BF, Gurioli L, Cashman KV, Hammer JE, Hobden BJ (2010) Textural studies of vesicles in volcanic rocks: an integrated methodology. J Volcanol Geotherm Res 190:271–289

Shemesh H (2016) Endodontic instrumentation and root filling procedures: effect on mechanical integrity of dentin. Endod Topics 33(1):43–49

Shilo M, Reuveni T, Motiei M, Popovtzer R (2012) Nanoparticles as computed tomography contrast agents: current status and future perspectives. Nanomedicine (Lond) 7:257–269. doi:10.2217/nnm.11.190

Shishkina O (2014) Experimental and modelling investigations of structure-property relations in nanoreinforced cellular materials, PhD Thesis. KU Leuven

Soete J, Claes S, Claes H, Cnudde V, Huysmans M, Swennen R (Submitted) Lattice Boltzmann simulations of gas flow in continental carbonate reservoir rocks and in upscaled rock models generated with multiple point geostatistics. AAPG Bulletin

Soete J, Kleipool LM, Claes H, Claes S, Hamaekers H, Kele S, Özkul M, Foubert A, Reijmer JJG, Swennen R (2015) Acoustic properties in travertines and their relation to porosity and pore types. Marine Pet Geol 59:320–335

Sonnaert M et al (2015) Multifactorial optimization of contrast-enhanced nanoCT for quantitative analysis neo-tissue formation in tissue engineering constructs. PLoS One (submitted—under revision)

Straumit I, Hahn C, Winterstein E, Plank B, Lomov SV, Wevers M (2016) Computation of permeability of a non-crimp carbon textile reinforcement based on X-ray computed tomography images. Compos A 81:289–295. doi:10.1016/j.compositesa.2015.11.025

Straumit I, Lomov SV, Wevers M (2015) Quantification of the internal structure and automatic generation of voxel models of textile composites from X-ray computed tomography data. Compos A 69:150–158. doi:10.1016/j.compositesa.2014.11.016

Sun J, Eidelman N, Lin-Gibson S (2009) 3D mapping of polymerization shrinkage using X-ray micro-computed tomography to predict microleakage. Dent Mater 25:314–320. doi:10.1016/j.dental.2008.07.010

Swain MV, Xue J (2009) State of the art of Micro-CT applications in dental research. Int J Oral Sci. 1(4):177–188

Tan Y, Kiekens K, Kruth J, Voet A, Dewulf W (2011) Material dependent thresholding for dimensional X-ray computed tomography. Int Symp Digit Ind Radiol Comput Tomogr 4.3:3–10

Taylor AM, Satterthwaite JD, Ellwood RP, Pretty IA (2010) An automated assessment algorithm for micro-CT images of occlusal caries. Surgeon 8(6):334–340

Terzopoulos D, Metaxas D (1990) Dynamic 3D models with local and global deformations: Deformable superquadrics. Proce Third Int Conf Comput Vision 13(7):606–615

Thi TBN, Morioka M, Yokoyama A, Hamanaka S, Yamashita K, Nonomura C (2015) Measurement of fiber orientation distribution in injection-molded short-glass-fiber composites using X-ray computed tomography. J Mater Process Technol 219:1–9. doi:10.1016/j.jmatprotec.2014.11.048

Timmins NE et al (2007) Three-dimensional cell culture and tissue engineering in a T-CUP (Tissue Culture Under Perfusion). Tissue Eng 13(8):2021–2028

Van Ende A, De Munck J, Van Landuyt KL, Poitevin A, Peumans M, Van Meerbeek B (2013) Bulk-filling of high C-factor posterior cavities: effect on adhesion to cavity-bottom dentin. Dent Mater 29:269–277. doi:10.1016/j.dental.2012.11.002

Van Eyndhoven G, Batenburg KJ, Kazantsev D, Van Nieuwenhove V, Lee PD, Dobson KJ, Sijbers J (2015) An iterative CT reconstruction algorithm for fast fluid flow imaging. IEEE Trans Image Process 24(11):4446–4458

van Lenthe GH et al (2007) Nondestructive micro-computed tomography for biological imaging and quantification of scaffold-bone interaction in vivo. Biomaterials 28(15):2479–2490

Van Marcke P (2008) Development of a pore network model to perform permeability computations on X-ray computed tomography images. Unpublished Ph.D. thesis, KU Leuven

Van Stappen J, De Kock T, Boone MA, Olaussen S, Cnudde V (2014) Pore-scale characterisation and modelling of CO_2 flow in tight sandstones using X-ray micro-CT: Knorringfjellet Formation of the Longyearbyen CO2 Lab, Svalbard. Norwegian J Geol 94(2–3):201–215

Vanaerschot A, Cox BN, Lomov SV, Vandepitte D (2016) Experimentally validated stochastic geometry description for textile composite reinforce-ments. Compos Sci Technol 122:122–129. doi:10.1016/j.compscitech.2015.11.023

Verboven P, Kerckhofs G, Mebatsion HK et al (2008) Three-dimensional gas exchange pathways in pome fruit characterized by synchrotron X-ray computed tomography. Plant Physiol 147:518–527

Verdonschot N, Fennis W, Kuijs R, Stolk J, Kreulen C, Creugers N (2001) Generation of 3-D finite element models of restored human teeth using micro-CT techniques. Int J Prosthodont 14:310–315

Vereecken E, Roels S (2014) A comparison of hygric performance of interior insulation systems: A hot box-cold box experiment. Energy Build 80:37–44

Verleye B, Croce R, Griebel M, Klitz M, Lomov SV, Morren G, Sol H, Verpoest I, Roose D (2008) Permeability of textile reinforcements: simulation; influence of shear, nesting and boundary conditions; validation. Compos Sci Technol 68(13):2804–2810. doi:10.1016/j.compscitech.2008.06.010

Versiani MA, Souza E, De-Deus G (2015) Critical appraisal of studies on dentinal radicular microcracks in endodontics: methodological issues, contemporary concepts, and future perspectives. Endod Topics 33(1):87–156

Verstrynge E, Adriaens R, Elsen J, Van Balen K (2014a) Multi-scale analysis on the influence of moisture on the mechanical behavior of ferruginous sandstone. Constr Build Mater 54:78–90

Verstrynge E, Pyka G, Van Balen K (2014) The influence of moisture on the mechanical behaviour of sandstone assessed by means of micro-computed tomography. In: Paper presented at the 9th International Masonry Conference, Guimaraes, Portugal

Verstrynge E, Van Steen C, Andries J, Van Balen K, Vandewalle L, Wevers M (2016) Experimental study of failure mechanisms in brittle construction materials by means of X-ray microfocus computed tomography. In: Saouma V, Bolander J, Landis E (eds) Proceedings of the ninth international conference on fracture mechanics of concrete and concrete structures—FraMCoS-9. Berkeley, California, USA, 29 May-1 June 2016 (art.nr. 92)

Verwer K, Eberli G, Baechle G, Weger R (2008) Effect of carbonate pore structure on dynamic shear moduli. Geophysics 75(1):E1–E8. doi:10.1190/1.3280225

Vinegar HJ, Wellington SL (1986) Tomographic imaging of three-phase flow experiments. Rev Sci Instrum 58:96–107

Vonlanthen P, Rausch J, Ketcham RA, Putlitz B, Baumgartner LP, Grobéty B (2015) High-resolution 3D analyses of the shape and internal constituents of small volcanic ash particles: the contribution of SEM micro-computed tomography (SEM micro-CT). J Volcanol Geotherm Res 29:1–12

Vranic DV, Saupe D (2001) 3D shape descriptor based on 3D Fourier transform. In Proceedings of the EURASIP conference on digital signal processing for multimedia communications and services (ECMCS 2001), (September), pp 1–4

Weger RJ, Eberli GP, Baechle GT, Massaferro JL, Sun YF (2009) Quantification of pore struucture and its effect on sonic velocity and permeability in carbonates. AAPG Bulletin 93 (10):1297–1317

Weissenbock J, Amirkhanov A, Li WM, Reh A, Groller E, Kastner J, Heinzl C, Ieee (2014) FiberScout: an interactive tool for exploring and analyzing fiber reinforced polymers. In: IEEE Pacific visualization symposium. pp 153–160. doi:10.1109/PacificVis.2014.52

Wirjadi O (2007) Survey of 3D image segmentation methods, ITWM

Wong MD et al (2012) A novel 3D mouse embryo atlas based on micro-CT. Development 139(17):3248–3256

Xie L et al (2010) Nondestructive assessment of sGAG content and distribution in normal and degraded rat articular cartilage via EPIC-mu CT. Osteoarthritis Cartilage 18(1):65–72

Youssef S, Dechamps H, Dautriat J, Rosenberg E, Oughanem R, Maire E, Mokso R (2013) 4D imaging of fluid flow dynamics in natural porous media by ultra-fast X-ray microtomography. In: International symposium of the SCA, Napa Valley, California

Zeiger DN, Sun J, Schumacher GE, Lin-Gibson S (2009) Evaluation of dental composite shrinkage and leakage in extracted teeth using X-ray microcomputed tomography. Dent Mater 25:1213–1220. doi:10.1016/j.dental.2009.04.007

Zhang Y, Sailer I, Lawn BR (2013) Fatique of Dental Ceramics. J Dent 41:1135–1147. doi:10.1038/nature13314.A

Zou W, Hunter N, Swain MV (2011) Application of polychromatic μCT for mineral density determination. J Dent Res 90(1):18–30

Zou W, Gao J, Jones AS, Hunter N, Swain MV (2009) Characterization of a novel calibration method for mineral density determination of dentine by X-ray micro-tomography. Analyst. 134(1):72–79

http://www.volumegraphics.com/en/products/vgstudio-max/fiber-composite-material-analysis/
http://biomedical.materialise.com/mimics

Chapter 9
Applications of CT for Dimensional Metrology

Andrea Buratti, Judith Bredemann, Michele Pavan, Robert Schmitt and Simone Carmignato

Abstract X-ray computed tomography (CT) has an increasing number of industrial applications in the field of dimensional metrology. In the first part of this chapter, an overview of the general measurement tasks that can be performed by using industrial CT is provided. Subsequently, specific measurement examples are given for different manufacturing fields (including casting, forming, machining, injection moulding and additive manufacturing). Then, trends and industrial demands are analysed to determine the current challenges of industrial CT and the future improvements. Finally, a special case study of CT for dimensional measurement in the medical field is presented.

A. Buratti (✉) · J. Bredemann · R. Schmitt
Chair for Production Metrology and Quality Management, Laboratory for Machine Tools and Production Engineering, RWTH Aachen University, Steinbachstraße 19, 52074 Aachen, Germany
e-mail: A.Buratti@wzl.rwth-aachen.de

J. Bredemann
e-mail: J.Bredemann@wzl.rwth-aachen.de

R. Schmitt
e-mail: R.Schmitt@wzl.rwth-aachen.de

M. Pavan
Materialise NV, Technologielaan 15, 3001 Louvain, Belgium
e-mail: Michele.Pavan@materialise.be

M. Pavan
Mechanical Engineering Department, KU Leuven, Celestijnenlaan 300, 3001 Louvain, Belgium

S. Carmignato
Department of Management and Engineering, University of Padova, Stradella San Nicola 3, 36100 Vicenza, Italy
e-mail: simone.carmignato@unipd.it

© Springer International Publishing AG 2018
S. Carmignato et al. (eds.), *Industrial X-Ray Computed Tomography*,
https://doi.org/10.1007/978-3-319-59573-3_9

333

9.1 CT Measurements in the Industrial Practice

Two important global trends are nowadays characterising the evolution and the challenges of manufacturing industry. Product variety and complexity are exponentially growing to match customers' specific needs. At the same time, customers demand for cheaper products. To comply with these demands, industry requires cost and time efficient solutions for product development and production. Recent advancements in manufacturing technology make possible to manufacture also parts with complex geometry and a multitude of features, opening new paths of product development. However, this creates new challenges while testing conformity of product characteristics. New measurement and testing techniques are required to inspect parts with, for instance, free-form surfaces or non-accessible features. Industrial demands require performing tolerance and geometrical quality control of complex parts, like additive manufacturing or injection moulding products (De Chiffre et al. 2014; Kruth et al. 2011).

In this context, industrial Computed Tomography (CT) is a powerful non-destructive technique, enabling the inspection and measurement of the full geometry of a product—including inner surfaces and non-accessible features—without altering or damaging it (with few exceptions, e.g. some biomaterials). CT generates a complete volumetric model of the inspected part that can be used for performing several quality control tasks, as introduced in Chap. 1. As shown in Fig. 9.1, the main tasks for industrial CT are visualisation, non-destructive testing, digitisation, and dimensional metrology:

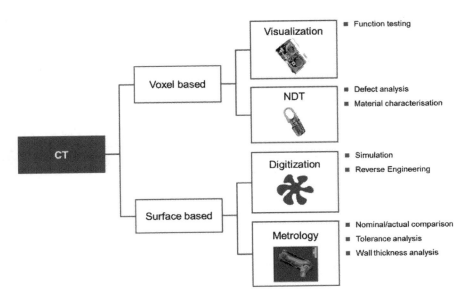

Fig. 9.1 Fields of application for industrial CT

- Visualisation is a qualitative test to inspect e.g. conformity and functioning of assembled parts.
- Non-destructive testing (NDT) is a broad category of qualitative testing methods including defect analysis and material characterisation. Defect analysis aims at identifying defects in manufactured parts, e.g. pores, voids, inclusions, foreign bodies. Material characterisation investigates the properties of the workpiece material.
- Digitisation deals with creating and evaluating virtual geometrical models of objects from CT scans and includes both simulation and reverse engineering.
- Finally, metrology addresses the challenge of performing dimensional measurements with CT. This includes any common metrological task, e.g. wall thickness measurements, nominal-actual comparison, and tolerance analysis (VDI/VDE 2010).

Going form visualization towards metrology, the complexity of the task increases and the metrological traceability becomes more important (see also Chap. 7).

In the following, the chapter will focus mainly on the dimensional metrology tasks (including tolerance analysis, nominal-actual comparison and wall thickness analysis). In tolerance analysis, features are measured in order to check whether they respect a given tolerance or not. Typical tasks include checking dimensional, form and position tolerances, and determining compensating elements, regular geometry and sculptured surfaces. Nominal-actual comparison allows visualising the geometric deviations between a nominal model (e.g. a computed aided design (CAD) model, or a reference workpiece) and an actual geometry (CT volume). A color-coded representation of the model is provided to visualise the deviations of the geometry. Measuring the wall thickness of a workpiece means analysing and quantifying the thickness of the volumetric model.

CT coordinate measurements are characterised by a paradigm change in the way measurements occur. Whereas conventional coordinate measurements (e.g. obtained

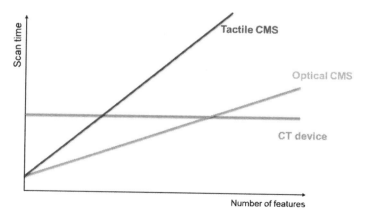

Fig. 9.2 Scanning time as a function of the number of features for different coordinate measuring systems (CMSs)

using a tactile coordinate measuring system, CMS) are performed directly on the workpiece surface, CT measurements are performed on a virtual model of the workpiece. This means that a large number of features can be analysed at the same time without compromising the scanning time. Figure 9.2 shows the relationship between the scanning and the number of analysed features for different CMS.

Dimensional CT measurements provide manufacturers with useful information for quality control during the whole development and manufacturing cycle of a product. There are several reasons why CT has increasingly become popular in the manufacturing industry. First, parts and products are characterised by an increasing geometrical complexity, e.g. involving the presence of a large number of complex features (including inner features), which may be difficult or impossible to be

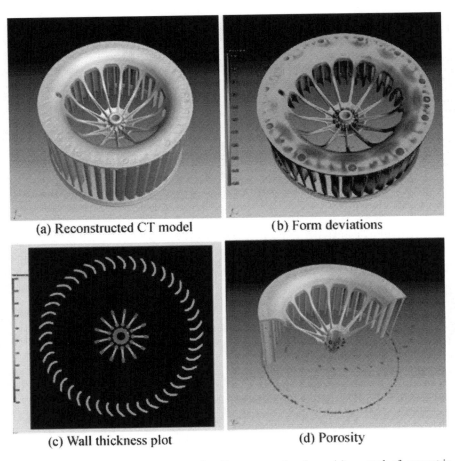

(a) Reconstructed CT model   (b) Form deviations

(c) Wall thickness plot   (d) Porosity

Fig. 9.3 Quality control of a car inlet fan. Reconstructed volume (**a**), control of geometric deviations (**b**), wall thickness (**c**) and material porosity (**d**) of car inlet fan (Courtesy of Nikon-Metrology/X-Tek)

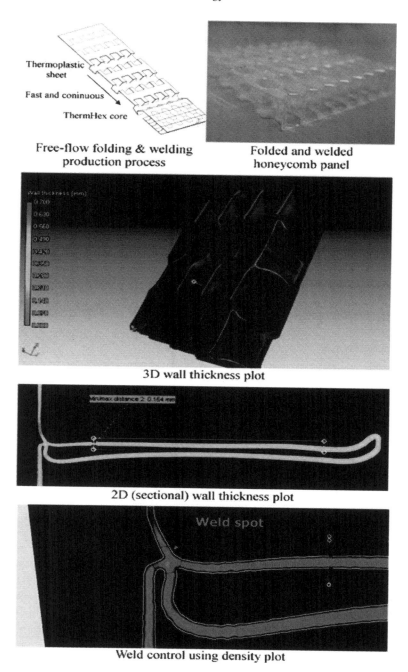

Fig. 9.4 Multiple quality checks for a honeycomb panel (Kruth et al. 2011)

accessed by means of the conventional measuring methods (optical and tactile CMSs). Second, in most of the cases, quality control of complex assemblies can be performed through optical and tactile CMS only in the non-assembled state. However, this does not ensure the assembled product is without mismatches (e.g. defective gaps, deformations due to wrong assembly); hence it is also necessary to inspect the assembly in a non-destructive way, calling for CT. Third, production methods like multimaterial injection moulding (e.g. 2K injection moulding, injection moulding of plastic parts with metal inserts) or additive manufacturing are capable of manufacturing complex features which require CT capability of inspecting quality in a non-destructive way.

Finally, CT allows performing dimensional measurements and material testing on the same CT data set, meaning a faster and holistic quality check. Figure 9.3 shows an example of combined quality check for a car inlet fan using the same volume data: form and geometrical deviations, wall thickness measurements, material testing. Figure 9.4 depicts 3D wall thickness measurements, cross-section wall thickness measurements, and welding quality control for a honeycomb panel produced in a continuous flow sheet embossing, folding and welding process.

In the next sections, it is shown through application examples how CT can play a key role in quality control of casted and formed products, machined products, additively manufactured products, injection moulding products, and complex assemblies.

9.2   Application Examples of CT in the Manufacturing Field

CT plays a key role in dimensional metrology as an alternative to conventional coordinate-measuring systems, such as optical and tactile CMSs. This Section provides an overview of state-of-the-art industrial applications for dimensional CT measurements.

9.2.1   Cast and Formed Products

Since CT is able to show the inner material structure of a part or product, it is very promising for detecting pores, inclusions, and cavities in the foundry and metal forming industry. Moreover, CT can be also used to perform dimensional quality control, also including hollow parts. An application example for this is shown in Fig. 9.5, where a nominal-actual comparison of a cast iron duct is shown. This workpiece features an inner geometry that can be hardly accessed by means of a tactile probe.

In the transportation industry, hollow components have been increasingly used as a means to reduce weight and thus fuel consumption. Quality control of the inner, non-accessible geometry is a crucial aspect for the manufacturing of these parts. Sterzing et al. showed that CT scans can be useful to perform quality inspection of hydro-formed hollow camshafts and hollow-constructed camshafts (Sterzing et al. 2013).

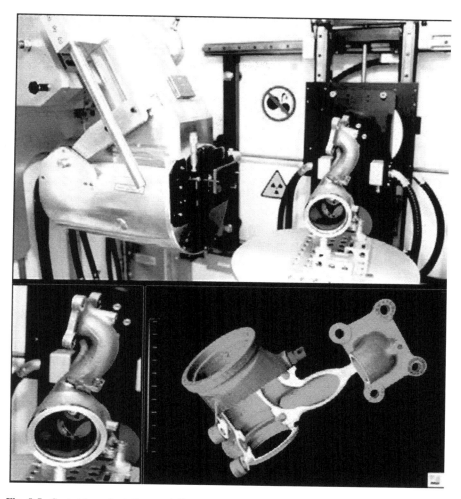

Fig. 9.5 Casted iron duct (*bottom left*) measured on a CT scanner (*top*) and compared with its nominal CAD model (*bottom right*) (De Chiffre et al. 2014)

Another example for CT-based quality control of casted products is shown in Fig. 9.6. Here, an aluminium casted cylinder head segment with reference spheres is analysed. This object can be used as a reference object for accuracy verification of CT measurements on castings (Bartscher et al. 2008).

CT measurements can be also used as a mean to optimise casting processes. In particular, twin-roll casting produces strips with an inhomogeneous material structure that can be affected by macrosegregation. This often leads to premature fracture and diminished strength (Slapakova et al. 2016). Microstructural analysis is typically performed by means of light optical microscopy or scanning electron microscopy. However, those methods are destructive and can be used to inspect

(a)   **(b)**

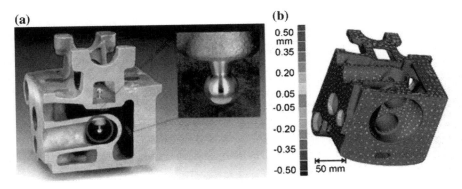

Fig. 9.6 Cast aluminium cylinder head segment with reference spheres (Kruth et al. 2011)

small volumes of material. On the other hand, CT measurements can reveal the spatial distribution of segregations in a non-destructive way and provide useful information to optimise twin-roll casting processes. Figure 9.7 depicts the CT reconstruction of centre line segregation in an experimental aluminium alloy from AA3003 series obtained by twin-roll casting.

This scan confirmed the assumption that segregations are oriented in the rolling direction and located in the central part of the strip.

Turbine blades also call for CT measurements, since they feature complex inner geometries that can be difficult to be probed with tactile sensors. An example is shown in Fig. 9.8. CT measurements were performed by using a Nikon Metrology 450 kV CT scanner with a linear detector. The quality control task involved wall thickness measurements for several cross-sections to perform a global fail/pass test of the blades.

CT can also be applied for inspection of combustion engines. Figure 9.9 shows the CT reconstruction of a three-cylinder head consisting of an aluminium block and steel bushings. Goal of the inspection was to perform reverse engineering and redesign the part to reduce overheating and cracking of the head. CT measurements involved several steps. First, the cylinder head was scanned by using a 10 meV CT scanner and 2D cross-sections were reconstructed. Then, the 3D model was reconstructed and meshed into the STL file. Subsequently, segmentation of different material parts occurred. Finally, both a re-meshing for finite-element method (FEM) and computational fluid dynamics (CFD) analysis and a CFD calculation occurred. It was thus possible to design again the cylinder head by means of a CAD software.

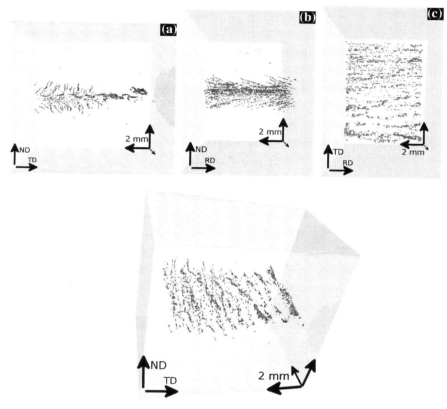

Fig. 9.7 CT scan of centre segregation in an aluminium alloy (*bottom*); central segregation viewed in rolling (*a*), transverse (*b*), and normal (*c*) direction (Slapakova et al. 2016)

9.2.2   Machined Products

In the industrial practice, machined parts are typically inspected by means of conventional CMSs equipped with tactile or optical probes. In this way, it is possible to access and accurately measure externally visible features on the workpiece surface. However, inner features sometimes require CT capabilities to be inspected. This is e.g. the case of narrow and long inner channels obtained through drilling, boring, tapping or laser machining. Figure 9.10 shows a CT scan of an aluminium part as well as its inner channels, which are not accessible by tactile or optical sensors.

Manufacturing of small cavities through e.g. milling or electrical discharge machining (EDM) also calls for CT inspection, since their size cannot be assessed by traditional means. Figure 9.11 depicts a nominal-actual comparison of a

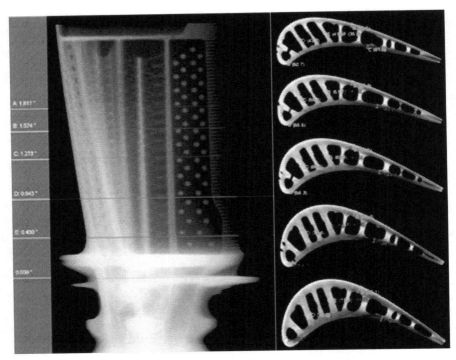

Fig. 9.8 Quality control of turbine blades: cross-sections (*left*); tolerance verification of sections (*right*, *green*/*red* = within/out of tolerance) (Courtesy of Nikon-Metrology/X-Tek)

hydraulic manifold of a race car. One can notice the small cavities that are hardly accessible by tactile probes, thus making CT inspection advantageous.

CT measurements can be also useful to provide feedback information for numerical simulation of machining processes. For example, Kersting et al. showed that it is possible to analyse geometrical deviations in micromilled parts and investigate the cause of defects by combining process simulation and CT measurements (Kersting et al. 2015). Micromilling simulation models cutting forces and conditions to evaluate tool deflections and, in turn, the effect on the undeformed chip. Moreover, it computes the local tool surface errors. Simulated results can be compared with CT measurements to identify the cause of defects. Figure 9.12 shows both CT reconstruction and simulation of a micromilled workpiece. Combining these techniques, it was possible to identify a defect due to transportation, one due to tool deflection during the finishing process, and one due to tool deflection in the roughing process.

Machining is at the same time a precise manufacturing process and very demanding in terms of accuracy of geometrical and dimensional measurements. However, the accuracy of CT measurements is affected by phenomena (e.g. beam hardening, scattering, and grey-value edge thresholding for surface determination)

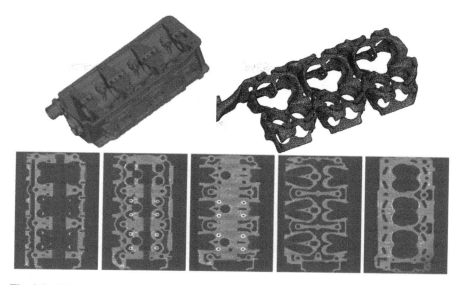

Fig. 9.9 Cylinder head showing aluminium block, steel bushings and cooling jacket (*top left*); extracted STL/FEM mesh of cooling jacket (*top right*); single slices for re-design (*bottom*) (Courtesy of Materialise NV)

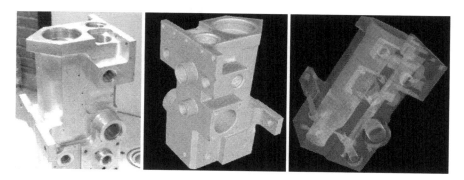

Fig. 9.10 CT scan of an aluminium workpiece (120 × 120 × 220 mm) (Courtesy of KU Leuven)

that dramatically limit CT performances, see Chap. 5. The influence of beam hardening on measurements can be seen in Fig. 9.13, which depicts a precision pin (diameter equal to 4 mm) inserted in a hollow step cylinder consisting of stainless steel. This assembly was scanned with a 225 kV CT scanner (Nikon Metrology MCT225), whereas the hollow step cylinder was calibrated with a CMS (Mitutoyo FN904). The cylinder was manufactured with a precision lathe (Mori Seiki

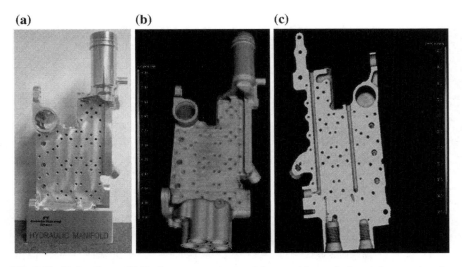

Fig. 9.11 Hydraulic manifold of a race car: picture of the part (**a**), nominal-actual comparison for the outer geometry (**b**), and for the inner geometry (**c**) (Courtesy of Nikon-Metrology/X-Tek)

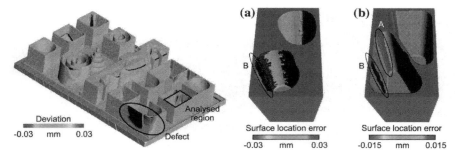

Fig. 9.12 Nominal-actual comparison between a CT scan and the CAD model of a micromilled part, showing a macroscopic defect (*left*). Two simulations allowed showing defects on the analysed region, regarding finishing (**a**) and roughing (**b**) (Courtesy of University of Padova, Italy, and TU Dortmund, Germany)

NL2000Y/500) and the precision pin had manufacturing tolerances of ±1 μm (Tan et al. 2013). One can notice from Fig. 9.13b that using techniques for beam hardening correction reduces the non-systematic error on the inner pin diameter from 7 μm to 2 μm. At the same time, it was possible to reduce the average offset of the outer cylinder by 3 μm (from 5 μm to −2 μm) and of the inner hole diameter by 8 μm (from −13 μm to −5 μm). In conclusion, beam hardening correction reduced the total error, i.e. both systematic and non-systematic errors, to within 5 μm.

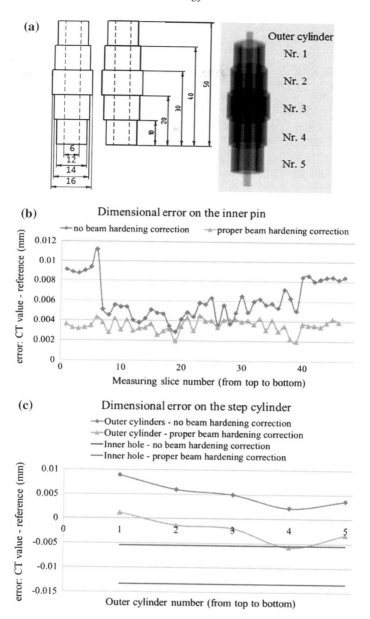

Fig. 9.13 CT measurements of a precision pin inserted in a stainless steel step cylinder: geometry and sizes of the assembly (**a**), measurement error of the inner pin diameter (**b**), measurement error of the step cylinder diameters (**c**) (De Chiffre et al. 2014)

9.2.3 Laser Cut Products

Laser cutting is a manufacturing technology that is often used as a substitute to conventional machining methods. Advantages of laser cutting include easier workpiece holding, lower contamination of the workpiece and absence of tool wear. CT can play a major role for inspecting dimensions and microcavities of laser cut products. An example is provided in Fig. 9.14, which depicts a CT scan of a drug-eluting stent, with a detailed view of a microcavity. In this example, CT scans were used to measure the volume of microcavities on drug-eluting stents (Carmignato et al. 2011). Laser-cut microcavities are manufactured to contain and release drug towards vessel walls. For predicting the maximum drug dosage that can be loaded on the stent, it is necessary to measure the volume of cavities. To this end, 2D optical measurements could be performed, but these measurements are destructive, time consuming, and can be inaccurate due to microcavities form errors. CT measurements were introduced to overcome the limitations of 2D optical measurements. Carmignato et al. were also able to assess measurement uncertainty of volumetric measurements by comparing measurements on the stent to those on a reference sample (Fig. 9.15) manufactured through micro-electrical discharge machining (EDM) milling. This reference object consists of the same material of the stent and features similar microcavities with negligible form errors. Thus, it was possible to perform optical calibration. Reference measurements were used to assess measurement uncertainty through the experimental procedure adapted from ISO 15530-3 (ISO 15530-3 2011). The overall uncertainty amounted 5×10^{-4} mm^3, i.e. 18% of microcavity volume.

9.2.4 Additive Manufactured Products

CT is increasingly used for characterizing additive manufactured (AM) parts, for dimensional metrology purposes, as well as for characterizing the defects present in the printed parts and for surface texture measurements of inner features. AM

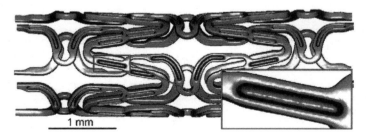

Fig. 9.14 CT reconstruction of a drug-eluting stent, with detailed view of a micro-cavity (Carmignato et al. 2011)

Fig. 9.15 SEM image of the reference sample (Carmignato et al. 2011)

processes allow producing complex geometries normally not obtainable with traditional manufacturing techniques. This design freedom is often useful to produce optimised parts meant to improve the performances of the products and reduce their weight. Many of these designs present internal features which need to be verified as all the other parts produced by traditional manufacturing techniques. From a metrological point of view, these objects represent a challenge because of the internal features which are not measurable by traditional dimensional measurement techniques, like tactile CMSs. In many cases X-ray CT represents the only option available to inspect and measure complex AM designs, with important implication in the verification of the geometrical conformity of the AM parts, which is a main issue for high quality demanding sectors like medical and aerospace.

An application example of CT for quality control of an additively manufactured injection nozzle is shown in Fig. 9.16. The nozzle (Fig. 9.16a) features a helical cooling channel that can be inspected only using CT (Fig. 9.16b). Deviations from the nominal geometry were computed by means of a nominal-actual comparison (Fig. 9.16c).

Figure 9.17 depicts customised 3D printed insoles with lightweight structures used to locally tailor the mechanical properties. Dimensional measurements are performed on the lightweight structure, in order to verify that the diameter of the lightweight structures is within the specification.

CT voxel models of AM parts are often compared to original CAD files, in order to assess the dimensional deviation from the original design and point out possible problems that happened during the process. An example of this approach is shown in Fig. 9.18, where a local deviation is visible in correspondence of the junction between two melting regions, which is probably due to local stress build up.

CT is also used to characterise the structure of the material, by analysing size, morphology and distribution of the internal defects. It is important to distinguish between designed porosity, like the one present in the lightweight structures, and the non-designed one, which refers to the pores in the material structure, e.g. resulting form consolidation limitations of the AM processes. In order to perform an accurate porosity measurement, there are a number of influencing factors that have to be taken into account. As for dimensional measurement, also for porosity analyses the quality of the CT datasets (intended as contrast-to-noise-ratio and signal-to-noise-ratio) is determinant in order to obtain an accurate result. Another important step is the choice of the thresholding algorithm which is used to define the surface between material and pores. Thresholding algorithms are divided in two

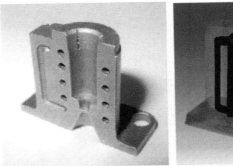

(a) Cross section of nozzle   (b) X-ray image of nozzle

(c) Deviation from nominal geometry

Fig. 9.16 Injection nozzle (Courtesy of KU Leuven)

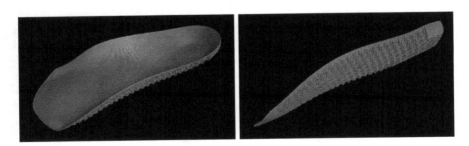

Fig. 9.17 CT scan of customised 3D printed insoles (Courtesy of RS print)

main families: the adaptive ones, which apply a local threshold by assessing for each voxel the grey values of the surrounding voxels, and the global ones, which apply a unique threshold to all the voxels in the dataset. In general, adaptive

(a)   **(b)**

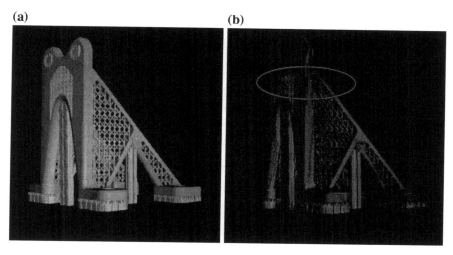

Fig. 9.18 **a** Voxel model of an engine bracket obtained by X-ray CT scan and **b** comparison between the CT voxel model and CAD model, which shows a local deviation (*red circle*) in correspondence of the junction of two islands, which is probably caused by local stress build up (Courtesy of Materialise NV)

thresholding methods show to be stable and with good performances, but the results are strongly dependent on the parameters chosen by the operator and they are computationally more demanding. Global thresholding can be overall less accurate, but more stable in term of results output (Iassanov et al. 2009).

The detection and measurement of the porosity in the structure strongly depends on the minimal pore size that is supposed to be measured and the voxel size achieved during the scan. CT-based porosity measurements are often compared with other porosity measurements like those according to Archimedes' principle (Spierings and Schneider 2010; Wits et al. 2016). Typically, Archimedes density measurements are overall more accurate than CT measurements, especially when the size of the sample increases, since the errors in the measurements induced by the surface effects and the weight measurements reduce their magnitude, while in the case of CT the increase in size leads to an increase of the voxel size, which reduces the resolution and accuracy of the measurement. However, CT-based porosity measurements have competitive advantages compared with other techniques, since it enables to assess the defects morphology and distribution within the part, as well as to point out the presence of critical pores (e.g. due to their size, shape or position), which could strongly affect the mechanical properties of the parts. Literature cases show the capabilities of CT to point out defects with a size that critically affects the fatigue properties of metal parts produced using AM processes (Leonard et al. 2012). CT is also used to assess the influence of the scanning strategies in parts produced using both metal and polymer AM processes (Tammas-Williams et al. 2015; Dewulf et al. 2016).

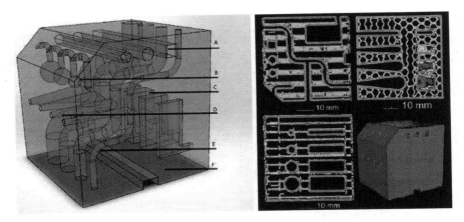

Fig. 9.19 Additive manufactured reference workpiece for characterisation of AM processes using CT (Courtesy of University of Huddersfield)

A current problem in additive manufacturing is lacking reproducibility and repeatability. As a consequence, parts produced at high rate may require scrapping or rework. In this sense, CT can provide useful information to close the quality control loop and improve the process. Shah et al. developed an artefact consisting of different polymers specifically for characterisation of AM processes using CT (Shah et al. 2016). The design took inspiration from the draft ISO 10360 standard for CT systems verification (see Chap. 6) and AM benchmarking artefacts as a basis and features complex inner structures and porosities that can be measured (Fig. 9.19).

AM plays also an important role in the aerospace industry. Figure 9.20 depicts an additive manufactured part from the company Premium Aerotec GmbH. It is made of a Titanium alloy and its size is $100 \times 55 \times 30$ mm. CT was used to perform inspection tasks, and in particular defect analysis, wall thickness measurements, and nominal-actual comparisons (Lübbehüsen 2014).

9.2.5 Injection Moulded Products

Inspecting polymeric parts is a very common task for CT, since polymeric materials typically feature a very good penetrability to X-rays, which in turns determines low artifacts due e.g. to beam hardening or scattering, see Chap. 5. Therefore, CT is an efficient tool for the holistic quality control of injection moulded parts. In the manufacture of those parts, common tasks include nominal-actual comparisons for trouble-shooting and run-in (Fig. 9.21). Typically, the comparison of the geometry of the moulded part occurs with a CAD model, or between single cavities, among different moulders, different materials, or after undergoing a heat treatment or wear (Stolfi 2017).

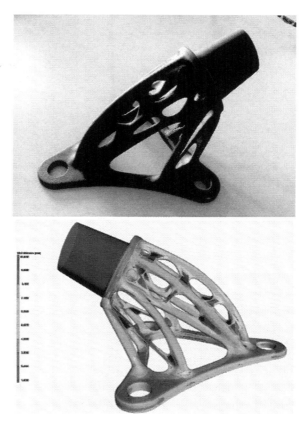

Fig. 9.20 Aerospace component produced through AM (*top*), and nominal-actual comparison of its CT scan (*bottom*) (Courtesy of Premium Aerotec GmbH)

Fig. 9.21 Nominal-actual comparison of an injection moulded cartridge holder (Courtesy of Technical University of Denmark)

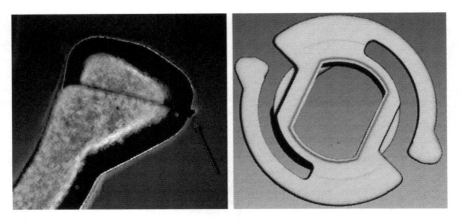

Fig. 9.22 Microscopic scan of injection moulded part featuring a flash (*left*), not visible with a CT scan (*right*) (Courtesy of Novo Nordisk A/S)

However, applying CT for inspecting injection moulded parts still involves several challenges. Measurement results strongly depend on the method used for aligning the measured geometry to the CAD model, especially when the reference points used for alignment (datum) include defects in the measured geometry. Limitation in CT resolution is also a problem, since it is not possible to recognise flash (Fig. 9.22) and mould lines, to resolve rough surfaces, rounded corners, and large variations in wall thickness.

A major industrial issue is mould approval or validation. In the industrial practice, injection tools are validated by verifying critical features of parts from different cavities by means of tactile CMSs. However, companies have started considering CT measurements as a tool for mould validation (De Chiffre et al. 2014).

For instance, in 2012, Novo Nordisk A/S (DK) reported that 90% of the use of CT in the Prefilled Device Quality Control Department was related to mould approval, whereas 10% was related to other special tasks (Sørensen 2012). Similarly, the verification department at LEGO A/S (DK) is responsible for quality control of LEGO pieces from new production moulds (CIA-CT 2013). Those products are inspected by means of conventional CMSs or manual measuring devices. In 2013, around 75% of features were still inspected manually. Some of these products, however, have high aspect ratios or hidden features, and cannot be inspected through conventional measuring methods. Also in this case, CT is applied to perform quality control of these parts.

A challenging task for CT inspection is measuring a large number of different parts. An example is provided in Fig. 9.23, where a set-up with many parts for CNC measurements on a tactile CMS and a multiple part holder for CT measurements are shown. Commonly, multiple parts are inspected for insert moulding (when plastic is injected around an insert, e.g. a metal part), outsert moulding (when a metal part surrounds injected plastic), and overmoulding (when a plastic part is injected

Fig. 9.23 Set-up with multiple parts for CNC measurements on a tactile CMS (*top*), multiple part holder for gripping different parts in CT measurements (Courtesy of Technical University of Denmark)

around another plastic part). Multimaterial aspects represent a critical issue, when the different parts feature a strongly different absorption behaviour (e.g. in case of a metal part and of a plastic part) or a similar behaviour (e.g. in case of two plastic parts), see Sect. 9.2.6.

Microinjection moulding is another application field for CT inspection (Ontiveros et al. 2012). However, it is advisable to use specific calibration artefacts to inspect microparts (Carmignato et al. 2009; Marinello et al. 2008).

Polymeric prosthetic joint components can be also inspected with CT, in particular for analysing the geometry of worn-bearing surfaces and assessing wear volumes. It is possible to achieve lower measurement uncertainties with respect to CMS measurements (Carmignato et al. 2011, 2014; Carmignato and Savio 2011; Spinelli et al. 2009).

Finally, metal injection moulding is also an application field for CT. Here, CT measurements are performed to e.g. detect defects, as shown in the example in Fig. 9.24.

Fig. 9.24 Porosity measurements from CT scans of an aluminium stator box, showing cavities (Courtesy of Grundfos A/S)

Fig. 9.25 CT scan of a watch assembly (Courtesy of Nikon-Metrology/X-Tek)

9.2.6 Assembled Products

CT has a great potential as a tool for inspecting assemblies. Even if quality of the assembly components is ensured by means of conventional measurement methods (e.g. tactile CMS), it is not guaranteed in the assembled state, since deformations or components misalignment may occur during the assembly process. On the other hand, CT allows visualising the single parts in the assembled state.

Application examples are shown in Figs. 9.25 and 9.26. Figure 9.25 depicts the CT scan of a watch assembly, showing the different components. Figure 9.26 shows an assembly consisting of a plastic actuator, an insert, and the area of

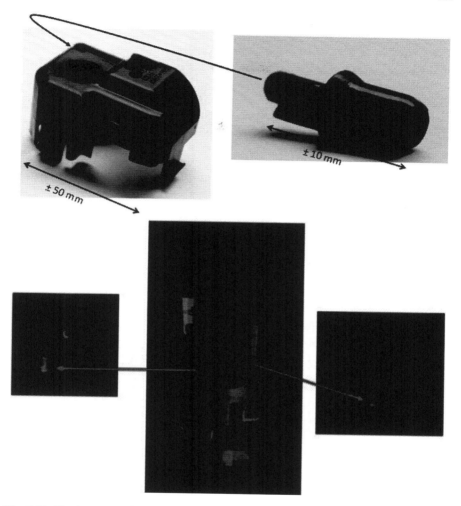

Fig. 9.26 Plastic actuator (*top left*), insert (*top right*), and the related fit analysis (*bottom*). The insert is coloured purple, the actuator is coloured *grey*, and the region of overlap is *green* (De Chiffre et al. 2014)

overlap. Here, the goal of CT inspection was to analyse several combinations of plastic actuators and inserts that should fit together and investigate failures in assembly.

CT can be also used for inspecting weld joints. Figure 9.27 depicts the CT scan of an electron-beam weld joint that holds together a titanium-aluminide wheel with the steel shaft of a turbine wheel (Baldo et al. 2016). Baldo et al. also investigated the influence of CT settings on the image quality of the joint area. It was possible to distinguish clearly the weld joint area between the Ni-based superalloy wheel and steel shaft of the shaft-wheel assembly.

Figure 9.28 shows an example of inspection of the assembly of two threaded components. In this case, CT inspection capabilities are insufficient to distinguish the two different parts of the assembly, since they are made of the same material. This may represent a typical problem when inspecting an assembly.

A special mention has to be made for multimaterial assemblies. Inspecting complex multimaterial assemblies is a common CT quality control task in the industrial practice. However, major challenges need to be addressed when inspecting these objects. Parts with similar absorption coefficients represent a common problem while performing the segmentation of each part. Figure 9.29

Fig. 9.27 CT scan of a weld joint area belonging to a turbine shaft-wheel assembly (Courtesy of Federal University of ABC and CERTI, Brazil)

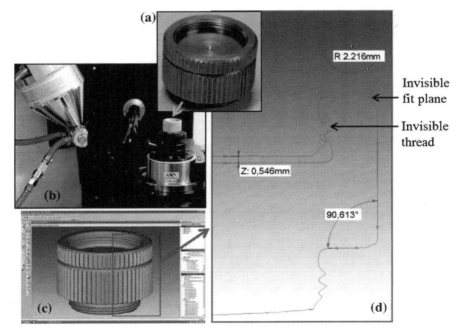

Fig. 9.28 Assembly consisting of two threaded parts made by the same material (**a**), CT measurement (**b**), reconstructed CT volume (**c**), cross-section cutting the right part of the CT model (**d**) (Courtesy of Nikon-Metrology/X-Tek)

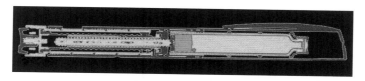

Fig. 9.29 Multimaterial assembly of an insulin pen, consisting of components of different polymeric materials (Courtesy of Novo Nordisk A/S)

shows the scan of a complex multi-material assembly of an insulin pen which consists of several polymeric materials. Here the main challenge is to segment the different material parts, since they have similar absorption coefficients.

Another issue arises when scanning parts with dramatically different absorption coefficients. Scanning assemblies which consist of both a high-absorbing part (e.g. made of metal) and a lowly absorbing one (e.g. made of plastic) is problematic in terms of identifying the suitable scan parameters. On one hand, high-absorbing parts need high-energy X-rays to ensure full penetration and thus show the inner structure. On the other hand, low-energy X-rays are necessary to reveal small details or to provide enough contrast for low-absorbing parts. To properly set up the CT scanner, a CT user needs to make a compromise between these two needs. Depending on the specific assembly that has to be inspected, this may not always be possible and therefore strategies like dual-energy CT (DECT) need to be applied. DECT scans a multimaterial assembly with two different X-rays spectra, in order to better resolve lowly and high-absorbing parts (Heinzl et al. 2008).

There are two ways to perform DECT: dual exposure and dual detectors techniques. The dual exposure technique involves scanning the object with two separate scans, one with high tube voltage and one with low tube voltage. An application example is shown in Fig. 9.30, where metal clippers with a plastic grip and its projection images with different voltages are depicted. CT projections with different energies are then combined to provide an improved reconstructed volume.

A method for combining multi-energy image stacks in dimensional metrology is shown in Fig. 9.31 (Krämer et al. 2010). It considers the high-energy image stack and searches for the outshined projection pixels. Information from a low-energy image stack is considered and fused with the previous stack.

Another important challenge that CT measurements of multimaterial assemblies need to address is surface determination. The common procedure for determining the surface is to identify the iso-surface, thus defining material surface with a single grey value. However, using this method for multimaterial workpieces can generate a fictitious material layer at the interface between two different materials. Figure 9.32 shows an example of this phenomenon: the scanned object consists of both aluminium and steel. By determining the iso-surface, a fictitious layer of aluminium appears at the interface between steel and background air.

Surface determination is particularly challenging in case of parts with similar density. An example is shown in Fig. 9.33, which depicts a multi-material workpiece consisting of layers of steel and ZrO_2 end gauges. The grey-value histogram

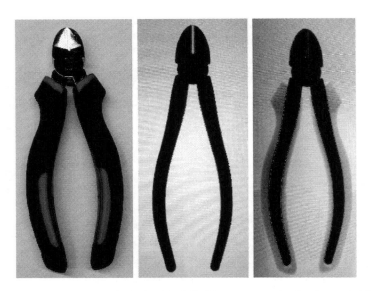

Fig. 9.30 Metal clippers with plastic handle (*left*), high-energy projection image (*middle*), and low-energy projection image (*right*)

Fig. 9.31 Workflow for multi-energy stack fusion. Adapted from Krämer et al. (2010)

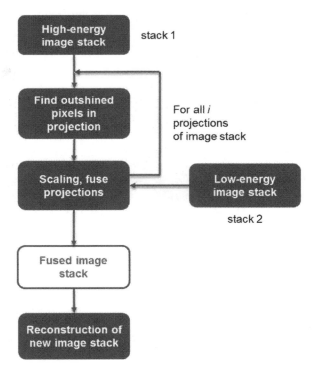

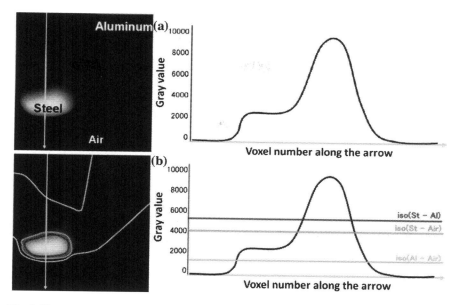

Fig. 9.32 Reconstructed volume containing steel, aluminium and air (*left*), grey value profiles along the *yellow line* (*right*), showing also the fictitious layer (*green*) (Haitham Shammaa et al. 2010)

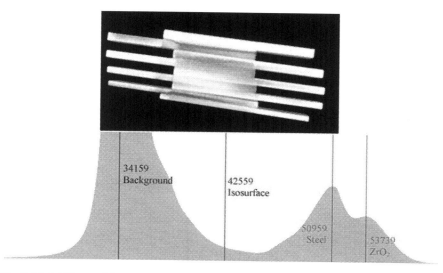

Fig. 9.33 Multi-material measurements: alternating steel and ZrO_2 end gauges; gauges (*top*) and gray levels and isosurface (*bottom*) (Kruth et al. 2011)

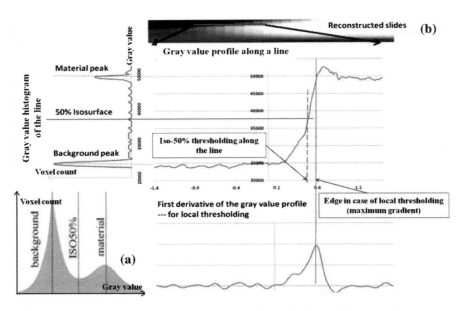

Fig. 9.34 iso-50 global thresholding method (**a**); local adaptive thresholding method applied to a reconstructed cross-section (**b**) (Courtesy of KU Leuven and Lessius University College)

shows three increasing grey levels for air, for steel and for ZrO_2. However, the optimal threshold to distinguish ZrO_2 from steel could be different from the optimal threshold to distinguish ZrO_2 from air. This effect is shown in Fig. 9.33 where the threshold is set at the iso-50 surface (middle between air and steel), resulting in a different thickness for the ZrO_2 gauges depending on the surrounding material.

A method to determine surfaces in the reconstructed volumes of multimaterial parts was suggested by Krämer and Weckenmann (2010a, b). This method identifies an initial surface defined by means of a single grey value and then determines the direction of the steepest change in the grey values within its neighbourhood. The object surface is determined locally by considering the highest variation of grey values. The application of this principle to a reconstruction slice is shown in Fig. 9.34.

In order to compare performances of different laboratories and companies across the world in the field of CT measurements on multimaterial assemblies, the "Interaqct Comparison on Assemblies" was performed in 2016 (Stolfi et al. 2016). For this comparison, two multimaterial workpieces were developed as reference objects (Fig. 9.35). The first one consists of a cylindrical step gauge made of aluminium and a tube made of glass with two fastening caps; this assembly features several uni- and bidirectional measurands, as well as different multimaterial measurands. The second object consists of an industrial assembly provided by Novo Nordisk A/S. It is a biomaterial component from a commercial insulin pen made of polyoxymethylene and ABS-polycarbonate produced through injection moulding.

(a)   **(b)**

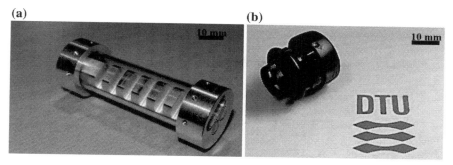

Fig. 9.35 Multimaterial assemblies that were investigated during the Interaqct Comparison on assembly: aluminium step gauge with glass tube and fastening claps (**a**), and insulin pen component made of two different plastics (**b**) (Courtesy of technical University of Denmark)

Measurands include diameters, and related roundness and concentricity. Reference measurements were provided by using tactile CMSs with a measurement uncertainty not higher than 5 μm (Stolfi et al. 2016).

Even if the state of the art is still characterised by several problems in the inspection of multimaterial assemblies (i.e. selection of suitable scan parameters and surface determination), industrial CT has already shown its great potential as a powerful tool for performing dimensional measurements on multimaterial assemblies.

9.3 CT Measurements for Medicine: A Case Study

Over the past forty years CT has become the most important imaging technology in medicine. With a mean of 131.8 per 1000 population performed exams it is widely used in the Organisation for Economic Cooperation and Development countries (OECD 2011). It has especially been applied for preventive medicine and diagnosis. Traditionally, a purely visual evaluation of the CT data is performed for the described applications by the treating doctor. The aim is for example, to detect a tumour or a fracture. With this form of evaluation, the image quality is most important which means that the images of the anatomical structures should be of high-contrast and low-noise. Nevertheless, the application of medical imaging technologies has developed over the last decades and is not limited to the use for diagnosis purposes any more.

In the field of image-based navigated surgery, CT data that are recorded pre-operatively are used to plan a surgical intervention in detail, especially to define possible access paths. In order to perform a surgical procedure more precisely, image guided systems are used to transfer the planning into the operating situs. This means that image guided systems are used to link the preoperative three dimensional image data of the patient's anatomy to the patient's position in the operating room and to guide surgical tools by projecting them onto the preoperative acqui-sitioned data in almost real time.

In the described context, CT data are used for three-dimensional measurements from the metrological point of view. The results of a CT scan form the basis for the guided movement of surgical tools based on the planned coordinates in three-dimensional space. This application corresponds to the use of CT scans as a coordinate measuring machine (Pollmanns 2014). The concrete measurement task consists of dimensionally stable imaging of anatomical structures in terms of their absolute dimensions and position in a fixed coordinate system. The following example illustrates the described importance of CT measurements in medicine.

9.3.1 CT Based Planning of Procedures at the Lateral Skull Base

The challenges of minimal-invasive approaches to the lateral skull base arise from the region's anatomy: critical neurovascular structures such as the jugular vein and the facial nerve are embedded in bone, only a few millimeters apart. Therefore, small canals have to be drilled to a desired target with sub-millimetric accuracy to pass the anatomical structures without damage.

In order to plan three collision free trajectories for example to the internal acustic meastus, critical structures need to be segmented manually. Then triangle meshes are extracted by the marching cubes algorithm resulting in a 3D model of the patients' temporal bone. Afterwards the planning of the canals is performed in a medical planning software.

The minimum distance of the various trajectories to the critical anatomical procedures can be determined by using the planning software. The surgeon selects three trajectories which have the widest distance to the critical structures. The feasibility of the intervention depends primarily on whether the selected reserve is large enough in order to compensate the uncertainties along the surgical process from the planning to the drilling of the planned boreholes.

To explain the importance of dimensional measurements for the planning of the access trajectories, the problem is discussed for the trajectory that is planned through the upper semicular canal (Fig. 9.36). For this trajectory, the determination of the minimal distance between the inner surface of the upper semicular canal and the outer diameter of the planned trajectory is important. In this context, their radial distance is critical.

From the metrological point of view, the task of the imaging is to visualize the position and the contour of the upper semicular canal as accurate as possible. Accordingly, the inspection characteristic for the measurement task is the position of a surface point on the inside of the upper semicular canal which defines the minimal distance to the outer diameter of the planned bore trajectory. However, due to the centrical position of the trajectory it is possible that not only the position of a

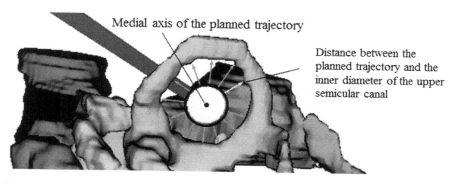

Medial axis of the planned trajectory

Distance between the planned trajectory and the inner diameter of the upper semicular canal

Fig. 9.36 Detailed view of a canal that leads through the upper semicular canal (Pollmanns 2014)

single point but a row of surface points along the inside of the canal need to be considered for the measurement of the closest distance between the canal and the trajectory (Pollmanns 2014).

Due to uncertainties that afflict the imaging processes and the following segmentation, there are deviations in the measurement of the upper semicular canal. Depending on the choice of grey level limit, the visualized surface points show a larger or smaller distance from the planned trajectory compared to the real anatomical structures. In addition, there are measurement deviations which result from a positional deviation of the entire anatomical structure related to a fixed origin. In the event that the set voxel size is significantly different from the actual size, the total volume is subject to a scaling error. In terms of an uncertainty estimation, both types of deviations need to be considered. (Pollmanns 2014).

9.3.2 Estimating the Uncertainty of Medical CT Measurements

In the described scenario the data is used for navigation purposes and the uncertainties in the imagining and image processing lead to false conditions for surgery planning. Therefore, the uncertainties of the CT measurements result in a higher risk of injuring vital structures. That is why methods for the uncertainty estimation of the CT-imaging process and the image processing are needed (Pollmanns 2014).

Pollmanns (2014) transferred the approach of ISO 15530-3 (explained in Chap. 7) to the medical domain. Within this chapter, the design of a special test specimen, the general experimental procedure and the determination of the uncertainty contributions of imaging and image processing are described more detailed.

For the design of the of the test specimen, the normative requirements need to be interpreted with regard to the medical scenario. Regarding the requirements that are defined within ISO 15530-3, the similarity of the test specimen to the human

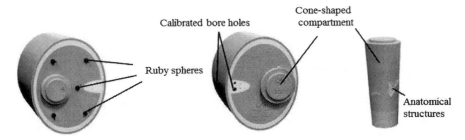

Fig. 9.37 Test specimen with cone-shaped compartment and integrated anatomical structures, ruby spheres (reference structures) and calibrated borehole

anatomy of the region of interest and to the radiometrical properties of the human skull are of great importance. Furthermore, there need to be structures/features that are assessable for tactile measurements for calibration purposes.

Based on these requirements, Pollmanns developed an elliptical test specimen (Fig. 9.37) which consists of different material layers that are equivalent to skin tissue, brain tissue, soft tissue, cortical and cancellous bone. An insert with artificial petrous bone and cylindrical boreholes are also integrated into the specimen. The diameter of the boreholes were chosen in accordance with the dimensions of the cavities in the cancellous bone of the human skull. Furthermore, a cone-shaped compartment with anatomic petrous including the relevant structures for minimally invasive surgery at the inner ear can be inserted into the test specimen. Five ruby spheres serve as reference structures for the definition of a coordinate system that is, together with the calibrated boreholes, important for the assessment of the systematic errors of the measurement process (Pollmanns 2014).

Since the measurements for the uncertainty estimation of the imaging process need to be performed in the same way as the actual CT measurements for the surgical planning are carried out, all relevant process steps and factors that influence the measurement results need to be considered. The process of the CT imaging contains the following process steps:

- positioning of the patient in the CT measurement volume,
- record of projections and reconstruction of the projection data,
- manual segmentation of relevant anatomical structures,
- transformation of volume data sets into surface data and
- postprocessing of the data, e.g. filtering.

Pollmanns performed 20 CT scans with a medical CT (Somatom Defintion AS, Siemens). The scan parameters were set in compliance with a scan protocol that is usually used for thin-layer CT of the petrous bone. The test specimen was repositioned after every CT scan. Therefore, the deviations of the measurements caused by the positioning of the test specimen with regard to the rotary axis of the CT can be determined.

For the metrological assessment of the uncertainty contributions VG Studio Max (Volume Graphics, Heidelberg) was used. Since it is impossible to calibrate the anatomical structures of the cone-shaped insert, the assessment of the systematic deviations and the standard uncertainty of the CT measurement process was determined according to the substitution method proposed by the ISO 15530-3. Based on this method, systematic and random errors of the measurement are detected at the same time by measuring two different features (Pollmanns 2014).

The assessment of the systematic deviations was based on the measurement of the diameters of the calibrated boreholes. According to the medical scenario, the boreholes with a diameter of 2 mm were most important, because their diameter is similar to the one of the cavities in the human bone. Since the calibrated boreholes are on the lateral and on the front face of the test specimen, different directions of penetration are considered. This is important in order to cover the uncertainties imposed by different bore directions, since the cavities in the human bone are curved and are therefore penetrated in different directions, too. The resulting standard uncertainty caused by the systematic deviation was calculated to $u_b = -0.054$ mm.

The uncertainty estimation was performed for the inner diameter of the upper semicircular canal (Fig. 9.36) and it's uncertainty was determined to $u_p = 0.162$ mm (Pollmanns 2014). In addition to u_p and u_b, the uncertainties resulting from the calibration process and from the test specimen need to be estimated. While u_{cal} can be taken from the calibration certificate, u_w is estimated based on the determined process uncertainty for CT-measurements. According to the opinion of different experts the following estimation is reasonable (Pollmanns 2014):

$$u_w = \frac{2}{3} u_p \tag{9.1}$$

Combining all components of uncertainty for the imaging and multiplying it with a coverage factor of $k = 3.579$ (level of confidence = 99.8%) results in $U_{img} = 0.724$ mm. (Pollmanns 2014)

Generally, this uncertainty needs to be considered for the planning of minimally invasive surgery at the inner ear. Depending on the planned boreholes, other measurement tasks are important for the assessment of the patient's risk. That is why it is necessary to analyse other measurement tasks that is important for the planning of minimally invasive surgery to the lateral skull base.

Moreover, there are several other applications where CT scans are used for preoperative planning of minimally invasive procedures. In these cases, the CT data are used for dimensional measurements and the knowledge of the corresponding uncertainty is important to plan high-precision and safe access paths. For this reason, more strategies for the evaluation of medical CT measurements are needed.

9.4 Challenges of CT Dimensional Measurements

Several trends are characterising the development of industrial CT. As regards general metrology, industry demands for faster cycle time of measurements with respect to the manufacturing cycle time. Moreover, industry requires more cost-effectiveness of measurement processes. This aspect is critical for CT, since the cost of CT scanners is considerably high and measurement times are long. Another important trend is the need for better operability for workers. This concerns both the CT set-up process and post-processing evaluation of data volumes. Finally, the range of applications of a measurement device tends to become wider. In case of CT, tendency is to explore new fields of application.

Considering industry's demands and the state of the art, there are several critical aspects, including accuracy of CT measurements that need to be improved (Carmignato 2012). Measurement uncertainties in the range of a micrometre, or even less for features in the micro- or nanometre range, are necessary, especially while comparing CT technology with tactile and optical coordinate measurements. Secondly, it is necessary to extend CT inspection capabilities to larger workpieces, workpieces consisting of high-absorbing materials—machine parts, crank cases, gear boxes, engine blocks, etc.—and huge objects like trucks or freight containers (Wenzel et al. 2009).

Improvement should come from several CT components and from the software side. X-ray tubes with higher current and smaller focus spot will allow inspecting larger parts and/or to achieve more precise measurement results at a lower exposure time (Gruhl 2010). More efficient detectors will ensure a lower integration time and faster measurements. Improved, task-specific reconstruction algorithms may allow acquiring fewer projection images to reconstruct the object, and thus speed up measurements. Lower reconstruction times and more accurate results may be achieved by using new techniques (e.g. algebraic reconstruction, iterative reconstruction) or a priori knowledge about the workpiece (e.g. geometrical information from a CAD model of the workpiece or from the manufacturing process) (Maass et al. 2010; Katsevich 2003). Assisting users in the choice of suitable CT scan parameters will contribute to reduce setup time and measurement uncertainty (Giedl-Wagner et al. 2012; Weckenmann et al. 2009). In this sense, a promising approach is to use expert system to help inexperienced users in setting up CT scanners (Schmitt et al. 2012). Determining measurement uncertainty will allow traceability of measurements and thus more reliable measurements (Krämer and Weckenmann 2010b; Kruth et al. 2011; Krämer et al. 2011; VDI/VDE 2013), see Chap. 7. CT cost-effectiveness as well as a larger application range may be achieved by adopting task-specific CT setups, e.g. specific combination of X-ray tube and detector, non-circular trajectories for tubes and detectors (by using robot CT or other similar technologies) (Sauerwein 2010; Fuchs et al. 2010), and special correction techniques for industrial CT (Baer et al. 2010). Finally, improvement in measuring multimaterial assemblies may be achieved by adopting multi-spectral CT, specific imaging and reconstruction techniques (Maass et al. 2011; Chen et al. 2000; Krämer and Weckenmann 2010a, b), see Chap. 3.

References

Baer M, Hammer M, Knaup M et al (2010) Scatter correction methods in dimensional CT. In: Conference on industrial computed tomography (ICT), Wels, Austria

Baldo C, Coutinho T, Donatelli G (2016) Experimental study of metrological CT system settings for the integrity analysis of turbine shaft-wheel assembly weld joint. In: 6th conference on industrial computed tomography (iCT 2016), Wels, Austria

Bartscher M, Hilpert U, Fiedler D (2008) Determination of the measurement uncertainty of computed tomography measurements using a cylinder head as an example. Tech Mess 75:178–186

Carmignato S (2012) Accuracy of industrial computed tomography measurements: experimental results from an international comparison. CIRP Ann 61(1):491–494. doi:10.1016/j.cirp.2012.03.021

Carmignato S, Savio E (2011) Traceable volume measurements using coordinate measuring systems. CIRP Ann 60(1):519–522. doi:10.1016/j.cirp.2011.03.061

Carmignato S, Dreossi D, Mancini L et al (2009) Testing of X-ray Microtomography systems using a traceable geometrical standard. Measure Sci Tech 20:084021. doi:10.1088/0957-0233/20/8/084021

Carmignato S, Spinelli M, Affatato S et al (2011) Uncertainty evaluation of volumetric wear assessment from coordinate measurements of ceramic hip joint prostheses. Wear 270(9–10): 584–590. doi:10.1016/j.wear.2011.01.012

Carmignato S, Balcon M, Zanini F (2014) Investigation on the accuracy of CT measurements for wear testing of prosthetic joint components. In: Proceedings of international conference on industrial computed tomography (ICT). Wels, Austria

Chen SY, Carroll JD (2000) 3-D coronary reconstruction and optimization of coronary interventions. IEEE Trans Med Imag 19(4):318–336

CIA-CT Project Newsletter NR 8 (2013) Technical University of Denmark

De Chiffre L, Carmignato S, Kruth J-P et al (2014) Industrial applications of computed tomography. CIRP Ann 63(2):655–677. doi:10.1016/j.cirp.2014.05.011

Fuchs T, Schön T, Hanke R (2010) A translation-based data acquisition scheme for industrial computed tomography. In: 10th European conference on non-destructive testing ECNDT, Moscow, Russia

Giedl-Wagner R, Miller T, Sick B (2012) Determination of optimal ct scan parameters using radial basis function neural networks. In: Conference on industrial computed tomography (ICT), Wels, Austria

Gruhl T (2010) Technologie der Mikrofokus-Röntgenröhren: Leistungsgrenzen und erzielte Fortschritte. Fraunhofer IPA Workshop F 207: Hochauflösende Röntgen-Computertomographie-Messtechnik für mikro-mechatronische Systeme

Haitham Shammaa M et al (2010) Segmentation of multi-material CT data of mechanical parts for extracting boundary surfaces. Comput Aided Des 42(2):118–128

Heinzl C et al (2008) Statistical analysis of multi-material components using dual energy CT. Vision Model Vis, Proc

Iassanov P, Gebrenegus T, Tuller M (2009) Segmentation of X-ray computed tomography images of porous materials—A crucial step for characterization and quantitative analysis of pore structures. Water Resour Res

ISO 15530-3 (2011) Geometrical product specifications (GPS)—coordinate measuring machines (CMM): technique for determining the uncertainty of measurement—Part 3: Use of calibrated workpieces or measurement standards

Katsevich A (2003) A general scheme for constructing inversion algorithms for cone beam CT. Int J Math Math Sci 21:1305–1321

Kersting P, Carmignato S, Odendahl S, Zanini F, Siebrecht T, Krebs E (2015) Analysing machining errors resulting from a micromilling process using CT measurement and process simulation. In: Proceedings of the 4M/ICOMM2015 conference, pp 137–140. doi:10.3850/978-981-09-4609-8_034

Krämer P, Weckenmann A (2010a) Multi-energy image stack fusion in computed tomography. Meas Sci Tech 21(4/045105):1–7

Krämer P, Weckenmann A (2010) Simulative Abschätzung der Messunsicherheit von Messungen mit Röntgen-Computertomographie. Conference on Industrial Computed Tomography (ICT), Wels, Austria

Krämer P, Weckenmann A (2011) Modellbasierte simulationsgestützte Messunsicherheitsbestimmung am Beispiel Roentgen-CT. VDI-Berichte 2149: 5. VDI-Fachtagung Messunsicherheit 2011—Messunsicherheit praxisgerecht bestimmen 2149:13–22

Dewulf W, Pavan M, Craeghs T and Kruth, J-P (2016) Using X-ray computed tomography to improve the porosity level of polyamide-12 laser sintered parts. CIRP Ann

Kruth J-P, Bartscher M, Carmignato S et al (2011) Computed tomography for dimensional metrology. CIRP Ann 60(2):821–842. doi:10.1016/j.cirp.2011.05.006

Leonard F, Tammas-Williams S, Pragnell PB, Todd I, Withers PJ (2012) Assessment by X-ray CT of the effects of geometry and build direction on defects in titanium ALM parts. In: Proceedings of international conference on industrial computed tomography (ICT), Wels, Austria

Lübbehüsen J (2014) Advances in automated high throughput fan beam CT for DICONDE-conform multi-wall turbine blade wall thickness inspection and 3D additive manufactured aerospace part CT inspection. In: 11th European conference on non-destructive testing (ECNDT), Prague, Czech Republic

Maass C, Knaup M, Sawall S et al (2010) ROI-Tomografie (Lokale Tomografie). IN: Proceedings of international conference on industrial computed tomography (ICT), Wels, Austria

Maass C, Meyer E, Kachelriess M (2011) Exact dual energy material decomposition from inconsistent rays (MDIR). Med Phys 38(2):691–700

Marinello F, Savio E, Carmignato S et al (2008) Calibration artefact for the micro scale with high aspect ratio: the fiber gauge. CIRP Ann 57(1):497–500. doi:10.1016/j.cirp.2008.03.086

Ontiveros S, Yague-Fabra JA, Jimenez R et al (2012) Dimensional measurement of micro-moulded parts by computed tomography. Meas Sci Tech 23(125401):9. doi:10.1088/0957-0233/23/12/125401

Organisation for Economic Co-operation and Development (2011) Health at a glance 2011. OECD indicators. OECD Publishing, Paris

Pollmanns S (2014) Bestimmung von Unsicherheitsbeiträgen bei medizinischen Computertomografiemessungen für die bildbasierte navigierte Chirurgie, 1. Aufl ed. Apprimus Verlag, Aachen, XI, 148 S

Sauerwein C (2010) Rekonstruktionsalgorithmen und ihr Potenzial. Fraunhofer IPA Workshop F 207: Hochauflösende Röntgen-Computertomographie-Messtechnik für mikro-mechatronische Systeme

Schmitt R, Isenberg C, Niggemann C (2012) Knowledge-based system to improve dimensional CT measurements. In: Conference on industrial computed tomography (ICT), Wels, Austria

Spierings A, Schneider, M (2010) Comparison of density measurement techniques for additive manufactured metallic parts. Rapid Prototyping J

Shah P, Racasan R, Bills P (2016) Comparison of different additive manufacturing methods using optimized computed tomography. In: Proceedings of international conference on industrial computed tomography (ICT), Wels, Austria

Šlapáková M, Zimina M, Zaunschirm S, Kastner J, Bajera J, Cieslar M (2016) 3D analysis of macrosegregation in twin-roll cast AA3003 alloy. Mater Charact 118:44–49

Sørensen T (2012) CT scanning in the medical device industry. In: Conference "industrial applications of CT scanning—possibilities and challenges in the manufacturing industry", Lyngby, Denmark

Spinelli M, Carmignato S, Affatato S et al (2009) CMM-based procedure for polyethylene non-congruous unicompartmental knee prosthesis wear assessment. Wear 267:753–756. doi:10.1016/j.wear.2008.12.049

Sterzing A, Neugebauer R, Drossel W-G (2013) Metal forming—challenges from a green perspective. In: Proceedings of international conference on competitive manufacturing (COMA 2013), Stellenbosch, South Africa, pp 19–24

Stolfi A (2017) Integrated quality control of precision assemblies. Technical University of Denmark

Stolfi A, De Chiffre L (2016) Selection of items for "InteraqCT comparison on assemblies". In: 6th conference on industrial computed tomography (iCT 2016), Wels, Austria

Tammas-Williams S, Zhao H, Leonard F, Derguti F, Todd I, Pragnell P (2015) XCT analysis of the influence of melt strategies on defect population in Ti-6Al-4V components manufactured by selective laser beam melting. Mater Charact 102:47–61

Tan Y, Kiekens K, Welkenhuyzen F et al (2013) Simulation-aided investigation of beam hardening induced errors in CT dimensional metrology. 11th ISMTII Symposium, Aachen, Germany

VDI/VDE 2630 Part 1.2 (2010) Computed tomography in dimensional measurement—Influencing variables on measurement results and recommendations for computed tomography dimensional measurements. VDI, Düsseldorf

VDI/VDE 2630 Part 2.1 (2013) Computertomografie in der dimensionellen Messtechnik—Bestimmung der Messunsicherheit und der Prüfprozesseignung von Koordinatenmessgeräten mit CT-Sensoren. VDI, Düsseldorf

Weckenmann A, Kraemer P (2009) Assessment of measurement uncertainty caused in the preparation of measurements using computed tomography. IMEKO XIX World Congress "fundamental and applied metrology", Lisbon, Portugal

Wenzel T, Stocker T, Hanke R (2009) Searching for the invisible using fully automatic X-ray inspection, Foundry Trade J Inst Cast Metals Eng 183:3666

Wits WW, Carmignato S, Zanini F, Vaneker THJ (2016) Porosity testing methods for the quality assessment of selective laser melted parts. CIRP Ann Manuf Tech 65(1):201–204. doi:10.1016/j.cirp.2016.04.054

22841900R00213

Made in the USA
San Bernardino, CA
17 January 2019